COLLEGE ALGEBRA

COLLEGE ALGEBRA

MUSTAFA A. MUNEM

WILLIAM TSCHIRHART

JAMES P. YIZZE

MACOMB COUNTY COMMUNITY COLLEGE

WORTH PUBLISHERS, INC.

COLLEGE ALGEBRA

COPYRIGHT © 1974 BY WORTH PUBLISHERS, INC.

444 PARK AVENUE SOUTH, NEW YORK, NEW YORK 10016

ALL RIGHTS RESERVED.

NO PART OF THIS PUBLICATION MAY BE REPRODUCED,

STORED IN A RETRIEVAL SYSTEM, OR TRANSMITTED,

IN ANY FORM OR BY ANY MEANS, ELECTRONIC,

MECHANICAL, PHOTOCOPYING, RECORDING, OR OTHERWISE,

WITHOUT THE PRIOR WRITTEN PERMISSION OF

THE PUBLISHER.

PRINTED IN THE UNITED STATES OF AMERICA

LIBRARY OF CONGRESS CATALOG CARD NO. 73-85130

ISBN: 0-87901-026-6

SECOND PRINTING DECEMBER, 1974

DESIGN BY MALCOLM GREAR DESIGNERS

PREFACE

PURPOSE: This book is designed to prepare students for further study in mathematics and other disciplines by providing them with a thorough grounding in the skills of algebra. The book is intended for use in a three-hour one-semester course or a four-hour one-quarter course.

PREREQUISITES: It is assumed that the students who use this text will have had the equivalent of at least (a) one year of high school geometry and (b) one year of high school algebra or a college course in beginning algebra.

APPROACH: We have the greatest success with our own students when we are careful to gear the presentation to the beginner's experience and when we make use of frequent detailed explanations and worked-out examples, many of them illustrated by geometric interpretations. Because this approach works, we have endeavored to emphasize techniques, drills, and applications throughout the text. Definitions and properties are carefully stated throughout the book. Review problem sets at the end of each chapter help students to gauge their understanding of the material. Answers to selected problems are provided in the back of the book so that students can test themselves.

CONTENT: Chapters 1 and 2 review the fundamental operations with polynomials, fractions, exponents, and radicals. These two chapters present more completely the material normally introduced in simpler form in an intermediate algebra course. Chapters 3, 4, 5, and 6 use the function concept—with emphasis on graphing—to present topics in college algebra. Linear, quadratic, rational, exponential, and logarithmic functions are explored in some detail. In Chapter 7 we present sequences, progressions, and a brief introduction to probability. Chapter 8 includes an elementary discussion of the solution of linear systems. The standard topics in analytic geometry, including the circle, the parabola, the ellipse, and the hyperbola, are covered in Chapter 9.

ADDITIONAL AIDS: A semi-programmed *Study Guide* for students conforms with the arrangement of topics in the book. It contains a great many carefully graded fill-in statements and problems broken down into simpler units, chapter reviews, and tests for each chapter. All answers are provided in the *Study Guide* to encourage self-testing at each student's own pace.

ACKNOWLEDGMENTS: This text owes a great deal to the special assistance of Professors Charles H. Ainley of Spokane Falls Community College, B. D. Arendt of the University of Missouri at Columbia, Rodney Chase of Oakland Community College, Lorraine T. Foster of California State University at Northridge, Adam J. Hulin of Louisiana State University at New Orleans, Calvin A. Lathan of Monroe Community College, Henry Nace at Lawrence Institute of Technology, Charles V. Peele at Marshall University, Carl W. Richards, formerly at UCLA, and George W. Schultz of St. Petersburg Junior College, the Clearwater Campus. We are also indebted to our colleagues at Macomb County Community College, especially Professor John von Zellen, for their helpful suggestions. Robert C. Andrews of Worth Publishers, who coordinated this project, merits special thanks.

Mustafa A. Munem
William Tschirhart
James P. Yizze

Warren, Michigan
January 1974

CONTENTS

CHAPTER	1	**Sets and Polynomials**	1
Section	1	Introduction	3
Section	2	Sets	3
Section	3	Real Numbers and Polynomials	14
Section	4	Field Axioms for Real Numbers	22
Section	5	Algebraic Fractions	33
Section	6	Operations on Rational Expressions	43
Section	7	Exponents	51
Section	8	Rational Exponents	58
Section	9	Radicals	63
CHAPTER	2	**Equations, Order Relations and the Cartesian Plane**	77
Section	1	Introduction	77
Section	2	First-Degree Equations	79
Section	3	Geometry of Line	87
Section	4	Absolute Value Equations and Inequalities	105
Section	5	Cartesian Coordinate System and the Distance Formula	117
CHAPTER	3	**Relations and Functions**	133
Section	1	Introduction	135
Section	2	Relations	135
Section	3	Functions	142
Section	4	Even and Odd Functions	160
Section	5	Increasing and Decreasing Functions	165

CHAPTER	4	**Linear and Quadratic Functions**	173
Section	1	Polynomial Functions	175
Section	2	Linear Functions	175
Section	3	Systems of Linear Equations	196
Section	4	Quadratic Functions	208
Section	5	Quadratic Inequalities	224
Section	6	Quadratic Forms and Radical Equations	231
CHAPTER	5	**Roots of Polynomials and Complex Numbers**	241
Section	1	Introduction	243
Section	2	Division of Polynomials	243
Section	3	Graphs of Polynomial Functions of Degree Greater Than 2	251
Section	4	Rational Functions	257
Section	5	Complex Numbers	264
Section	6	Complex Zeros of Polynomial Functions	273
CHAPTER	6	**Exponential and Logarithmic Functions**	281
Section	1	Introduction	283
Section	2	Inverse Functions	283
Section	3	Exponential Functions and Their Properties	291
Section	4	Logarithmic Functions and Their Properties	295
Section	5	Properties of Logarithms	299
Section	6	Common Logarithms	304
CHAPTER	7	**Sequences, Mathematical Induction, and the Binomial Theorem**	323
Section	1	Introduction	325
Section	2	Sequences	325

Section	3	Progressions	329
Section	4	Mathematical Induction	340
Section	5	Combinations and Permutations	346
Section	6	Binomial Theorem	353
Section	7	Probability	359

CHAPTER	8	**Linear Systems: Matrices and Determinants**	375
Section	1	Introduction	377
Section	2	Matrices and Row Reduction	377
Section	3	Determinants	386
Section	4	Cramer's Rule	394

CHAPTER	9	**Analytic Geometry**	401
Section	1	Introduction	403
Section	2	Circle	404
Section	3	Parabola	409
Section	4	Ellipse	418
Section	5	Hyperbola	428
Section	6	Systems with Second-Degree Equations	438

Appendix

Table	I	Common Logarithms	448
Table	II	Powers and Roots	450
Table	III	Natural Logarithms	451

Answers to Selected Problems   453

Index   511

CHAPTER 1

Sets and Polynomials

CHAPTER 1

Sets and Polynomials

1 Introduction

Our objective in this chapter is to discuss some important concepts considered to be necessary for the study of algebra. The topics include the symbolism of sets and the properties of real numbers, polynomials, rational expressions, exponents, and radicals. We will begin by surveying the basic terminology of sets.

2 Sets

Intuitively, we think of a *set* as a collection of objects. Thus, one might speak of a particular set of golf clubs, a specific chess set, the set of books in a library, or the set of counting numbers: 1, 2, 3, and so on. The objects in a given set are called the *elements* or *members* of the set. Notationally, capital letters, A, B, C, D, E, and so on, are usually used to denote sets, whereas lowercase letters, a, b, c, d, and so on, are used to represent elements of sets. For example, the set A, which has as its elements b, c, d, and e, and no other elements, can be written $A = \{b, c, d, e\}$. The symbol used to show that an element "belongs to" or is a "member of" a set is $\in$. In our example, $b \in A$, $c \in A$, $d \in A$, and $e \in A$. To indicate that a particular element "does not belong" to a set, we use the notation $\notin$. For instance, $f \notin A$ and $g \notin A$.

We call the method of representing a set by listing its members, *enumeration*. For example, we can use enumeration to write the set B of all integers from 1 to 7 as

$$B = \{1, 2, 3, 4, 5, 6, 7\}$$

If a set is so defined that it does not have any elements, we call that set the *null set* (or *empty set*). For example, the set of all Presidents of the United States who died before their thirty-fifth birthday is the empty set, since the President of the United States must be at least 35 years old. The symbol for the null set is $\emptyset$ or $\{\ \}$. It is important to realize that 0 and $\emptyset$ do not mean the same thing. That is, $\emptyset$ is not equal to 0 or $\{0\}$, since $\{0\}$ is a set with the one element, the number 0, while $\emptyset$ (or $\{\ \}$) is the set with no elements. The order in which the elements of a set appear when enumeration is used to describe the set does not matter. Hence, the set $A = \{b, c, d, e\}$ could also be described as

$$A = \{e, c, b, d\} \quad \text{or} \quad A = \{d, e, c, b\}$$

or in any other possible order.

Besides enumeration there is another method used to describe sets. This method consists of describing some identifying property of the set. A standard notation, called *set builder notation*, is used for this description as follows. $A = \{x \mid x \text{ has property } p\}$, which is read "$A$ is the set of all elements x such that x has the property p."

For example, the set B of all positive integers between 1 and 100 could be described as follows. Let a letter, say x, represent an element of the set; then set B can be described as "the set of all x such that x is a positive integer between 1 and 100." That is,

$$B = \{x \mid x \text{ is a positive integer between 1 and 100}\}$$

A set is said to be *finite* if its members can be counted out by the numbers up to some counting number; a set that is neither finite nor empty is an *infinite* set. For example, if A is the set of all 20 students in a particular class, A is a finite set since *all* its elements can be enumerated. On the other hand, if C is the set of all counting numbers, C is an infinite set since all its members cannot be enumerated. It should be noted, however, that infinite sets whose elements form a general pattern can be described by displaying "the pattern" that the members satisfy. For example, the set C of all counting numbers can be described as

$$C = \{1, 2, 3, 4, \ldots\}$$

where the three dots mean the same as "etc."

Also, the set B of all positive counting numbers from 1 to 100 can be described as

$$B = \{1, 2, 3, \ldots, 100\}$$

where the three dots are used to represent the elements of B between 3 and 100.

EXAMPLES

Describe each of the following sets and indicate which set is finite and which is infinite.

1 A, the set of all positive odd counting numbers.

SOLUTION

$$A = \{x \mid x \text{ is a positive odd counting number}\}$$

All the members of A cannot be enumerated because A is an infinite set; however, the known pattern of the elements of A suggests that A can also be written as $A = \{1, 3, 5, \ldots\}$.

2 E, the set of all even counting numbers from 2 to 500.

SOLUTION

$$E = \{x \mid x \text{ is an even counting number from 2 to 500}\}$$

Since all the members of set E can be enumerated as $E = \{2, 4, 6, \ldots, 500\}$, it is a finite set.

2.1 Set Relations

Suppose that F is the set of all Ford automobiles and that M is the set of all motor vehicles. Clearly, all the members of F are also found in M. We say that F is a subset of M or, symbolically, $F \subseteq M$, which is read "F is contained in M." The set of all girls in a biology class is a subset of the set of all students in the class. In general, set A is a *subset* of set B, written $A \subseteq B$, if every element of A is an element of B. *The empty set is considered to be a subset of every set.*

EXAMPLES

1 Let $A = \{1, 2, 3, 4, 5, 6\}$ and $B = \{2, 5, 3\}$; then $B \subseteq A$.

2 If $A = \{1, 2, 3\}$ and $B = \{x \mid x \text{ is a counting number less than 4}\}$, then $A \subseteq B$ and $B \subseteq A$.

Example 2 illustrates the definition of *equality of sets;* for, if $A \subseteq B$ and $B \subseteq A$, we consider A and B to be different names for the same sets and we write $A = B$.

In Example 1, we have $B \subseteq A$, but $B \neq A$. B is an example of a proper subset of A. In general, B is said to be a *proper subset* of a set A, written $B \subset A$ (notice that the horizontal bar is left off), if all members of B are in A and A has at least one member not in B; that is, $B \subseteq A$ but $B \neq A$.

EXAMPLES

1 If $A = \{1, 2, 3\}$, $B = \{2, 3, 1, 7, 9\}$, and $C = \{2, 3, 1\}$, then $A \subseteq B$,

$C \subseteq B$, $A \subseteq C$, and $C \subseteq A$. More informatively, $A \subset B$, $C \subset B$, and $A = C$.

2. List all the subsets of $\{a, b, c\}$.

   SOLUTION. $\{a\}$, $\{b\}$, $\{c\}$, $\{a, b\}$, $\{a, c\}$, $\{b, c\}$, $\{a, b, c\}$, and $\emptyset$ are the subsets of $\{a, b, c\}$. Note that all the subsets, with the exception of $\{a, b, c\}$ itself, are proper subsets.

3. If $A = \{2, 3, 4\}$, $B = \{1, 2, 3, 4, 7, 8\}$, and $C = \{7, 8\}$, then $A \subset B$ and $C \subset B$.

In Example 3 we see that A and C have no members in common. The set of all girls taking biology has no member in common with the set of all boys taking the same class. Such sets are called disjoint sets. In general, two sets that have no members in common are *disjoint*. For example, the sets $\{1, 2, 3\}$ and $\{4, 8, 10\}$ are disjoint sets.

Suppose that $A = \{1, 2, 3\}$ and $B = \{2, 8, 9\}$. Clearly, $A \nsubseteq B$ and $B \nsubseteq A$. (Why?) Also, A and B are *not* disjoint because $2 \in A$ and $2 \in B$. We say that A and B are overlapping. In general, sets A and B *overlap* if there is at least one member common to A and B and if each set contains at least one member not found in the other. For example, if $A = \{2, 3, 5, 9\}$ and $B = \{3, 5, 10, 11, 12\}$, A and B overlap because $3 \in A$ and $3 \in B$, $2 \in A$ and $2 \notin B$, and $11 \in B$ but $11 \notin A$.

When the selection of elements of subsets is limited to a fixed set, the limiting set is called a *universal* set or a *universe*. A universal set represents the complete set or the largest set from which all other sets in that same discussion are formed. The choice of the universal set is dependent upon the problem being considered. For example, in one case it may be the set of all people in the United States, and in another case, it may be the set of all people in Michigan.

EXAMPLE

Describe set A where $A = \{x \mid x$ is a number greater than 2 and x is a member of universal set $U\}$.

a) $U = \{1, 2, 3, \frac{4}{3}, \frac{1}{8}\}$
b) U is the set of all counting numbers.
c) $U = \{0, 1, 2\}$

SOLUTION

a) $A = \{3\}$
b) $A = \{x \mid x$ is a counting number greater than 2$\}$ or, equivalently, $A = \{3, 4, 5, \ldots\}$
c) $A = \emptyset$

Subsets can be represented pictorially by drawings called *Venn diagrams*. These diagrams often help in understanding set concepts. If we let U be the universal set, an arbitrary set $A \subseteq U$ can be represented as another closed region within the closed region representing U (Figure 1). Each of the four set relations discussed above can be represented by one of four Venn diagrams (Figure 2a, b, c, and d).

Figure 1

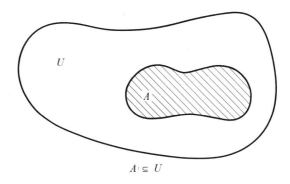

$A \subseteq U$

Figure 2

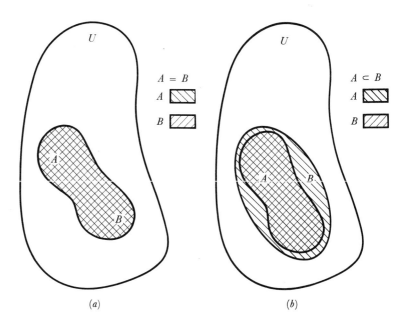

Figure 2 continues on the next page.

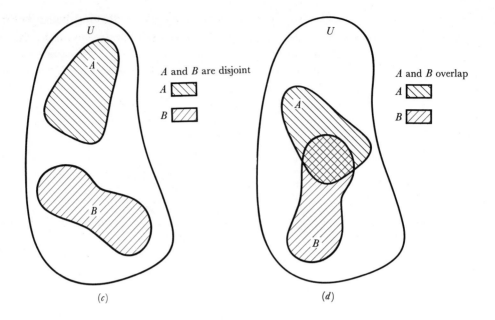

(c)   (d)

EXAMPLE

Let N be the set of counting numbers and assume that

$$A = \{x \mid x = 3n, n \in N\} \quad \text{and} \quad B = \{y \mid y = 4m, m \in N\}$$

Use a Venn diagram to illustrate the set relationship between A and B.

SOLUTION. $A = \{3, 6, 9, 12, \ldots\}$ and $B = \{4, 8, 12, 16, \ldots\}$ are infinite sets. $3 \in A$ but $3 \notin B$; $4 \in B$ but $4 \notin A$; however, $12 \in A$ and $12 \in B$. Hence, A and B are overlapping subsets of N (Figure 3).

Figure 3

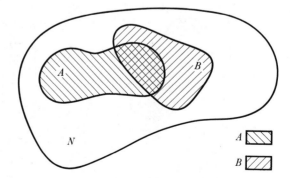

2.2 Set Operations

Consider a universal set $U = \{1, 2, 3, 4, 5, 6, 7, 8\}$. From U we can form $A = \{1, 2, 3, 4\}$ and $B = \{1, 3, 7\}$. How can sets A and B be used to form other sets? One way is simply to combine all the elements of A and B to form $\{1, 2, 3, 4, 7\}$. The operation suggested by this example is that of *set union*.

A union B, written $A \cup B$, and represented by the entire shaded region in Figure 4, is defined as

$$A \cup B = \{x \mid x \in A \quad \text{or} \quad x \in B \text{ (or both)}\}$$

Figure 4

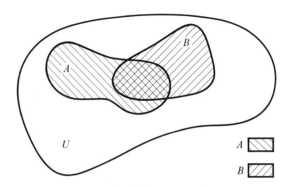

$A \cup B$ is represented by the entire shaded region.

Hence, in the example,

$$\{1, 2, 3, 4\} \cup \{1, 3, 7\} = \{1, 2, 3, 4, 7\}$$

Another way to use A and B to form another set is to form set $\{1, 3\}$, the set of all elements common to A and B. This is an example of *set intersection*.

A intersect B, written $A \cap B$, and represented by the shaded region in Figure 5, is defined as

$$A \cap B = \{x \mid x \in A \quad \text{and} \quad \text{(simultaneously)} \ x \in B\}$$

Figure 5

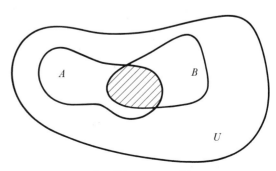

$A \cap B$ is represented by the shaded region

For example,

$$\{1, 2, 3, 4\} \cap \{1, 3, 7\} = \{1, 3\}$$

If $A = \{1, 2, 3, 4\}$ and $B = \{5, 6, 7\}$, then $A \cap B = \emptyset$ and A and B are disjoint sets. In general, A and B are disjoint sets whenever $A \cap B = \emptyset$.

The union of two sets, then, is simply the set that results when all the elements of the two sets are combined; the intersection is merely the set of all elements common to the two sets. Note that when the union of two sets containing common elements is described, the common elements are *not* listed twice; hence, $\{2, 3, 3\} \cup \{1, 4, 8\}$ is *not* written as $\{2, 3, 4, 1, 4, 8\}$ but as $\{2, 3, 4, 1, 8\}$, since the listing of 4 twice is superfluous.

EXAMPLES

1. Determine $A \cup B$ and $A \cap B$ if $A = \{1, 2, 3, 4, 5\}$ and $B = \{2, 5, 6, 7\}$.

    SOLUTION

    $$A \cup B = \{1, 2, 3, 4, 5, 6, 7\} \quad \text{and} \quad A \cap B = \{2, 5\}$$

2. Let $A = \{x \mid x \text{ is a counting number}\}$ and let $B = \{x \mid x \text{ is an even counting number}\}$; that is, $B = \{2, 4, 6, 8, \ldots\}$. Form $A \cup B$ and $A \cap B$.

    SOLUTION. Note that $B \subset A$ and U is the set of all counting numbers (Figure 6).

    Figure 6

    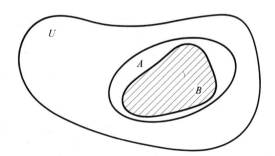

    $A \cup B = \{x \mid x \text{ is a counting number or } x \text{ is an even counting number (or both)}\}$

Therefore,

$A \cup B = \{x \mid x \text{ is a counting number}\} = A$
$A \cap B = \{x \mid x \text{ is a counting number and (simultaneously) } x \text{ is an even counting number}\}$

Therefore,

$$A \cap B = \{x \mid x \text{ is an even counting number}\} = B$$

3 Use Venn diagrams to illustrate that $A \cap (B \cup C)$ and $(A \cap B) \cup (A \cap C)$ are equal sets.

SOLUTION. The shaded area of Figure 7a represents $B \cup C$, and the shaded area of Figure 7b represents $A \cap (B \cup C)$. The shaded areas of

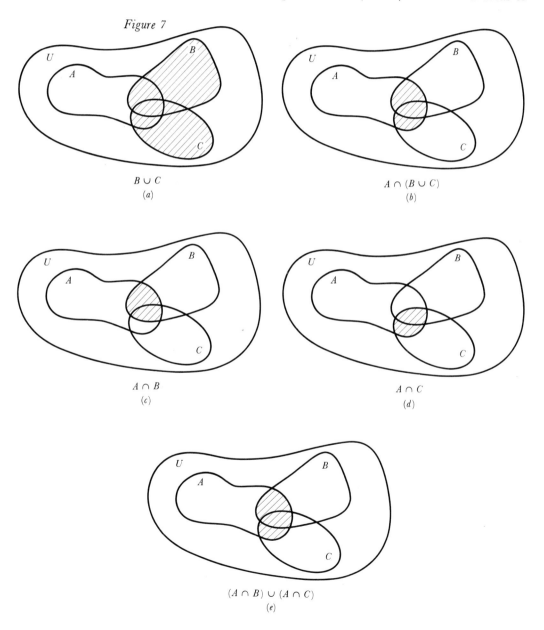

Figure 7

Figures 7c and 7d represent $A \cap B$ and $A \cap C$, respectively, and the shaded area of Figure 7e represents $(A \cap B) \cup (A \cap C)$. Clearly, parts b and e have the same shaded areas, so that the Venn diagrams illustrate the fact that

$$A \cap (B \cup C) = (A \cap B) \cup (A \cap C)$$

PROBLEM SET 1

1. Let $A = \{4, 5, 6, 7, 8, 9, 10\}$, $B = \{3, 6, 8, 9, 10\}$, and $C = \{5, 6, 8, 12\}$. Insert in the following blanks the correct symbol—$\in$ or $\notin$.

5 _____ A	5 _____ B	5 _____ C
10 _____ A	10 _____ B	10 _____ C
12 _____ A	12 _____ B	12 _____ C

2. Let A be the set of numbers greater than 1 but less than 9. Describe A by enumeration if the universal set from which A is formed is
   a) $\{2, 3, 5, 7, 8\}$
   b) $\{7, 3, 10, 2, 15\}$
   c) $\{5, 7, 9, 10\}$
   d) $\{11, 13, 15\}$
   e) The set of counting numbers

3. Decide which sets are finite and which sets are infinite.
   a) The set of students at Macomb College
   b) The set of baseball players who have a college degree
   c) The set of counting numbers
   d) The set of even numbers
   e) The set of odd numbers
   f) The set of the letters of the English alphabet
   g) The set $A = \{1, 3, 5, 7, \ldots\}$
   h) The set of cars in Detroit
   i) The set $A = \{5, 6, 7, 8, 9, 10, \ldots, 15\}$
   j) The set of grains of wheat in the United States

4. Use set builder notation, $\{x \mid x \text{ has property } p\}$, to describe each of the following sets. Indicate which of the sets are finite and which are infinite.
   a) The set of letters in the word "maximum"
   b) The set of even counting numbers
   c) The set of counting numbers divisible by 5, that is, all counting numbers that have a zero remainder when divided by 5
   d) The set of counting numbers greater than 2 and less than 13

5. Which of the following statements are true and which are false?
   a) $3 \in \{3, 4\}$
   b) $\{a\} \subseteq \{a, \{a\}, \{\{a\}\}\}$
   c) $\{3\} \subset \{3, 4\}$
   d) $\{2\} \in \{2, \{2\}\}$

e) $\{3\} \subseteq \{3, 4\}$
f) $\{1, 3\} \subseteq \{1, 5, 15\}$
g) $\{0\}$ is empty
h) $\{a, b\} = \{b, a\}$

6 Indicate which of the set relations (proper subset, equal, disjoint) hold between each pair of the following sets.
   a) The set of all cigarettes; the set of all pipes
   b) The empty set; $\{2, 3, 4\}$
   c) The set of all counting numbers greater than 3; the set of all counting numbers less than 3
   d) The set of all counting numbers greater than 3; the set of all counting numbers greater than 8
   e) The set of all positive even counting numbers less than 11; $\{2, 4, 6, 8, 10\}$

7 List all the subsets of each of the following sets. Indicate which of the subsets are proper subsets.
   a) $\{2\}$
   b) $\{2, 3\}$
   c) $\{a, b, c\}$
   d) $\{\emptyset, \{0\}\}$
   e) $\{5, 6, 7, 8\}$

8 Tabulate the number of different subsets of a set having
   a) 0 elements
   b) 1 element (see Problem 7a)
   c) 2 elements (see Problem 7b)
   d) 3 elements (see Problem 7c)
   e) 4 elements (see Problem 7e)

   Can you generalize your result? That is, if a set has n elements, how many subsets can be formed?

9 Use $A = \{1, 2, 4\}$, $B = \{2, 3, 5, 7\}$, and $C = \{6, 3, 5, 8\}$ to form each of the following sets.
   a) $A \cap B$
   b) $B \cup A$
   c) $B \cap C$
   d) $B \cup C$
   e) $A \cap C$
   f) $C \cup B$
   g) $C \cap A$
   h) $A \cap \emptyset$
   i) $C \cup A$
   j) $B \cup \emptyset$
   k) $A \cap (B \cup C)$
   l) $A \cup (B \cap C)$

10 a) Why is it true that any set is a subset of itself?
   b) *Transitive law.* If $A \subseteq B$ and $B \subseteq C$, then $A \subseteq C$. Give two examples to illustrate the law.
   c) Is $\emptyset \subseteq \emptyset$? (Explain) yes
   d) Is $\emptyset \in \emptyset$? (Explain) No
   e) For any set A, is $\emptyset \subseteq A$?

3 Real Numbers and Polynomials

The language of sets will be used to describe these number sets.

1 *Positive Integers* (N): The *natural numbers* or *counting numbers*, 1, 2, 3, ..., form the fundamental number set of algebra for both historical and logical reasons. This set, which we will designate as set N, is most often referred to as the set of *positive integers*. Thus,

$$N = \{1, 2, 3, \ldots\} \text{ is the set of positive integers}$$

2 *Negative Integers* (I_n): The set of *negative integers* consists of negatives of all the positive integers. Thus, if we denote the set of negative integers by I_n, then

$$I_n = \{-1, -2, -3, \ldots\}$$

3 *Integers* (I): The set of *integers* is the set formed by the union of the positive integers, the negative integers, and the number zero. If we use I to denote the set of integers, we have

$$I = \{1, 2, 3, 4, \ldots\} \cup \{-1, -2, -3, -4, \ldots\} \cup \{0\}$$
$$= \{\ldots, -4, -3, -2, -1, 0, 1, 2, 3, 4, \ldots\}$$

4 *Rational Numbers* (Q): A *rational number* is any number that can be written as the ratio of two integers, where the denominator is different from zero. For example, 3, $2\frac{1}{2}$, and 43 percent are rational numbers since each can be written as the ratio of two integers, as follows:

$$3 = \tfrac{3}{1}$$
$$2\tfrac{1}{2} = \tfrac{5}{2}$$
$$43\% = \tfrac{43}{100}$$

More formally, if Q is used to denote the set of rational numbers, we have

$$Q = \left\{ q \mid q = \frac{a}{b},\, a \in I,\, b \in I,\, b \neq 0 \right\}$$

Elements of I can be expressed in the form a/b, $b \neq 0$. For example, $3 = \tfrac{3}{1}$, $-2 = -\tfrac{2}{1}$, and $0 = \tfrac{0}{1}$.

In general, for $a \in I$, $a = a/1$ and $a/1 \in Q$; hence, $I \subset Q$.

The relations among the sets N, I, and Q can be illustrated by a Venn diagram (Figure 1).

The rational numbers can also be described by investigating their decimal representations. For example, let us consider the specific rational

Figure 1

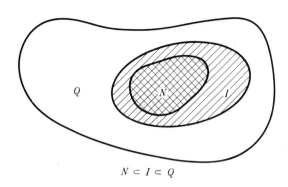

$N \subset I \subset Q$

number $\frac{3}{7}$. The number $\frac{3}{7}$ can be interpreted as $3 \div 7$, as shown in Figure 2. Notice the pattern. In each step within the division, the remainder must be either 0, 1, 2, 3, 4, 5, or 6. Therefore, if enough zeros are annexed after the decimal of the dividend (this does not affect the value of the dividend) and the division by 7 is performed more than seven times, one of the remainders must reoccur; but as soon as a remainder appears again (in this example it is 3), the digits in the quotient repeat. In this example

$$\frac{3}{7} = 0.428571428571\overline{428571}$$

where the bar identifies the block of digits that repeats infinitely often.

Figure 2

```
          0.428571428571 · · ·
        ┌─────────────────────
    7   │ 3.0000000
          2 8
          ───
            2 0
            1 4
            ───
              6 0
              5 6
              ───
                4 0
                3 5
                ───
                  5 0
                  4 9
                  ───
                    1 0
                     7
                    ───
                  ┌─────┐
                  │ 3 0 │
                  │ 2 8 │
                  │ ─── │
                  │  2  │
                  └─────┘
```

This concept can be generalized; for if a/b is a rational number, where a is an integer and b is a positive integer,

$$b \overline{)a.0000\ldots}$$

can be performed until a remainder repeats [there are only b remainders (Why?)]. When the remainder repeats, the digits in the quotient repeat. Hence, it follows that *every rational number can be represented by an eventually repeating decimal.* The converse of this statement also holds. That is, *every eventually repeating decimal represents a rational number.*

EXAMPLES

Show the repeating block for each of the following rational numbers.

1. $2 = \frac{2}{1} = 2.\overline{0}$
2. $\frac{2}{3} = 0.666\overline{6}$
3. $\frac{1,310}{99} = 13.23\overline{23}$

Rational numbers, such as 3 and $\frac{12}{4}$, in which the repeating block is the digit 0, are sometimes called *terminating decimals*.

In summary, a rational number is a number that can be considered from two different viewpoints: either as a ratio of two integers, or as a repeating decimal.

EXAMPLES

1. Show that each of the following numbers is a rational number by examining the decimal representation of the number.
   a) $\frac{6}{3}$ b) 16 percent
   c) $\frac{10}{3}$ d) $7.142845\overline{845}$

   SOLUTION

   a) $\frac{6}{3} = 2.0$ is a terminating decimal (the repeating block is the digit 0).
   b) 16 percent $= 0.\overline{16}$ is a terminating decimal.
   c) $\frac{10}{3} = 3.33\overline{3}$ has a repeating block (the digit 3).
   d) The number $7.142845\overline{845}$ is a rational number since $\overline{845}$ is a repeating block.

2. Express each of the following rational numbers as the ratio of two integers.
   a) 1.07   b) $0.777\overline{7}$
   c) $0.3131\overline{31}$ d) $1.3528\overline{28}$

   SOLUTION

   a) $1.07 = 1 + \frac{7}{100} = \frac{107}{100}$

b) Let $x = 0.77\overline{7}$. Multiplying both sides of the equation by 10 moves one of the repeating blocks of decimals to the left of the decimal point, so that

$$10x = 7.77\overline{7}$$

and

$$x = 0.77\overline{7}$$

Subtracting the corresponding sides of the equations, we have

$$9x = 7 \quad \text{or} \quad x = \tfrac{7}{9}$$

c) Let $x = 0.3131\overline{31}$; then
$$100x = 31.31\overline{31}$$
$$-x = -0.31\overline{31}$$
$$\overline{99x = 31}$$

so that

$$x = \tfrac{31}{99}$$

d) Let $x = 1.352828\overline{28}$; then
$$10{,}000x = 13{,}528.28\overline{28}$$
$$-100x = -135.28\overline{28}$$
$$\overline{9{,}900x = 13{,}393}$$

so that

$$x = \tfrac{13{,}393}{9{,}990}$$

5   *Irrational Numbers* (L): So far, we have seen that the set of rational numbers is the set of numbers represented by repeating decimals and that the set of repeating decimals represents rational numbers. But there are decimals that do not repeat, for example, the decimal 1.01001000100001 ..., where there is one more "0" after each "1" than there is before the "1." Also $\sqrt{2}$ is not rational, for it can be shown that $\sqrt{2}$ cannot be represented as a ratio of two integers. The numbers $1 + \sqrt{2}$, $3 - \sqrt{2}$, $3\sqrt{2}$, $3 + \sqrt{3}$, $\sqrt[3]{3}$, π and $\sqrt{5}$ are also *irrational numbers*.

6   *Real Numbers* (R): The set of *real numbers* is, in a sense, the set of all numbers that can be written as decimal numbers. Consequently, the set of real numbers consists of two sets of numbers—the rational numbers,

which represent repeating decimals, and the irrational numbers, which are represented by nonrepeating decimals. If we use R to denote the set of real numbers, we can express R by using set notation as follows:

$$R = Q \cup L \quad \text{(note that } Q \cap L = \emptyset\text{)}$$

where L is the set of irrational numbers (Figure 3).

Figure 3

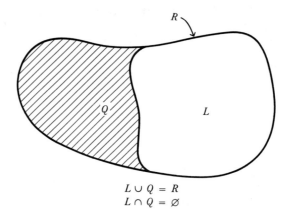

$L \cup Q = R$
$L \cap Q = \emptyset$

3.1 Polynomials

Arithmetic entails the study of the rules and techniques that govern the addition, subtraction, multiplication, and division of *real numbers*. In algebra we are concerned with generalizing the rules of arithmetic operations to enable us to add, subtract, multiply, and divide *expressions that represent real numbers*. Expressions of the form

$$5x + 7, \quad 3x - 9, \quad \frac{2x + 1}{5x - 4}, \quad \text{and} \quad -8x$$

are called *algebraic expressions* (where the symbol x is called a *variable*). Unless otherwise stated, we will assume henceforth that variables represent real numbers.

One particular type of algebraic expression that we will study extensively in algebra is the *polynomial*. Before defining polynomials it will be necessary to review the use of *exponents* to represent repeated multiplication.

$a \cdot a \cdot a \cdot a \cdot a$ can be written in shorthand notation as a^5, where a is called the *base* and 5 the exponent or *power*.

$$5 \cdot 5 \cdot 5 = 5^3$$
$$(xy)(xy)(xy)(xy) = (xy)^4$$

and

$$(2t)(2t) = 4t^2$$

are other examples of the use of the exponents. In general, if n is a positive integer,

$$\underbrace{a \cdot a \cdot a \cdots a}_{n \text{ factors}} = a^n$$

where a^n is called the nth *power of a.*

In Section 7 we will study in greater depth the algebra of exponents but for our purpose here the above definition suffices.

The algebraic expressions $4x$, $3x - 5$, $y^2 + 4y + 7$ and $-z^5 - 3z^3 + z$ are examples of polynomials in one variable.

In general, a *polynomial in one variable*, say x, is any algebraic expression formed from the real numbers and x by using only the operations of addition, subtraction, and multiplication. Those parts of the polynomials separated by either the "$+$" or "$-$" sign are called *terms* of the polynomial. If a polynomial contains only one term, it is called a *monomial*. A polynomial of two terms is called a *binomial;* a polynomial of three terms is called a *trinomial*. Thus, $4x$ is a monomial, $3x - 5$ is a binomial, and $y^2 + 4y + 7$ is a trinomial. An important feature of a polynomial in one variable is the highest power of the variable it contains. We call this number the *degree* of the polynomial. For example, the degree of $4x$ is 1, the degree of $y^2 + 4y + 7$ is 2, and the degree of $-z^5 - 3z^3 + z$ is 5. A nonzero real number, such as 8, is called *a polynomial of degree zero*. The real number zero is not assigned a degree, it is described as the *zero polynomial*.

We can extend the notion of polynomials to include algebraic expressions of several variables. For example, $3xy + 5$ is a polynomial of two variables, $x^2y^3z + x^3y^2z^4 - 4$ is a polynomial of three variables, and $xy + yzw - 3$ is a polynomial of four variables. The degree of a polynomial of several variables is the largest sum of exponents in any one term of the polynomial. For example, the degree of $3xy + 5$ is 2, the degree of $x^2y^3z + x^3y^2z^4 - 4$ is 9, and the degree of $xy + yzw - 3$ is 3.

The *coefficient* of any of the variables of a term in a polynomial is the product of the remaining factors. For example, in $3xyz$, the coefficient of yz is $3x$, the coefficient of z is $3xy$, the coefficient of $3z$ is xy, and so on. From now on, we will use the word "coefficient" to refer to the *numerical factor*, unless otherwise stated. In the above example, then, the coefficient of the term $3xyz$ is 3. The coefficient of y in the second term of the binomial $8x + 4xyz$ is $4xz$, whereas the coefficient of the second term is 4.

EXAMPLES

1 Determine the degree of each of the following polynomials.
   a) $z^2 + 1$
   b) $2x^4 - 3x^3 + 7x^5 - 5$
   c) $3x^3y - 2x^2y + 7xy^2 - y^3$

SOLUTION

a) 2, since the highest power of the variable z is 2.
b) 5, since the largest exponent of the variable x is 5.
c) For $3x^3y - 2x^2y + 7xy^2 - y^3$:
   first term, $3 + 1 = 4$
   second term, $2 + 1 = 3$
   third term, $1 + 2 = 3$
   fourth term, 3
   Since the highest sum of exponents is 4, the degree of the polynomial is 4.

2 Which of the following expressions is not a polynomial in one variable?
   a) $3x^4 - 2x^3 + \frac{1}{2}x + 1$ b) $2x + (1/x^3) + 2$

SOLUTION

a) $3x^4 - 2x^3 + \frac{1}{2}x + 1$ is a polynomial since it is formed by adding, subtracting, and multiplying real numbers and x.
b) $2x + (1/x^3) + 2$ is not a polynomial because the term $1/x^3$ does not comply with the definition of a polynomial in that it is formed by dividing 1 by x^3.

PROBLEM SET 2

1 Use N, I, Q, L, and R as defined in this section to indicate the set relation between each pair of the following sets.
   a) N, I b) N, Q
   c) N, L d) N, R
   e) I, Q f) I, L
   g) I, R h) Q, L
   i) Q, R j) L, R

2 Using N, Q, L, and R as defined in this section, describe each of the following sets of numbers by the use of a Venn diagram.
   a) $N \cup Q$ b) $L \cap Q$
   c) $I \cap Q$ d) $I \cap L$
   e) $Q \cup L$

3. Express each of the following rational numbers in decimal representation.
   a) $\frac{3}{8}$ b) $\frac{2}{9}$ c) $\frac{4}{13}$ d) $\frac{1}{7}$ e) $\frac{17}{9}$

4. Indicate which of the following real numbers are rational and which are irrational.
   a) $\dfrac{3 - \sqrt{2}}{4}$
   b) $\dfrac{\sqrt{5}}{5}$
   c) $\dfrac{\pi}{2}$
   d) $0.34567567\overline{567}$
   e) $1.1234567891011121314\ldots$
   f) $\dfrac{33}{1.2}$

5. Express each of the following rational numbers as a ratio of two integers.
   a) 2.05
   b) $0.499\overline{9}$
   c) $7.36\overline{262}$
   d) $5.33\overline{3}$
   e) $0.77\overline{7}$
   f) $0.464\overline{646}$

6. Use decimal representations to construct rational numbers between each of the following pairs of real numbers.
   a) 3.2 and 3.6
   b) $\sqrt{2}$ and $\sqrt{3}$
   c) 5.1 and 5.7
   d) 1.21 and 1.27

7. Identify the given polynomial as a monomial, binomial, trinomial, or other. Find the degree of the polynomial and list the coefficients.
   a) $x^2 - 5x + 6$
   b) $3x^2 - 5x + 17$
   c) $4x^5 - 21$
   d) $-x^4 - x^2 + 13$
   e) $7x^2$
   f) $3x^3 + 5x + 7$
   g) $2x^7 - 3$
   h) $-2x^4 - x$
   i) $4x^6 - 13$
   j) $-16x$
   k) -3
   l) -25

8. Which of the following are not polynomials? Indicate the degree of each expression that is a polynomial.
   a) $7x^2 + 13x^2y^3 + 9$
   b) $\dfrac{2x^3}{y^2 + 1}$
   c) $5x^2y + 16$
   d) $\dfrac{x^2y^2}{z + 13}$
   e) $\dfrac{5x^2y^3 + 1}{y^2}$
   f) $3x^2 + 5xyz - 1$
   g) $x + \dfrac{y^2}{5}$
   h) $6xyz + 13x - \frac{1}{2}$
   i) $x^2 - 5xy + 2z^2$
   j) $x^3 - 4xy - 3z$
   k) $3x^3 - 2x^2y^2 + 5$
   l) $3x^4 - 4xy + 5xz^4$
   m) $x^4 - 3x^2 + 5$
   n) $4x^2 - 3xy^2 + 5xy^2z$

4 Field Axioms for Real Numbers

The rules that govern the addition, subtraction, multiplication, and division of polynomials in algebra are the same as those for the arithmetic of real numbers. This is one reason for referring to algebra as generalized arithmetic.

Our purpose here is to *survey* the rules—*field axioms* for the arithmetic of real numbers—so that the operations of adding, subtracting, and multiplying polynomials can be introduced. The division of polynomials, however, will be delayed until Chapter 5.

We will begin by listing the *equality axioms* for the real numbers.

Suppose that a, b, and c are real numbers; then the equality, "$=$," relationship satisfies the following axioms.

AXIOM 1 REFLEXIVE LAW

$a = a$

AXIOM 2 SYMMETRIC LAW

If $a = b$, then $b = a$.

AXIOM 3 TRANSITIVE LAW

If $a = b$ and $b = c$, then $a = c$.

AXIOM 4 SUBSTITUTION PROPERTY

If $a = b$, then a can be substituted for b in any statement involving b without affecting the truthfulness of the statement. For example, if $a = b$ and $b + 3 = 10$, then $a + 3 = 10$. If a, b, and c represent real numbers, then the operations of addition and multiplication satisfy the following *field axioms*.

AXIOM 5 THE CLOSURE LAWS

i $a + b$ is a real number.
ii $a \cdot b$ is a real number.

AXIOM 6 THE COMMUTATIVE LAWS

i $a + b = b + a$
ii $a \cdot b = b \cdot a$

AXIOM 7 THE ASSOCIATIVE LAWS

i $(a + b) + c = a + (b + c)$

ii $(a \cdot b) \cdot c = a \cdot (b \cdot c)$

AXIOM 8 THE DISTRIBUTIVE LAWS

i $a \cdot (b + c) = a \cdot b + a \cdot c$
ii $(a + b) \cdot c = a \cdot c + b \cdot c$

AXIOM 9 THE IDENTITY ELEMENTS

i There exists a real number zero, denoted by 0, such that for any real number a,

$$a + 0 = 0 + a = a$$

ii There exists a real number one, denoted by 1, such that for any real number a,

$$a \cdot 1 = 1 \cdot a = a$$

AXIOM 10 THE INVERSE ELEMENTS

i For each real number a, there exists a real number, the *additive inverse of a*, denoted by $-a$ such that

$$a + (-a) = (-a) + a = 0$$

ii For each real number a, where $a \neq 0$, there exists a real number, the *multiplicative inverse or reciprocal*, denoted by $1/a$ such that

$$a \cdot \frac{1}{a} = \frac{1}{a} \cdot a = 1$$

Axiom 6 states that the order in which we add or multiply two real numbers does not affect the sum or product. Thus,

$$3 + 2 = 2 + 3 \quad \text{and} \quad 3 \times 2 = 2 \times 3$$

Axiom 7 asserts that three real numbers in a sum or in a product may be associated in either of two ways without altering the result. For example,

$$2 + 3 + 4 = (2 + 3) + 4 = 2 + (3 + 4)$$

and

$$2 \times 3 \times 4 = (2 \times 3) \times 4 = 2 \times (3 \times 4)$$

EXAMPLES

1 State the axiom that justifies each of the following equalities.
   a) $3 + 4 = 4 + 3$
   b) $(x)(y) + z = (y)(x) + z$
   c) $3(x + y) = 3x + 3y$
   d) $(x + y)(a + b) = (x + y)a + (x + y)b$
   e) $(x + y) + 0 = (x + y)$
   f) $(2 \cdot 3)5 = 2(3 \cdot 5)$

SOLUTION

a) $3 + 4 = 4 + 3$, commutative law of addition
b) $(x)(y) + z = (y)(x) + z$, commutative law of multiplication
c) $3(x + y) = 3x + 3y$, distributive law
d) $(x + y)(a + b) = (x + y)a + (x + y)b$, distributive law
e) $(x + y) + 0 = x + y$, identity element of addition
f) $(2 \cdot 3)5 = 2(3 \cdot 5)$, associative law of multiplication

2 Show that $x + [y + (z + w)] = [(x + y) + z] + w$

PROOF

$$x + [y + (z + w)] = (x + y) + (z + w) \quad \text{(associative law of addition)}$$
$$= [(x + y) + z] + w \quad \text{(why?)}$$

3 Show that $(x + y)(z + w) = xz + xw + yz + yw$.

PROOF

$$(x + y)(z + w) = x(z + w) + y(z + w) \quad \text{(why?)}$$
$$= xz + xw + yz + yw \quad \text{(distributive law)}$$

Notice that the field axioms do not involve the operations of subtraction or division. This is because these two operations are definable in terms of addition and multiplication.

The *difference* of two real numbers a and b, denoted by $a - b$, is defined by $a - b = a + (-b)$. That is, to subtract b from a, add the additive inverse of b to a. For example, to subtract 3 from 7, add -3 to 7, thus $7 - 3 = 7 + (-3) = 4$.

The *quotient* of real numbers a and b, for $b \neq 0$, denoted by $a \div b$ or a/b, is defined by $a/b = a \cdot 1/b$. That is, to *divide* a by b, where $b \neq 0$, multiply a by the multiplicative inverse of b. For example, $12 \div 4 = 12 \cdot \frac{1}{4} = 3$.

The field axioms can be used to prove some familiar properties of arithmetic, as illustrated in the following examples.

EXAMPLES

1 Use the field axioms to prove that if a, b, and c are real numbers and

a) if $a + c = b + c$, then $a = b$.
b) if $ac = bc$ and $c \neq 0$, then $a = b$.

(These are referred to as the *cancellation laws* of addition and multiplication, respectively.)

PROOF. Part a) can be verified as follows. Since

$$a + c = b + c$$

we have, after adding $-c$ to each side,

$$a + c + (-c) = b + c + (-c)$$

so that $a + (c + (-c)) = b + (c + (-c))$, or

$$a + 0 = b + 0 \qquad \text{(additive inverse)}$$

Therefore,

$$a = b \qquad \text{(identity element)}$$

Now we shall verify part b). Since $ac = bc$ and c has inverse $1/c$, we have, after multiplying each side by $1/c$,

$$ac \frac{1}{c} = bc \frac{1}{c}$$

so that

$$a \left(c \cdot \frac{1}{c} \right) = b \left(c \cdot \frac{1}{c} \right) \qquad \text{(associative law)}$$

or

$$a \cdot 1 = b \cdot 1 \qquad \text{(multiplicative inverse)}$$

Therefore,

$$a = b \qquad \text{(identity element)}$$

2 Prove that $a \cdot 0 = 0 \cdot a = 0$, where a is a real number.

PROOF. We can write $a \cdot 0 = a \cdot (0 + 0) = a \cdot 0 + a \cdot 0$ (distributive property). But also $a \cdot 0 = a \cdot 0 + 0$ (identity element), so that $a \cdot 0 + 0 = a \cdot 0 + a \cdot 0$. Hence, using the cancellation law, we have $a \cdot 0 = 0$. Finally, by the commutative law, $a \cdot 0 = 0 \cdot a$, so that $a \cdot 0 = 0 \cdot a = 0$.

3. Prove that $-(-a) = a$ for any real number a.

PROOF. Now $(-a) + [-(-a)] = 0$ (inverse element). Also,

$$(-a) + a = 0 \quad \text{(inverse element)}$$

so that

$$(-a) + [-(-a)] = (-a) + a$$

By the cancellation law

$$-(-a) = a$$

4.1 Addition, Subtraction, and Multiplication of Polynomials

The field axioms that serve as the basis for the addition, multiplication, subtraction and division of real numbers also apply to polynomials. This means that the four operations on polynomials are "governed" by the same axioms. We will now use the axioms to review the operations of addition, subtraction, and multiplication of polynomials (division will be delayed until Chapter 5).

Suppose that we are to add the polynomials

$$(3x^3 + 2x^2 + 4) + (5x^2 + 7x^3 + 8)$$

By using the commutative and associative laws of addition we can arrange and regroup the terms as follows:

$$(3x^3 + 2x^2 + 4) + (5x^2 + 7x^3 + 8)$$
$$= (3x^3 + 7x^3) + (2x^2 + 5x^2) + (4 + 8)$$

(By the distributive law, we can simplify the right-hand side.)

$$= (3 + 7)x^3 + (2 + 5)x^2 + (4 + 8)$$
$$= 10x^3 + 7x^2 + 12$$

Often the rearrangement and reordering of the terms is done by using the "vertical scheme."

$$
\begin{array}{r}
\; 3x^3 + 2x^2 + 4 \\
(+) \; 7x^3 + 5x^2 + 8 \\
\hline
10x^3 + 7x^2 + 12
\end{array}
$$

Let us now consider an example of polynomial subtraction:

$$(3x^3 - 7x^2 + 4) - (5x + 8x^2 - 7)$$

SECTION 4 FIELD AXIOMS FOR REAL NUMBERS

We have

$$(3x^3 - 7x^2 + 4) - (5x + 8x^2 - 7)$$
$$= 3x^3 + [(-7x^2) - 8x^2] - 5x + [4 - (-7)]$$
$$= 3x^3 + (-7 - 8)x^2 - 5x + [4 - (-7)]$$
$$= 3x^3 - 15x^2 - 5x + 11$$

Using the vertical scheme,

$$\begin{array}{r} 3x^3 - 7x^2 + 4 \\ (-) + 8x^2 + 5x - 7 \\ \hline 3x^3 - 15x^2 - 5x + 11 \end{array}$$

EXAMPLES

1 Perform the following additions using the vertical scheme.
   a) $(4x^3 + 7x - 3) + (2 + x^3 - x)$
   b) $(7 - 3x^2 + x) + (2x^2 + 8 + 13x) + (3x - 7 + 2x^2)$

SOLUTION

a)
$$\begin{array}{r} 4x^3 + 7x - 3 \\ (+) x^3 - x + 2 \\ \hline 5x^3 + 6x - 1 \end{array}$$

b)
$$\begin{array}{r} -3x^2 + x + 7 \\ 2x^2 + 13x + 8 \\ (+) 2x^2 + 3x - 7 \\ \hline x^2 + 17x + 8 \end{array}$$

2 Perform the following subtractions.
   a) $(3x - 7) - (2x + 1)$
   b) $(4x + 7x^3) - (2x - x^3 + 5)$

SOLUTION

a) $(3x - 7) - (2x + 1) = (3x - 7) + (-2x - 1)$
$$= x - 8$$
b) $(4x + 7x^3) - (2x - x^3 + 5) = (4x + 7x^3) + (-2x + x^3 - 5)$
$$= 2x + 8x^3 - 5$$

3 Subtract $x^2 - 5x + 8$ from the sum of $x^2 - 3x + 4$ and $2x^2 - x + 2$.

SOLUTION. First,

$$\begin{array}{r} x^2 - 3x + 4 \\ (+) 2x^2 - x + 2 \\ \hline 3x^2 - 4x + 6 \end{array}$$

Then

$$\begin{array}{r} 3x^2 - 4x + 6 \\ (-) \quad x^2 - 5x + 8 \\ \hline 2x^2 + x - 2 \end{array}$$

In order to multiply two polynomials, we use the distributive law and combine similar terms. For example, to find the products

1. $4xy(2x^2 - 3xy + 4y^2)$
2. $(x + y)(2x - 3y)$
3. $(x^2 + 2xy - y^2)(x + xy + y)$

we use the distributive law and combine like terms to get

1. $4xy(2x^2 - 3xy + 4y^2) = (4xy)(2x^2) + (4xy)(-3xy) + (4xy)(4y^2)$
$= 8x^3y - 12x^2y^2 + 16xy^3$

2. $(x + y)(2x - 3y) = (x + y)2x + (x + y)(-3y)$
$= 2x^2 + 2xy - 3xy - 3y^2$
$= 2x^2 - xy - 3y^2$

3. $(x^2 + 2xy - y^2)(x + xy + y)$
$= (x^2 + 2xy - y^2)x + (x^2 + 2xy - y^2)xy + (x^2 + 2xy - y^2)y$
$= x^3 + 2x^2y - xy^2 + x^3y + 2x^2y^2 - xy^3 + x^2y + 2xy^2 - y^3$
$= x^3 + 3x^2y + xy^2 + x^3y + 2x^2y^2 - xy^3 - y^3$

Since in many multiplications involving polynomials the actual computations can become quite tedious, a device to help perform the computations and at the same time simplify the work is desirable. One such device, the *vertical scheme*, will be illustrated.

EXAMPLES

1. Find the product of $2x - 1$ and $x^3 + x^2 - 2x - 1$ by using the vertical scheme.

SOLUTION

$$\begin{array}{l} x^3 + x^2 - 2x - 1 \\ 2x - 1 \\ \hline 2x^4 + 2x^3 - 4x^2 - 2x \qquad [(2x)(x^3 + x^2 - 2x - 1)] \\ \quad - x^3 - x^2 + 2x + 1 \qquad [(-1)(x^3 + x^2 - 2x - 1)] \\ \hline 2x^4 + x^3 - 5x^2 \qquad + 1 \qquad \text{(product)} \end{array}$$

2. Use the vertical scheme illustrated above to multiply each of the following.

a) $(x + 4)(2x - 3)$
b) $(3x^3 - 5xy + 7y^2)(x - y)$

SOLUTION

a)
$$
\begin{array}{r}
x + 4 \\
2x - 3 \\
\hline
2x^2 + 8x \\
 - 3x - 12 \\
\hline
2x^2 + 5x - 12
\end{array}
$$

b)
$$
\begin{array}{r}
3x^3 - 5xy + 7y^2 \\
x - y \\
\hline
3x^4 - 5x^2y + 7xy^2 \\
5xy^2 - 3x^3y - 7y^3 \\
\hline
3x^4 - 5x^2y + 12xy^2 - 3x^3y - 7y^3
\end{array}
$$

Certain products occur so frequently in algebra that they deserve special recognition. These products are listed below.

Special Product 1

The square of the binomial $a + b$,

$$
\begin{aligned}
(a + b)^2 &= (a + b)(a + b) \\
&= (a + b)a + (a + b)b \\
&= a^2 + ba + ab + b^2 \\
\boxed{(a + b)^2 &= a^2 + 2ab + b^2}
\end{aligned}
$$

For example,

$$
\begin{aligned}
(x + 2y)^2 &= (x)^2 + 2(x)(2y) + (2y)^2 \\
&= x^2 + 4xy + 4y^2
\end{aligned}
$$

Special Product 2

The square of the binomial $a - b$. We can find this product by applying special product 1, since $a - b = a + (-b)$. Hence,

$$
\begin{aligned}
(a - b)^2 &= [a + (-b)]^2 \\
&= a^2 + 2a(-b) + (-b)^2 \\
\boxed{(a - b)^2 &= a^2 - 2ab + b^2}
\end{aligned}
$$

For example,

$$
\begin{aligned}
(2x - 5)^2 &= (2x)^2 - 2(2x)(5) + 5^2 \\
&= 4x^2 - 20x + 25
\end{aligned}
$$

Special Product 3

$$(a + b)(a - b)$$

$$\begin{aligned}
(a + b)(a - b) &= (a + b)a + (a + b)(-b) \\
&= a^2 + ba + a(-b) + b(-b) \\
&= a^2 + ab - ab - b^2
\end{aligned}$$

$$\boxed{(a + b)(a - b) = a^2 - b^2}$$

For example,

$$\begin{aligned}
(2x + 3y)(2x - 3y) &= (2x)^2 - (3y)^2 \\
&= 4x^2 - 9y^2
\end{aligned}$$

Special Product 4

$$(a + b)(a^2 - ab + b^2)$$

$$\begin{aligned}
(a + b)(a^2 - ab + b^2) &= (a + b)a^2 + (a + b)(-ab) + (a + b)b^2 \\
&= a^3 + a^2b - a^2b - ab^2 + ab^2 + b^3
\end{aligned}$$

$$\boxed{(a + b)(a^2 - ab + b^2) = a^3 + b^3}$$

For example,

$$\begin{aligned}
(2x + y)(4x^2 - 2xy + y^2) &= (2x)^3 + y^3 \\
&= 8x^3 + y^3
\end{aligned}$$

Special Product 5

$(a - b)(a^2 + ab + b^2)$. By replacing b by $-b$ in special product 4, we have

$$[a + (-b)][a^2 - a(-b) + (-b)^2] = a^3 + (-b)^3$$

$$\boxed{(a - b)(a^2 + ab + b^2) = a^3 - b^3}$$

For example,

$$\begin{aligned}
(3x - 2y)(9x^2 + 6xy + 4y^2) &= (3x)^3 - (2y)^3 \\
&= 27x^3 - 8y^3
\end{aligned}$$

Later these formulas will be used "from the other side" to see, for instance, that $a^3 + b^3$ may be factored into $a + b$ times $a^2 - ab + b^2$.

We will now conclude this section by examining other examples to see how the substitution axiom, together with some known products, can be used to find other products.

EXAMPLES

In each of the following expressions determine the product in part a, and then use the result to find the product in part b.

1 a) $(a + b)^3$ b) $(w + 2z)^3$

SOLUTION

a) $(a + b)^3 = (a + b)(a + b)^2$
$= (a + b)(a^2 + 2ab + b^2)$
$= a(a^2 + 2ab + b^2) + b(a^2 + 2ab + b^2)$
$= a^3 + 2a^2b + ab^2 + a^2b + 2ab^2 + b^3$

That is, we have,

Special Product 6

$$\boxed{(a + b)^3 = a^3 + 3a^2b + 3ab^2 + b^3}$$

b) If $a = w$ and $b = 2z$, we get
$(w + 2z)^3 = w^3 + 3w^2(2z) + 3w(2z)^2 + (2z)^3$
$= w^3 + 6w^2z + 3w(4z^2) + 8z^3$
$= w^3 + 6w^2z + 12wz^2 + 8z^3$

2 a) $(a - b)^3$ b) $(3w - 2u)^3$

SOLUTION

a) By replacing b in Example 1a by $-b$, we have

$$[a + (-b)]^3 = a^3 + 3a^2(-b) + 3a(-b)^2 + (-b)^3$$

That is, we have,

Special Product 7

$$\boxed{(a - b)^3 = a^3 - 3a^2b + 3ab^2 - b^3}$$

b) After substituting $3w$ for a and $2u$ for b, we get
$(3w - 2u)^3 = (3w)^3 - 3(3w)^2(2u) + 3(3w)(2u)^2 - (2u)^3$
$= 27w^3 - 3(9w^2)(2u) + 3(3w)(4u^2) - 8u^3$
$= 27w^3 - 54w^2u + 36wu^2 - 8u^3$

PROBLEM SET 3

1 Justify the following statements by giving the appropriate axiom or property. Assume that all letters represent real numbers.
   a) $x + 5$ is a real number.
   b) $a + (b + 3) = (a + b) + 3$
   c) $3 \cdot x = x \cdot 3$

d) $a(b + c)$ is a real number.
e) $6 + (-6) = 0$
f) $2(xy) = (2x)y$
g) $1 \cdot 4 = 4$
h) $5 + 0 = 5$
i) $0 \cdot x = 0$
j) If $2x = 2 \cdot 3$, then $x = 3$.

2 Prove each of the following properties, where a and b are real numbers.
a) If $a \cdot b = 0$, then $a = 0$ or $b = 0$, or both.
b) $(-a)b = a(-b) = -(ab)$
c) $(-a)(-b) = ab$
d) If $a + b = 0$, then $a = -b$.
e) $(-1) \cdot a = -a$

3 Perform the indicated operations for each of the following expressions.
a) $(3x^2 + 5x - 3) + (-2x^2 - 7x + 17)$
b) $(3x^2 - 7x) - (-3x^2 + 5x - 3)$
c) $-(-5x^2 + 4x + 9)$
d) $(x^4 - 3x^3 - x^2 - 1) - (-2x^4 + 2x^3 + x^2 + 2)$
e) $(5x^4 - 3x^3 - 5x^2 - x - 1) + (1 + 2x + x^2 + x^3 - x^4)$
f) $(x^2 - 2x + 3) + 2x$
g) $(5x^3 - 5x + 4) - (4x^3 - 2x - 3)$
h) $(10x^2 + 29x - 21) + (5x^2 - 3x + 7)$
i) $(x^4 + 5x^2 + 10) + (3x^4 - 7x^2 + 17)$
j) $(1 - x - 3x^2 - x^5) - (1 - 2x + x^2)$
k) $(y^4 + 4by^3 + 2b^2y^2) + (3y^4 - 6by^3 - 13b^2y^2)$
l) $(8x^2 + 1 - 4x) - (4x - 13 - 6x^2)$
m) $(x^2 + 5x - 16) + (3x^2 + 7x + 6) - (13x^2 + 7)$
n) $(8x^2y - 12xy^2 + 6xy) - (8x^2y + 10xy^2 + 7xy)$
o) $x^2 - 2x - 12 - 3(x^2 - x + 1)$
p) $(4x^2y^2 - 3xy + 7xy^2) - (5x^2y^2 + 5xy - 13xy^2)$

4 Determine the following products.
a) $(x - y)(2x^2y)$
b) $(x - 5)(x + 5)$
c) $(x + y)(-xy^2)$
d) $(x - 3)(2x + 3)$
e) $(5x + 3)(7x + 2)$
f) $(z + 6)(z^2 - 1)$
g) $(3x - 3)(4x - 4)$
h) $(x + \frac{1}{4})(x - \frac{1}{4})$
i) $(2x + 1)(-4 + 3x)$
j) $(3x + 5)(3x - 5)$
k) $(x + y + 3)(2x - y + 4)$
l) $(6x + 5)(2x + 7)$
m) $(3x - 1)(2 - x)(3x + 1)$
n) $[2x - (a + b)][2x + (a + b)]$
o) $(3x + x^2y - y^2)(x^2 - xy + y)$
p) $(3x + 17)(3x + 1)$
q) $(7 + x)(7 - x)(-7 + x)(-7 - x)$
r) $(5x - 23)(-x + 2)$
s) $(x^2 - y^2)(x - y)^2$
t) $(x + 2 + y)(x + 2 - y)$
u) $(3x + 5)(x - 1)(2x + 1)$

v) $(4x + 5)(4x - 1)3y^2$
w) $(7x + 3y)(9x + 5y)(3x + 1)$
x) $(6x^2 + 5y)(3x + 7y)$
y) $(x - 1)(x^2 + x - 2)$
z) $(3x + 1)(x - 1)^2$

5 Use the special products to find the product for each of the following.

a) $(3x + 5y)^2$
b) $\left(\dfrac{x}{2} - \dfrac{y}{3}\right)^2$

c) $(-2x - 4)^2$
d) $\left(\dfrac{x}{y} - \dfrac{y}{x}\right)^2$

e) $(4 - x)(4 + x)$
f) $(x^{n+1} - 6)(x^{n+1} + 6)$

g) $(x^n - 3)(x^{2n} + 3x^n + 9)$
h) $(3 - w - y - z)(3 - w + y + z)$

i) $(x - 2y)(x^2 + 2xy + 4y^2)$
j) $(x + 5y)(x^2 - 5xy + 25y^2)$

k) $(3 + x)(9 - 3x + x^2)$
l) $(2x - 1)(4x^2 + 2x + 1)$

m) $(x + 5y)^3$
n) $(2x - 2y)^3$

o) $(x + y + z)^3$
p) $(x - 3y)^3$

q) $(3x - 5y)^3$
r) $(x + 2y + z)^3$

5 Algebraic Fractions

Reviewing the methods for adding, subtracting, and multiplying polynomials, we may see that the result is always another polynomial. That is, the set of polynomials is closed under the operations of addition, subtraction, and multiplication. Hence, the sum, difference, and product of two polynomials always results in a polynomial. However, as is the case for integers, the set of polynomials is not closed under the operation of division. For example,

$$\frac{2x^2 + 1}{x} = 2x + \frac{1}{x}$$

is not a polynomial. In order to have a set containing the polynomials that is closed under division, it becomes necessary to introduce algebraic expressions of the form P/Q, $Q \neq 0$, where P and Q are polynomials. Since the set of algebraic expressions, which is closed under division, is similar in form to that of the set of rational numbers, they are called *rational expressions*. For example,

$$\frac{x^2 - 5x + 1}{x^2 + 7} \quad \text{and} \quad \frac{x^2 + 1}{3x + 2}$$

are rational expressions. As in arithmetic, rational expressions can be converted to simplified equivalent forms called *reduced fractions*. The processing of reducing fractions is based upon factoring both the numerator and denominator to find common factors.

For example, the fraction $\frac{12}{16}$ can be reduced as follows:

$$\frac{12}{16} = \frac{3 \cdot 4}{4 \cdot 4} = \frac{3}{4}$$

Here the key step in the process is the factorization of both 12 and 16 to find the common factor of 4.

Consequently, before we can apply this principle of reducing arithmetic fractions to reducing rational expressions, it is necessary to investigate methods of factoring polynomials in order to find common factors.

5.1 Factoring Polynomials

If a polynomial is written as a product of other polynomials, then each of the latter polynomials is called a *factor* of the original polynomial. The procedure of finding such a product is called *factoring*. For example, since $x^2 - 16 = (x - 4)(x + 4)$, then $x - 4$ and $x + 4$ are factors of $x^2 - 16$. The process of factoring an algebraic expression is similar to that of finding the factors of a composite number. This process is frequently performed by reversing the operation of multiplication. In fact, we shall have to depend on our experience with multiplication in order to recognize factors.

The simplest type of factoring polynomials is one in which each term of the polynomial contains a *common factor*. For example,

$$3x + 5x^2 = x(3 + 5x)$$
$$15x^3 + 4x^2y - x^2 = x^2(15x + 4y - 1)$$
$$25xy - 15x = 5x(5y - 3)$$
$$7xy + 4x^2y^2 + x^3y^3 = xy(7 + 4xy + x^2y^2)$$

The polynomial $x + 1$ is said to be *prime* even though it can be expressed as $1 \cdot (x + 1)$ or $-[-(x + 1)]$. Likewise, the polynomials $x^2 + 5$, $x^2 + 1$, and $x^2 + 2$ are said to be primes. A prime polynomial is a polynomial that only has 1, -1, and itself as its factors in the real number system.

To *factor completely* a polynomial means to express it as a product of its prime factors. Thus, $x^2 - 3x + 2 = (x - 1)(x - 2)$ is factored completely since $x - 1$ and $x - 2$ are prime. If we have a polynomial with rational coefficients such as $25x^2 - \frac{4}{9}$, then the factors $5x + \frac{2}{3}$ and $5x - \frac{2}{3}$ have rational coefficients. If the polynomial has integers as coefficients, then the factors have integers as coefficients. For example, $x^2 + 2x - 8 = (x - 2)(x + 4)$.

Since factoring will be performed largely from experience with multiplication, it is advisable to become familiar with the following products, which were developed in Section 4.1.

1. $a^2 + 2ab + b^2 = (a + b)^2$ PERFECT SQUARE TRINOMIAL
2. $a^2 - 2ab + b^2 = (a - b)^2$
3. $a^2 - b^2 = (a + b)(a - b)$
4. $a^3 + b^3 = (a + b)(a^2 - ab + b^2)$
5. $a^3 - b^3 = (a - b)(a^2 + ab + b^2)$
6. $a^3 + 3a^2b + 3ab^2 + b^3 = (a + b)^3$
7. $a^3 - 3a^2b + 3ab^2 - b^3 = (a - b)^3$

For example, we can use product 1 to perform the factoring

$$x^2 + 6x + 9 = x^2 + 2(3)x + 3^2 = (x + 3)^2$$

Other types of factoring will be considered in the following examples.

EXAMPLES

1. Use the product forms above to factor each of the following polynomials completely.
   a) $x^2 + 4x + 4$
   b) $16x^2 - 25y^2$
   c) $9x^4 - 49y^2$
   d) $a^3b^3 + 27$
   e) $16x^4 - 1$
   f) $4x^2 - 4xy + y^2$
   g) $27 - x^3$
   h) $x^3 - 6x^2 + 12x - 8$
   i) $27 + 27x + 9x^2 + x^3$

   SOLUTION

   a) Using product 1 with $a = x$ and $b = 2$, we have

   $$\begin{aligned} x^2 + 4x + 4 &= x^2 + 2(2)(x) + (2)^2 \\ &= (x + 2)^2 \end{aligned}$$

   b) By product 3,

   $$\begin{aligned} 16x^2 - 25y^2 &= (4x)^2 - (5y)^2 \\ &= (4x + 5y)(4x - 5y) \end{aligned}$$

   c) Again we use product 3 to get

   $$\begin{aligned} 9x^4 - 49y^2 &= (3x^2)^2 - (7y)^2 \\ &= (3x^2 + 7y)(3x^2 - 7y) \end{aligned}$$

d) By product 4, with a replaced by ab and b by 3, we have

$$a^3b^3 + 27 = (ab + 3)(a^2b^2 - 3ab + 9)$$

e) By product 3,

$$16x^4 - 1 = (4x^2)^2 - 1^2$$
$$= (4x^2 - 1)(4x^2 + 1)$$

and, again by product 3,

$$4x^2 - 1 = (2x + 1)(2x - 1)$$

so that

$$16x^4 - 1 = (2x + 1)(2x - 1)(4x^2 + 1)$$

f) $4x^2 - 4xy + y^2 = (2x)^2 - 2(2x)y + y^2$
$= (2x - y)^2$ (product 2)

g) $27 - x^3 = 3^3 - x^3$
$= (3 - x)(9 + 3x + x^2)$ (product 5)

h) By product 7, with $a = x$ and $b = 2$, we have

$$x^3 - 6x^2 + 12x - 8 = x^3 - 3x^2(2) + 3x(2)^2 - 2^3$$
$$= (x - 2)^3$$

i) Using product 6, with $a = 3$ and $b = x$, we get

$$27 + 27x + 9x^2 + x^3 = (3)^3 + 3(3)^2x + 3(3)x^2 + x^3$$
$$= (3 + x)^3$$

2 Factor each of the following completely.
  a) $(x - y)^2 - (a - b)^2$
  b) $a^2 - 2ab + b^2 - 4x^2 + 12xy - 9y^2$
  c) $(a - b)^3 - (c - d)^3$
  d) $x^4 + 2x^2y^2 + 9y^4$
  e) $4x^4 + 1$

SOLUTION

a) By product 3,

$$(x - y)^2 - (a - b)^2 = [(x - y) + (a - b)][(x - y) - (a - b)]$$

b) $a^2 - 2ab + b^2 - 4x^2 + 12xy - 9y^2$
$= (a^2 - 2ab + b^2) - (4x^2 - 12xy + 9y^2)$
$= (a - b)^2 - (2x - 3y)^2$
$= [(a - b) + (2x - 3y)][(a - b) - (2x - 3y)]$ (Why?)

c) By product 5,

$$(a - b)^3 - (c - d)^3$$
$$= [(a - b) - (c - d)][(a - b)^2 + (a - b)(c - d) + (c - d)^2]$$

d) The trinomial would be a perfect square if the middle term were $6x^2y^2$. Thus, we have

$$\begin{aligned}
x^4 + 2x^2y^2 + 9y^4 &= x^4 + 2x^2y^2 + 9y^4 + 4x^2y^2 - 4x^2y^2 \\
&= x^4 + 6x^2y^2 + 9y^4 - 4x^2y^2 \\
&= (x^2 + 3y^2)^2 - 4x^2y^2 \quad \text{(product 1)} \\
&= [(x^2 + 3y^2) - 2xy][(x^2 + 3y^2) + 2xy] \\
&\quad \text{(product 3)}
\end{aligned}$$

e) Since $(2x^2 + 1)^2 = 4x^4 + 4x^2 + 1$ and $4x^4 + 1 = 4x^4 + 4x^2 + 1 - 4x^2$, we have $4x^4 + 1 = (2x^2 + 1)^2 - (2x)^2$. Using the formula for factoring the difference of two squares, which is product 3, that is,

$$a^2 - b^2 = (a - b)(a + b)$$

we have

$$4x^4 + 1 = (2x^2 + 1 - 2x)(2x^2 + 1 + 2x)$$

Suppose that we are given a polynomial of the form $ax^2 + bx + c$, where a, b, and c are integers; then the factors, if they exist must be the product of two first-degree polynomials. As before, our skill in this process depends on how well we can multiply first-degree polynomials. For example, $3x^2 + x = x(3x + 1)$ is a simple factorization, whereas $3x^2 + x - 2 = (3x - 2)(x + 1)$ is a factorization which requires more insight into the combination of products of first-degree polynomials which might yield $3x^2 + x - 2$. We notice that the first term of the quadratic expression $3x^2$ can be obtained as the product of x and $3x$, which is the only way to obtain the coefficient 3 as a product of positive integers. Thus, our factors are $(x \quad)(3x \quad)$, where we still have to determine the constant terms of the linear factors. The only ways we can express the term -2 of the quadratic as a product of integers are -1 times 2 or 1 times -2. Hence, a list of possibilities for the factorization of $3x^2 + x - 2$ as the product of linear factors is as follows:

	Possible combinations	Products
1	$(3x - 1)(x + 2)$	$3x^2 + 5x - 2$
2	$(3x + 1)(x - 2)$	$3x^2 - 5x - 2$
3	$(3x - 2)(x + 1)$	$3x^2 + x - 2$
4	$(3x + 2)(x - 1)$	$3x^2 - x - 2$

We find the correct factorization of $3x^2 + x - 2$ from the table. That is,

$$(3x - 2)(x + 1) = 3x^2 + x - 2$$

This example illustrates that the method for factoring second-degree polynomials such as $3x^2 + x - 2$ involves testing various combinations of the *possible* factors.

Notice that the first term of the expression $3x^2 + x - 2$, namely $3x^2$, is the product of the first terms of the linear factors, and the last term of the expression $3x^2 + x - 2$, namely -2, is the product of the last terms of the linear factors. Finally, notice that the "middle term" of the quadratic expression namely x, is the sum of the cross products $(3x)(1)$ and $(-2)(x)$.

EXAMPLES

Factor each of the following polynomials completely.

1 $6x^2 + 5x - 14$

SOLUTION. The terms whose product is $6x^2$ are $6x$ and x, or $2x$ and $3x$. The factors of -14 are -14 and 1, or 14 and -1, or 7 and -2, or -7 and 2. Testing various possibilities, we find that

$$6x^2 + 5x - 14 = (6x - 7)(x + 2)$$

2 $6x^2 + x - 2$

SOLUTION. We can gain some insight into the most likely factors by comparing the coefficient of the x term with the possible factors of the coefficient of the x^2 term and the constant term. The factors of 6 are 6 and 1, or 3 and 2 (considering only positive factors here), and the factors of -2 are -2 and 1, or 2 and -1. Since the coefficient of the x term is 1, this suggests the first trial of 3 and 2 as factors of 6. (Why?) That is,

$$6x^2 + x - 2 = (3x + 2)(2x - 1)$$

By testing possible factors mentally, we can obtain the correct factorization rather quickly. We need only to test various possible choices of factors of the x^2 term and the constant term to see if their cross products will yield the x term.

EXAMPLE

Factor $12x^2 - 29x + 15$ completely.

SOLUTION. Factors of 12 are 12 and 1 or 6 and 2 or 4 and 3. Factors of 15 are 15 and 1 or 5 and 3 or -15 and -1 or -5 and -3. We now test mentally the cross products of these factors to find the coefficient -29. Since $4(-5) = -20$, $3(-3) = -9$, and $(-20) + (-9) = -29$, we have the following:

$$(4x - 3)(3x - 5) = 12x^2 - 29x + 15$$

5.2 Reducing Fractions

We say that two rational expressions (or fractions) are *equivalent* if they give equal real numbers for each assignment of values to their variables, except when such an assignment gives an undefined value to either (or both) expressions. For example, $x/2x = \frac{1}{2}$ except when $x = 0$. If two rational expressions are equivalent it will be denoted by writing an *equals sign* between them.

We shall assume throughout this text that the variables in any fraction may not be assigned values that will involve a division by zero. Accordingly, we can develop a method for determining equivalent rational expressions by considering equivalent fractions of the form $a/b = c/d$, where a, b, c, and d are real numbers and b, $d \neq 0$. We have from arithmetic that

$$\frac{a}{b} = \frac{c}{d} \quad \text{if and only if} \quad ad = bc$$

For example, $\frac{2}{3} = \frac{4}{6}$, since $(2)(6) = (3)(4)$. This property of equivalence also holds for rational expressions. That is, if P, Q, R, and S are polynomials, then

$$\frac{P}{Q} = \frac{R}{S} \quad \text{if and only if} \quad PS = QR$$

For example, $3x/6x^2 = 1/2x$, since $(3x)(2x) = (6x^2)(1)$; and

$$\frac{x + 3}{x^2 + 5x + 6} = \frac{1}{x + 2}$$

since $(x + 3)(x + 2) = (x^2 + 5x + 6)(1)$. On the other hand, if $3x = 5y$, then $\frac{3}{5} = y/x$ and $\frac{5}{3} = x/y$. (Why?)

We can use the concept of equal fractions to verify that rational expressions can be reduced to lowest terms in the same way that rational numbers are reduced. For example, $\frac{2}{4} = \frac{1}{2}$, since $2 \cdot 2 = 4 \cdot 1$. Also, since $\frac{2}{4} = (2 \cdot 1)/(2 \cdot 2)$, we can reduce this fraction by eliminating the factors common to both the numerator and the denominator. Thus, $\frac{2}{4} = (2 \cdot 1)/(2 \cdot 2) = \frac{1}{2}$. This approach can be applied to rational expressions. For example,

$$\frac{x^3 - xy^2}{x^3 - 2x^2y + xy^2} = \frac{x(x-y)(x+y)}{x(x-y)(x-y)} = \frac{x+y}{x-y}$$

The equality is verified by the following products:

$$(x^3 - xy^2)(x - y) = x^4 - x^3y - x^2y^2 + xy^3$$
$$(x^3 - 2x^2y + xy^2)(x + y) = x^4 - x^3y - x^2y^2 + xy^3$$

This process of reducing fractions is known as the *fundamental principle of fractions*, which is stated as follows. If a, b, and k are real numbers, where $b \neq 0$ and $k \neq 0$, then

$$\frac{a}{b} = \frac{ak}{bk}$$

EXAMPLES

Use the fundamental principle of fractions in each of the following problems.

1 Reduce each of the following fractions to an equivalent form.

a) $\frac{36}{44}$

b) $\frac{28x^3y}{21xy^2}$

c) $\frac{x^2 - y^2}{x + y}$

d) $\frac{30x^2y^2}{25x^2y - 20xy^2}$

SOLUTION

a) $\frac{36}{44} = \frac{9 \cdot 4}{11 \cdot 4} = \frac{9}{11}$

b) $\frac{28x^3y}{21xy^2} = \frac{(7xy)(4x^2)}{(7xy)(3y)} = \frac{4x^2}{3y}$

c) $\frac{x^2 - y^2}{x + y} = \frac{(x-y)(x+y)}{1 \cdot (x+y)} = \frac{x-y}{1} = x - y$

d) $\frac{30x^2y^2}{25x^2y - 20xy^2} = \frac{(5xy)(6xy)}{5xy(5x - 4y)} = \frac{6xy}{5x - 4y}$

2 Reduce the rational expression $\frac{x^2 - 4}{x^2 + x - 6}$ to an equivalent fraction.

SOLUTION

$$\frac{x^2 - 4}{x^2 + x - 6} = \frac{(x-2)(x+2)}{(x-2)(x+3)} = \frac{x+2}{x+3}$$

3 Convert $y/(2y - 1)$ to a fraction with a denominator of $4y^2 - 1$.

SOLUTION. Since $4y^2 - 1 = (2y - 1)(2y + 1)$, we can use the fundamental principle of fractions to multiply the numerator and denominator by $2y + 1$ as follows:

$$\frac{y}{2y - 1} = \frac{y(2y + 1)}{(2y - 1)(2y + 1)} = \frac{2y^2 + y}{4y^2 - 1}$$

PROBLEM SET 4

1. Factor the following completely by looking for the common factors.
   a) $4x^2 + 7xy$
   b) $ab^2 - ab + a$
   c) $17x^3 - 34x^2$
   d) $5x^3 - 10x^2y$
   e) $3x(2a + b) + 5y(2a + b)$
   f) $x(y + z) + 3(y + z)$
   g) $(a^2 + 5)(7x) + (a^2 + 5)(13a)$
   h) $a(x - y) - b(x - y)$
   i) $4xy^2z + x^2y^2z^2 - x^3y^3$
   j) $m(x - y) + (y - x)$

2. Factor the following completely if possible.
   a) $x^2 + 5x + 6$
   b) $6x^2 + 21x - 45$
   c) $40 - 3x - x^2$
   d) $6x^2 - 31x + 35$
   e) $x^2 - 3x - 4$
   f) $5x - 36 + x^2$
   g) $x^2 + 3x + 5$
   h) $x^2 + 10x + 25$
   i) $x^2 - 3x - 40$
   j) $3x^2 - 8x - 3$
   k) $16 - 6x - x^2$
   l) $x^2 - 9xy - 10y^2$
   m) $6x^2 - 13xy + 6y^2$
   n) $2x^2 + 5x - 12$
   o) $12x^2 - 14x - 10$
   p) $-2x^2 + 13x - 21$

3. Use the known products of this section to factor the following.
   a) $x^4 - 9y^2$
   b) $49x^2 - 98x + 49$
   c) $(x + y) - (x^2 - y^2)$
   d) $a^2x^2 - 1$
   e) $729x^3 - 8y^3$
   f) $(x^2 - y^2) + 4(x - y)$
   g) $1 - 81y^2$
   h) $64 + 8y^3x^3$
   i) $x^2 - x - y^2 - y$
   j) $9(x + y)^2 - 4(z + w)^2$
   k) $8x^3 - 12x^2 + 6x - 1$
   l) $x^2 - 9y^2 + x - 3y$
   m) $x^3 + 3x^2 + 3x + 1$
   n) $(x + y)^3 - 125y^3$
   o) $x^2 - y^2 - (9x - 9y)$

4. Factor completely each of the following.
   a) $x^2 - 5x - 84$
   b) $x^3 - 1$
   c) $y^2 - 16x^2$
   d) $2x^3 - 4x^2 - 30x$
   e) $18zx^2 - 2zy^2$
   f) $x^2 + 4x - 5$
   g) $25 - 10x + x^2$
   h) $27y^3 + 1$
   i) $2y^2 - 14y + 24$
   j) $9x(3c + d) - 8y(3c + d) + (x + y)(3c + d)$
   k) $625y^2 - (y^2 + 4y + 4)$
   l) $4x^2 - 11x + 6$
   m) $(1 - 5x)^2 - 121b^2$

n) $216x^6 - y^9$
p) $1 + 64z^3$
r) $6x^2 - 5xy - 6y^2$
o) $a^2 - 14a - 15$
q) $y^{16} - 1$
s) $(x + y)^2 - 81z^2$

5 Use the property "$a/b = c/d$ if and only if $ad = bc$" to decide whether or not each of the following pairs of fractions are equivalent.

a) $\frac{14}{147}$, $\frac{2}{21}$

b) $\frac{x^2 + 2x + 1}{x + 1}$, $\frac{3x + 3}{3}$

c) $\frac{x + 2}{x - 2}$, $\frac{-2 - x}{2 - x}$

d) $\frac{x + y}{1}$, $\frac{x^3 + y^3}{x^2 - xy + y^2}$

e) $\frac{a + b}{1}$, $\frac{ca + b}{c}$

f) $\frac{x^2 - 9}{x - 3}$, $\frac{x + 3}{1}$

6 Reduce each of the following fractions to an equivalent fraction by applying the fundamental principle of fractions.

a) $\frac{x^2 y}{xy^3}$

b) $\frac{-a^2 bc}{-ab^2 c}$

c) $\frac{15x^2 y^5 c^7}{45x^2 y^3 c^6}$

d) $\frac{x^2 + 2xy + y^2}{x^3 + y^3}$

e) $\frac{2x(-y)(-z)^2}{6x^3 y^3 (-z)^3}$

f) $\frac{18x^8 y^6}{24x^5 y^6 - 36x^6 y^5}$

7 Use the fundamental principle of fractions to determine the missing expression for each of the following fractions.

a) $\frac{3}{7} = \frac{?}{21}$

b) $\frac{x - 3}{x + 1} = \frac{x^2 - 4x + 3}{?}$

c) $\frac{x - y}{x + y} = \frac{x^2 - y^2}{?}$

d) $\frac{2xy}{-3a^2 bc} = \frac{?}{12a^3 b^2 c}$

e) $\frac{xy}{x^2 + xy + y^2} = \frac{?}{x^3 - y^3}$

f) $\frac{x^2 - y^2}{x^2 + xy} = \frac{?}{(x^2 - y^2)x}$

g) $\frac{x}{x^2 + 3x + 9} = \frac{?}{x^3 - 27}$

h) $\frac{x + 3}{x^2 - 9} = \frac{x^2 + 3x}{?}$

8 Write each of the following rational expressions in a reduced equivalent form if possible.

a) $\frac{x^2 + x}{x^2 - x}$

b) $\frac{x^3 - xy^2}{x^2 + 2xy - 3y^2}$

c) $\frac{x^2 + 2x - 3}{x^2 + 5x + 6}$

d) $\frac{x^2 - 1}{x^3 - 1}$

e) $\frac{7(9x^2 - y^2)}{14(3x + y)^2}$

f) $\frac{x^3 + y^3}{x^3 + x^2 y}$

g) $\frac{x^4 - y^4}{x^4 + 2x^2 y^2 + y^4}$

h) $\frac{x^3 + 1}{x^3 - 4x^2 - 5x}$

i) $\frac{x^2 + x - 12}{x^2 + 4x - 21}$

j) $\frac{x + x^2 - y - xy}{x^2 - 2xy + y^2}$

k) $\frac{x^2 - 9x + 18}{3x^2 - 5x - 12}$

l) $\frac{6x^4 - 48x}{5x^5 - 80x}$

m) $\dfrac{x^2 + xy - 2y^2}{x + 2y}$ n) $\dfrac{7x^2 - 5xy}{49x^3 - 25xy}$

o) $\dfrac{3x^2 + 21x}{4x^3 + 28x^2}$ p) $\dfrac{8x^4 - x}{4x^2 + 2x + 1}$

q) $\dfrac{15 + 19x - 10x^2}{3x + 8x^2 + 5x^3}$ r) $\dfrac{4 - x^4}{8 + x^6}$

s) $\dfrac{x^3 - 2x^2 + 5x - 10}{3x^5 + 15x^3 - x^2 - 5}$ t) $\dfrac{x^2 - 4 + 2y - xy}{2xy - 4y - x^2 + 4}$

u) $\dfrac{x^2 - y^2}{y^2 + xy - 2x^2}$ v) $\dfrac{x^2 - (y + 3)^2}{9 - (x + y)^2}$

6 Operations on Rational Expressions

The rules for adding algebraic expressions are the same as those for adding numerical fractions from arithmetic. First we will consider adding "like" fractions, that is, fractions with the same denominator. The rule is: *To add like fractions, add the numerator to find the numerator of the sum, and retain the common denominator as the denominator of the sum.* That is,

$$\frac{a}{b} + \frac{c}{b} = \frac{a + c}{b}$$

EXAMPLE

Determine each of the following sums.

a) $\dfrac{x}{x^2 + 1} + \dfrac{2x + 1}{x^2 + 1}$

b) $\dfrac{-x^2}{5 - x} + \dfrac{25}{5 - x}$

SOLUTION

a) $\dfrac{x}{x^2 + 1} + \dfrac{2x + 1}{x^2 + 1} = \dfrac{x + 2x + 1}{x^2 + 1} = \dfrac{3x + 1}{x^2 + 1}$

b) $\dfrac{-x^2}{5 - x} + \dfrac{25}{5 - x} = \dfrac{-x^2 + 25}{5 - x} = \dfrac{(-x + 5)(x + 5)}{5 - x} = x + 5$

The rule for adding "unlike fractions," that is, fractions with different denominators, is: *To add unlike fractions, change them to equivalent fractions having the same denominator, then follow the rule for adding like fractions.* That is,

$$\frac{a}{b} + \frac{c}{d} = \frac{ad}{bd} + \frac{bc}{bd}$$

$$= \frac{ad + bc}{bd}$$

EXAMPLE

Add each of the following fractions.

a) $\dfrac{3}{x-1} + \dfrac{x}{x+1}$

b) $\dfrac{2x}{x+7} + \dfrac{1}{x-5} + \dfrac{-12}{x^2+2x-35}$

SOLUTION

a) $\dfrac{3}{x-1} + \dfrac{x}{x+1} = \dfrac{3(x+1)}{(x-1)(x+1)} + \dfrac{x(x-1)}{(x+1)(x-1)}$

$= \dfrac{3(x+1) + x(x-1)}{(x-1)(x+1)}$

$= \dfrac{x^2 + 2x + 3}{(x-1)(x+1)}$

b) $\dfrac{2x}{x+7} + \dfrac{1}{x-5} + \dfrac{-12}{x^2+2x-35}$

$= \dfrac{2x(x-5)}{(x+7)(x-5)} + \dfrac{1(x+7)}{(x-5)(x+7)} + \dfrac{-12}{(x+7)(x-5)}$

$= \dfrac{2x(x-5) + (x+7) + (-12)}{(x+7)(x-5)}$

$= \dfrac{2x^2 - 10x + x + 7 + (-12)}{(x+7)(x-5)}$

$= \dfrac{2x^2 - 9x - 5}{(x+7)(x-5)}$

$= \dfrac{(2x+1)(x-5)}{(x+7)(x-5)} = \dfrac{2x+1}{x+7}$

Similarly, a general rule for subtraction of fractions is

$$\dfrac{a}{b} - \dfrac{c}{d} = \dfrac{ad - cb}{bd}$$

Thus, to add or subtract two fractions having different denominators, it is customary to find a common denominator by multiplying the numerator and the denominator of each of the given fractions by a suitable factor. For example,

$\dfrac{x}{x-2} - \dfrac{2}{x+2} = \dfrac{x(x+2)}{(x-2)(x+2)} - \dfrac{2(x-2)}{(x-2)(x+2)}$

$= \dfrac{x(x+2) - 2(x-2)}{(x-2)(x+2)}$

$= \dfrac{x^2 + 2x - 2x + 4}{(x-2)(x+2)} = \dfrac{x^2 + 4}{x^2 - 4}$

EXAMPLE

Perform each of the following subtractions.

a) $\frac{2}{3} - \frac{1}{2}$

b) $\dfrac{x+1}{x-1} - \dfrac{x-1}{x+1}$

SOLUTION

a) $\dfrac{2}{3} - \dfrac{1}{2} = \dfrac{2 \cdot 2 - 1 \cdot 3}{3 \cdot 2} = \dfrac{1}{6}$

b) $\dfrac{x+1}{x-1} - \dfrac{x-1}{x+1} = \dfrac{(x+1)(x+1) - (x-1)(x-1)}{(x-1)(x+1)}$

$= \dfrac{(x^2 + 2x + 1) - (x^2 - 2x + 1)}{(x-1)(x+1)}$

$= \dfrac{4x}{x^2 - 1}$

It is usually desirable, although not essential, to find the *least common denominator* (L.C.D.) in obtaining the algebraic sum of fractions. This can be accomplished by obtaining the prime factorization of each denominator of the fractions, then forming the product of the different prime factors, using the highest exponent that appears with each prime factor. This procedure is illustrated in the following examples.

EXAMPLE

1 Determine

$$\dfrac{3}{2x-2} + \dfrac{x}{3x+3} + \dfrac{1-x}{6}$$

SOLUTION. First, factor each denominator to obtain $2x - 2 = 2(x-1)$ and $3x + 3 = 3(x+1)$ and $6 = 3 \cdot 2$. The L.C.D. for the given fractions must contain factors 3, 2, $x - 1$, and $x + 1$. Thus, the L.C.D. is the product $3 \cdot 2(x-1)(x+1)$.

$\dfrac{3}{2x-2} + \dfrac{x}{3x+3} + \dfrac{1-x}{6}$

$= \dfrac{3}{2(x-1)} + \dfrac{x}{3(x+1)} + \dfrac{1-x}{3 \cdot 2}$

$= \dfrac{3 \cdot 3(x+1)}{2(x-1)3(x+1)} + \dfrac{x \cdot 2(x-1)}{3(x+1)2(x-1)}$

$\quad + \dfrac{(1-x)(x-1)(x+1)}{3 \cdot 2(x-1)(x+1)}$

$= \dfrac{9(x+1) + 2x(x-1) + (1-x)(x-1)(x+1)}{6(x-1)(x+1)}$

$= \dfrac{9x + 9 + 2x^2 - 2x + x^2 - 1 - x^3 + x}{6(x-1)(x+1)}$

$= \dfrac{-x^3 + 3x^2 + 8x + 8}{6(x-1)(x+1)}$

2 Determine

$$\frac{x}{x^2 + 5x + 6} - \frac{2}{3x^2 - 16x + 5} + \frac{3x}{x^2 - 3x - 10}$$

SOLUTION

$$\frac{x}{x^2 + 5x + 6} - \frac{2}{3x^2 - 16x + 5} + \frac{3x}{x^2 - 3x - 10}$$

$$= \frac{x}{(x + 2)(x + 3)} - \frac{2}{(3x - 1)(x - 5)} + \frac{3x}{(x - 5)(x + 2)}$$

The L.C.D. is $(x + 2)(x + 3)(3x - 1)(x - 5)$, so that

$$\frac{x}{x^2 + 5x + 6} - \frac{2}{3x^2 - 16x + 5} + \frac{3x}{x^2 - 3x - 10}$$

$$= \frac{x(3x - 1)(x - 5)}{(x + 2)(x + 3)(3x - 1)(x - 5)}$$

$$- \frac{2(x + 2)(x + 3)}{(x + 2)(x + 3)(3x - 1)(x - 5)}$$

$$+ \frac{3x(x + 3)(3x - 1)}{(x + 2)(x + 3)(3x - 1)(x - 5)}$$

$$= \frac{3x^3 - 16x^2 + 5x - 2x^2 - 10x - 12 + 9x^3 + 24x^2 - 9x}{(x + 2)(x + 3)(3x - 1)(x - 5)}$$

$$= \frac{12x^3 + 6x^2 - 14x - 12}{(x + 2)(x + 3)(3x - 1)(x - 5)}$$

$$= \frac{2(6x^3 + 3x^2 - 7x - 6)}{(x + 2)(x + 3)(3x - 1)(x - 5)}$$

The rule for multiplying fractions is: *The numerator of the product of two fractions is the product of the numerator of the factors, and the denominator of the product is the product of the denominator of the factors.* That is,

$$\frac{a}{b} \cdot \frac{c}{d} = \frac{a \cdot c}{b \cdot d}$$

EXAMPLES

1 Find $\dfrac{x}{x - 1} \cdot \dfrac{x^2 - 1}{x^3}$

SOLUTION

$$\frac{x}{x - 1} \cdot \frac{x^2 - 1}{x^3} = \frac{x(x^2 - 1)}{x^3(x - 1)} = \frac{x(x + 1)(x - 1)}{x \cdot x^2(x - 1)} = \frac{x + 1}{x^2}$$

2 Find $\dfrac{2x - 2}{x^2 + 2x} \cdot \dfrac{x^2 + 4x + 4}{2x^2 + 2x - 4}$

SECTION 6 OPERATIONS ON RATIONAL EXPRESSIONS

$$\frac{2x-2}{x^2+2x} \cdot \frac{x^2+4x+4}{2x^2+2x-4} = \frac{2(x-1)}{x(x+2)} \cdot \frac{(x+2)^2}{2(x+2)(x-1)} = \frac{1}{x}$$

The rule for dividing fractions can be stated as follows: *To divide fractions, multiply the first fraction by the reciprocal of the second fraction.* That is,

$$\frac{a}{b} \div \frac{c}{d} = \frac{a}{b} \cdot \frac{1}{c/d}$$

$$= \frac{a}{b} \cdot \frac{d}{c}$$

$$= \frac{ad}{bc}$$

EXAMPLES

1 Divide $\dfrac{x^2+5x+6}{x^2-4}$ by $\dfrac{x^2+4x+4}{x^2-4x+4}$

SOLUTION

$$\frac{x^2+5x+6}{x^2-4} \div \frac{x^2+4x+4}{x^2-4x+4} = \frac{(x^2+5x+6)(x^2-4x+4)}{(x^2-4)(x^2+4x+4)}$$

$$= \frac{(x+2)(x+3)(x-2)(x-2)}{(x-2)(x+2)(x+2)(x+2)}$$

$$= \frac{(x+3)(x-2)}{(x+2)(x+2)}$$

Divide $\dfrac{x^2-x}{y^2-y} \cdot \dfrac{y^2x-yx}{x-1}$ by $\dfrac{x^2}{x-1}$

SOLUTION

$$\frac{x^2-x}{y^2-y} \cdot \frac{y^2x-yx}{x-1} \div \frac{x^2}{x-1} = \frac{x^2-x}{y^2-y} \cdot \frac{y^2x-yx}{x-1} \cdot \frac{x-1}{x^2}$$

$$= \frac{x(x-1)}{y(y-1)} \cdot \frac{xy(y-1)}{x-1} \cdot \frac{x-1}{x^2} = x-1$$

So far we have considered fractions of the form P/Q, $Q \neq 0$, where P and Q are polynomial expressions. Now we shall consider fractions in which the numerator or denominator, or both, contain fractions. Fractions of this type are called *complex fractions*. For example, the fractions

$$\frac{1}{\frac{2}{3}}, \quad \frac{\frac{2}{3}}{7}, \quad \frac{\frac{1}{2}}{\frac{3}{5}}, \quad \frac{3}{1-x}, \quad \frac{\frac{2}{3}}{1+\frac{1}{x}}, \quad \text{and} \quad \frac{x-\frac{1}{x}}{\frac{2}{1+x}-\frac{x}{1-x}}$$

are complex fractions. Let us consider some examples in which complex fractions can be written in equivalent reduced fractions. In these cases the complex fractions are said to be *simplified*.

EXAMPLES

1. Simplify $\dfrac{2 + \dfrac{1}{x}}{x - \dfrac{1}{1-x}}$

SOLUTION

$$\dfrac{2 + \dfrac{1}{x}}{x - \dfrac{1}{1-x}} = \dfrac{\dfrac{2x+1}{x}}{\dfrac{x-x^2-1}{1-x}} = \dfrac{2x+1}{x} \div \dfrac{x-x^2-1}{1-x}$$

$$= \dfrac{2x+1}{x} \cdot \dfrac{1-x}{x-x^2-1} = \dfrac{(2x+1)(1-x)}{x(x-x^2-1)}$$

$$= \dfrac{2x - 2x^2 + 1 - x}{x^2 - x^3 - x} = \dfrac{-2x^2 + x + 1}{-x^3 + x^2 - x}$$

$$= \dfrac{2x^2 - x - 1}{x^3 - x^2 + x}$$

2. Simplify $\dfrac{x - \dfrac{1}{y}}{1 - \dfrac{x}{\dfrac{1}{y}}}$

SOLUTION. Consider only the term $\dfrac{-x}{1/y}$ in the denominator:

$$\dfrac{-x}{\dfrac{1}{y}} = -x \div \dfrac{1}{y} = -xy$$

Therefore,

$$\dfrac{x - \dfrac{1}{y}}{1 - \dfrac{x}{\dfrac{1}{y}}} = \dfrac{x - \dfrac{1}{y}}{1 - xy} = \dfrac{\dfrac{xy-1}{y}}{1-xy} = \dfrac{xy-1}{y} \div (1-xy)$$

$$= \dfrac{xy-1}{y} \cdot \dfrac{1}{1-xy} = \dfrac{-(1-xy)}{y} \cdot \dfrac{1}{1-xy} = \dfrac{-1}{y}$$

PROBLEM SET 5

1. Change each of the following fractions to an equivalent fraction having the same L.C.D.

a) $\frac{7}{12}, \frac{1}{10}, \frac{2}{45}$ b) $\frac{3x}{89z}, \frac{5}{12xz}, \frac{7z}{6xy}$

c) $\frac{x-2}{5xy^3}, \frac{4y+1}{25x^3y}$ d) $\frac{x}{x^2+3x+9}, \frac{x-2}{x^2-3x}, \frac{1}{x^3-27}$

e) $\frac{5}{12x^2}, \frac{7}{6x^2-24}, \frac{1}{12x-24}$

f) $\frac{2}{x+4}, \frac{x-3}{x^2-16}, \frac{x^2}{x^3+64}$

2 Combine each of the following fractions into a single fraction in reduced form.

a) $\frac{x}{y} + \frac{7x}{y}$ b) $\frac{3x^2+5x}{x} + \frac{3x-3x^2}{x}$

c) $\frac{x+x^2}{5x} + \frac{3x-x^2}{5x}$ d) $\frac{5}{x} + \frac{7}{x} + \frac{x-12}{x}$

e) $\frac{2x^2+5x^3}{x^2} + \frac{5x^2-5x^3}{x^2}$ f) $\frac{x^2-xy-2y^2}{x+y} + \frac{x^2-2xy-3y^2}{x+y}$

g) $\frac{x^2+5x+16}{x+1} - \frac{3x^2-2x+7}{x+1}$

h) $\frac{x^2-y^2}{x+y} - \frac{x^2+y^2}{x+y}$

i) $\frac{x^3+y^3}{x+y} - \frac{x^3-y^3}{x+y}$ j) $\frac{x^3-27}{x-3} - \frac{x^3+27}{x-3}$

3 Perform each of the following operations of addition and subtraction.

a) $\frac{4}{x^2} + \frac{1}{2x}$ b) $\frac{3}{5x} + \frac{2}{6x} + \frac{5}{7x}$

c) $\frac{51}{4x} - \frac{13}{5x}$ d) $\frac{1}{28xy} + \frac{1}{36yz} + \frac{1}{21xz}$

e) $\frac{x-2}{3} - \frac{x+2}{3}$ f) $\frac{1}{x-1} + \frac{1}{x+1}$

g) $\frac{1}{x-2} + \frac{3}{x+2}$ h) $\frac{a}{a-x} - \frac{x}{a+x} - \frac{a^2+x^2}{a^2-x^2}$

i) $\frac{x}{x^2-y^2} - \frac{1}{x+y}$ j) $\frac{x}{x-1} - \frac{1}{x^2-x}$

k) $\frac{x+1}{x-1} - \frac{(x^2+1)^2}{x^2-1}$ l) $\frac{x}{x^2-9} - \frac{x-1}{x^2-5x+6}$

m) $\frac{x}{3x-2} - \frac{2}{2x+3}$ n) $\frac{1}{x-y} - \frac{xy+2y^2}{x^3-y^3}$

o) $\frac{x}{x+y} - \frac{y}{x-y} + \frac{xy}{x^2-y^2}$

4 Perform each of the following operations of multiplication and division.

a) $\frac{2xy}{3z^2} \cdot \frac{x^2}{z}$ b) $\frac{5x^2y}{3x^2y} \cdot \frac{6y}{10x^2}$

c) $\frac{x+4}{x^2} \cdot \frac{y}{x-y}$ d) $\frac{15xyz}{16a^2} \cdot \frac{10y^3}{9x^2a} \cdot \frac{12xa^2}{25y^2z}$

e) $\dfrac{x^2 + xy}{xy} \cdot \dfrac{y}{x^2 - y^2}$

f) $\dfrac{x^3 + y^3}{x^2} \cdot \dfrac{4x^3}{(x+y)^3} \cdot \dfrac{x^2 + 2xy + y^2}{2x}$

g) $\dfrac{x^3 - 27y^3}{x + 3y} \div \dfrac{x^2 + 3xy + 9y^2}{x^2 - 9y^2}$

h) $\left(x - 1 - \dfrac{3}{x}\right) \div \left(x + 1 + \dfrac{x}{x - 3}\right)$

i) $\dfrac{x^2 - (y - z)^2}{x^4 - y^4} \cdot \dfrac{x^2 + y^2}{x - y + z} \div \dfrac{1}{x + y}$

j) $\left(x - 7 + \dfrac{24}{x + 3}\right) \div \left(x - 4 + \dfrac{9}{x + 3}\right)$

k) $\left(\dfrac{x}{x + 1} - \dfrac{x - 1}{x}\right) \div \left(\dfrac{x}{x + 1} + \dfrac{x - 1}{x}\right)$

5 Simplify each of the following complex fractions.

a) $\dfrac{\frac{2}{3}}{\frac{4}{3}}$

b) $\dfrac{\frac{17}{6}}{\frac{7}{3}}$

c) $\dfrac{\frac{11}{24}}{\frac{12}{9}}$

d) $\dfrac{\frac{1}{x}}{\frac{2}{x + 1}}$

e) $\dfrac{\frac{1}{x}}{1 - \frac{1}{x}}$

f) $\dfrac{1 - \frac{x}{3}}{1 - \frac{x}{5}}$

g) $\dfrac{\frac{1}{x + 1}}{1 + \frac{1}{x + 1}}$

h) $\dfrac{1}{1 + \dfrac{1}{1 + \frac{1}{x}}}$

i) $\dfrac{\frac{2}{x^3 - 1}}{\frac{x}{x^2 + x + 1}}$

j) $\dfrac{\frac{1}{x + 1} - 1}{\frac{x}{1 - \frac{1}{x}}}$

k) $\dfrac{\frac{3}{x} + \frac{x}{x + 1}}{\frac{3}{x + 1} - \frac{1 - x}{x}}$

l) $\dfrac{\frac{x^2 - y^2}{x + y}}{\frac{x/y + 1}{x/y - 1}}$

m) $\dfrac{x + \frac{1}{x^2 - 2x + 1}}{\frac{1}{x^2 - 1}}$

n) $\dfrac{x - \frac{x - 1}{1 + x}}{\frac{1 + x}{x} + \frac{1 - \frac{1}{x}}{1 - x}}$

7 Exponents

In Section 3.1 we discussed the meaning of a^n, where $a \neq 0$ and n is any positive integer. Accordingly, a^n means the product formed by taking a as a factor n times. That is,

$$a^n = a \cdot a \cdot a \cdots a \quad (n \text{ factors})$$

In this section we shall first consider the properties of positive integral exponents, then extend the definition of exponents to include the negative integers together with zero exponents, as well as the rational numbers.

7.1 Properties of Positive Integral Exponents

If a and b are real numbers and m and n are positive integers, then

i $a^m a^n = a^{m+n}$

ii $(a^m)^n = a^{mn}$

iii $(ab)^n = a^n b^n$

iv $\left(\dfrac{a}{b}\right)^n = \dfrac{a^n}{b^n} \quad b \neq 0$

and

v $\dfrac{a^m}{a^n} = \begin{cases} a^{m-n} & \text{if } m \text{ is greater than } n \\ 1 & \text{if } m \text{ is equal to } n \\ \dfrac{1}{a^{n-m}} & \text{if } m \text{ is less than } n \end{cases} \quad a \neq 0$

These properties can be verified by applying the definition of positive integral exponents. We shall verify Properties i and iii here and leave the verifications of ii, iv, and v to the reader as an exercise in Problem Set 6, Problem 2.

PROOF OF i

$$\begin{aligned} a^m a^n &= a^{m+n} \\ &= \underbrace{(a \cdot a \cdots a)}_{m \text{ factors}} \underbrace{(a \cdot a \cdots a)}_{n \text{ factors}} \quad \text{(definition)} \\ &= \underbrace{a \cdot a \cdot a \cdots a}_{m+n \text{ factors}} \\ &= a^{m+n} \quad \text{(definition)} \end{aligned}$$

PROOF OF iii

$$\begin{aligned}(ab)^n &= a^n b^n \\ &= \underbrace{(ab)(ab) \cdots (ab)}_{n \text{ factors}} \qquad &\text{(definition)} \\ &= \underbrace{(a \cdot a \cdot a \cdots a)}_{n \text{ factors}} \underbrace{(b \cdot b \cdot b \cdots b)}_{n \text{ factors}} \qquad &\text{(why?)} \\ &= a^n b^n \qquad &\text{(definition)}\end{aligned}$$

The above properties can be extended to properties such as $x^m x^n x^k = x^{m+n+k}$, $(xyz)^n = x^n y^n z^n$, and so on.

EXAMPLES

Use the above properties to simplify the given expressions.

1 $x^4 x^6$

SOLUTION

$$x^4 x^6 = x^{4+6} = x^{10} \qquad \text{(Property i)}$$

2 $(3x^2)^3$

SOLUTION

$$\begin{aligned}(3x^2)^3 &= 3^3 (x^2)^3 \qquad &\text{(Property iii)} \\ &= 27 x^{2 \cdot 3} \qquad &\text{(Property ii)} \\ &= 27 x^6\end{aligned}$$

3 $\dfrac{x^7}{x^3 x^2}$

SOLUTION

$$\dfrac{x^7}{x^3 x^2} = \dfrac{x^7}{x^{3+2}} = \dfrac{x^7}{x^5} \qquad \text{(Property i)}$$
$$= x^{7-5} = x^2 \qquad \text{(Property v)}$$

4 $\left(\dfrac{x^3 y^5}{2 x^4 y^2}\right)^7$

SOLUTION

$$\left(\dfrac{x^3 y^5}{2 x^4 y^2}\right)^7 = \dfrac{(x^3)^7 (y^5)^7}{(2)^7 (x^4)^7 (y^2)^7}$$

$$= \frac{x^{21}y^{35}}{128x^{28}y^{14}} \qquad \text{(why?)}$$

$$= \frac{y^{21}}{128x^{7}}$$

5. $\left(\dfrac{8xy^2}{2x^2z}\right)^4$

SOLUTION

$$\left(\frac{8xy^2}{2x^2z}\right)^4 = \frac{(8)^4 x^4 (y^2)^4}{(2)^4 (x^2)^4 z^4} = \frac{(2)^{12} x^4 y^8}{(2)^4 x^8 z^4} = \frac{(2)^8 y^8}{x^4 z^4} \qquad \text{(why?)}$$

$$= \frac{256 y^8}{x^4 z^4}$$

6. $\dfrac{3x^4}{4z^2} \cdot \dfrac{5x^2 y^2}{3z^4}$

SOLUTION

$$\frac{3x^4}{4z^2} \cdot \frac{5x^2 y^3}{3z^4} = \frac{15 x^4 x^2 y^3}{12 z^2 z^4} = \frac{15 x^6 y^3}{12 z^6} \qquad \text{(why?)}$$

$$= \frac{5 x^6 y^3}{4 z^6}$$

7.2 Zero and Negative Integral Exponents

Now we will extend the use of exponents to include the negative integers together with zero exponents, as well as the rational numbers, so that Properties i through v in Section 7.1 remain valid.

Let us assume that a is a nonzero real number. If we let $m = 0$ in Property i, we have

$$a^0 a^n = a^{0+n} \qquad \text{or} \qquad a^0 a^n = a^n$$

This statement is true if we assign the value 1 for a^0, since $1 \cdot a^n = a^n$. (Why?) Properties ii through v can also be verified for the choice of $a^0 = 1$. Hence, if $m = n$ in Property v, then

$$\frac{a^m}{a^n} = \frac{a^n}{a^n} = a^{n-n} = a^0$$

Again, it follows that a^0 must be assigned the value 1, since $a^n/a^n = 1$. Thus, $3^0 = 1$, $(\tfrac{1}{2})^0 = 1$, and $(-5)^0 = 1$. Notice that 0^0 is not defined.

Once again, let us consider Property i for negative integral exponents. If we let $m = -n$, where n is a positive integer, then

$$a^m a^n = a^{-n} a^n = a^0 = 1$$

Thus, $a^{-n}a^n = 1$. After dividing both sides of this equation by a^n, we have $a^{-n} = 1/a^n$. For example,

$$3^{-2} = \frac{1}{3^2} = \frac{1}{9}$$

$$(xy)^{-3} = \frac{1}{(xy)^3} = \frac{1}{x^3 y^3}$$

$$\left(\frac{1}{2}\right)^{-2} = \frac{1}{\left(\frac{1}{2}\right)^2} = \frac{1 \cdot 4}{\frac{1}{4} \cdot 4} = 4$$

This is formalized in the following definition.

DEFINITION

If a is any real number, $a \neq 0$, and n is a positive integer, then

i $\quad a^0 = 1$

ii $\quad a^{-n} = \dfrac{1}{a^n}$

From here on, Properties i through iv can be applied to the set of integers when used as exponents. Furthermore, we can simplify Property v to read

$$\frac{a^m}{a^n} = a^{m-n}$$

where m and n are integers, since m − n is an integer.

EXAMPLES

In Examples 1 to 4, eliminate negative exponents and simplify the expressions.

1 $\quad \dfrac{3ab^{-2}}{c^3 d^{-4}}$

SOLUTION

$$\frac{3ab^{-2}}{c^3 d^{-4}} = \frac{3ab^{-2} \cdot b^2 d^4}{c^3 d^{-4} \cdot b^2 d^4} \quad \text{(fundamental principle of fractions)}$$

$$= \frac{3ab^0 d^4}{c^3 d^0 b^2}$$

$$= \frac{3ad^4}{c^3 b^2}$$

2 $\quad \dfrac{x^{-4}(-3x)}{(-5x)^{-2}}$

SECTION 7 EXPONENTS 55

SOLUTION

$$\frac{x^{-4}(-3x)}{(-5x)^{-2}} = \frac{-3x^{-4+1}}{(-5)^{-2}x^{-2}} \quad \text{(Properties i and iii)}$$

$$= \frac{-3x^{-3}}{[1/(-5)^2]x^{-2}} \quad \text{(definition)}$$

$$= -\frac{3}{\frac{1}{25}} x^{-3-(-2)} = -75x^{-1}$$

$$= \frac{-75}{x}$$

3 $\dfrac{x^{-1}+y^{-1}}{(x+y)^{-1}}$

SOLUTION

$$\frac{x^{-1}+y^{-1}}{(x+y)^{-1}} = \frac{\dfrac{1}{x}+\dfrac{1}{y}}{\dfrac{1}{x+y}} \quad \text{(definition)}$$

$$= \frac{\dfrac{x+y}{xy}}{\dfrac{1}{x+y}} \quad \text{(why?)}$$

$$= \frac{(x+y)(x+y)}{xy} = \frac{(x+y)^2}{xy}$$

4 $\dfrac{x^{-2}+y^{-2}}{(xy)^{-1}}$

SOLUTION

$$\frac{x^{-2}+y^{-2}}{(xy)^{-1}} = \frac{\dfrac{1}{x^2}+\dfrac{1}{y^2}}{\dfrac{1}{xy}} \quad \text{(definition)}$$

$$= \frac{\dfrac{y^2+x^2}{x^2y^2}}{\dfrac{1}{xy}} \quad \text{(why?)}$$

$$= \frac{y^2+x^2}{xy}$$

5 Write the expression $(x^n y^{-3n+2}/y^{2n})^2$ as a product without fractions, in which each variable occurs once.

SOLUTION

$$\left(\frac{x^n y^{-3n+2}}{y^{2n}}\right)^2 = (x^n y^{-3n+2-2n})^2 = (x^n y^{-5n+2})^2 = x^{2n} y^{-10n+4}$$

6 Simplify $\dfrac{2^{3x+2}8^2}{2^{3x}2^6 - 2^{3x+2}2^3}$

SOLUTION

$$\dfrac{2^{3x+2}8^2}{2^{3x}2^6 - 2^{3x+2}2^3} = \dfrac{2^{3x+2}(2^3)^2}{2^{3x}2^2 2^4 - 2^{3x+2}2^3}$$

$$= \dfrac{2^{3x+2}2^3 2^3}{2^{3x+2}2^4 - 2^{3x+2}2^3}$$

$$= \dfrac{2^{3x+2}2^3 2^3}{2^{3x+2}2^3(2-1)}$$

$$= \dfrac{2^3}{2-1} = 8$$

PROBLEM SET 6

1 Use the properties of exponents to simplify each of the following expressions, where m and n are positive integers.

a) $x^7 x^3 x^5$ b) $x^{3n} x^7$

c) $x^{2n} x^n$ d) $\dfrac{x^{5n}}{x^5}$

e) $\dfrac{x^{5n}}{x^n}$ f) $(x^{4n})^3$

g) $\left(\dfrac{x^2}{y}\right)^3$ h) $\left(\dfrac{-2x}{3y^2}\right)^3 \left(\dfrac{3x^2}{4y}\right)^4$

i) $\left(\dfrac{3x}{3y}\right)^{2n}$ j) $\left(\dfrac{x^2 y}{x^3}\right)^{2n}$

k) $\left(\dfrac{-4a^2 b^3}{2a^4 b^6}\right)^4$ l) $\left(\dfrac{4x^2 y^3}{8x^3 y}\right)^7$

m) $\left(\dfrac{-a^2 xyz^5}{a^4 x^2 y^2 z^3}\right)^3$ n) $\dfrac{(x+y)^7}{(x+y)^2}$

o) $\dfrac{(x^{n+1} x^{2n+1})^3}{x^{3n}}$ p) $\dfrac{(x^{n+1} x^{n-1})^2}{x^{2n}}$

q) $\dfrac{(x^{7n} x^{3n} x^{2n})^3}{x^{5n}}$ r) $\dfrac{(x^2 x^3 x^7 x^{10})^4}{x^{13}}$

s) $\dfrac{(9^m)^{2m-1} \cdot 27^{m+2}}{81^{2m+1}}$ t) $\dfrac{(x^m)^n}{x^{m(m-5n)} \cdot x^{n(m-4n)}}$

u) $(-3xy^3)^4$ v) $(-5x^2 y^3 z^5)^2$

w) $(5a^2 b^3 z^2)^3$ x) $\left(\dfrac{x^2 y^3 z^4}{x^4}\right)^3$

2 Verify the following properties of exponents, where m and n are positive integers.

SECTION 7 EXPONENTS 57

a) $(a^m)^n = a^{mn}$

b) $\left(\dfrac{a}{b}\right)^n = \dfrac{a^n}{b^n} \quad b \neq 0$

c) $\dfrac{a^m}{a^n} = \begin{cases} a^{m-n} & \text{if } m \text{ is greater than } n \\ 1 & \text{if } m \text{ is equal to } n \\ \dfrac{1}{a^{n-m}} & \text{if } m \text{ is less than } n \end{cases} \quad a \neq 0$

3 Write each of the following expressions without negative exponents by applying the definition and the properties of exponents.

a) $5^{-1} \cdot 3^{-2}$
b) $5^{-2} \cdot 3^2$
c) $\left(\dfrac{3}{7}\right)^{-2}$
d) $\left(\dfrac{4}{5}\right)^{-2}$
e) $\dfrac{x^{-2}}{5^{-2}}$
f) $\dfrac{3^{-1} + 2^{-1}}{3 + 2}$
g) $3(x + y)^{-1}$
h) $x^{-3}y^{-2}$
i) $\left(\dfrac{3x^2}{y-2}\right)^{-1}$
j) $3x^{-2}y^{-4}$
k) $x^{-2}y^{-1}$
l) $\left(\dfrac{4x^{-2}}{3x}\right)^{-3}$
m) $x^{-2} + y^{-2}$
n) $(x^{-1} + y^{-1})^{-2}$
o) $(\tfrac{1}{2}x^3y^{-4})^2$
p) $5(x + y)^{-2}$
q) $\left(\dfrac{x^{-1}}{y-1}\right)^2$
r) $\left(\dfrac{x^0}{y-1}\right)^{-2}$

4 Eliminate the negative exponents and simplify.

a) $(2^{-3})^{-1} + 5^{-1}$
b) $4^{-1} \cdot 4^2$
c) $(2^{-3} \cdot 5^{-1})^{-2}$
d) $(3^{-2})^{-1} \cdot (2^3)^{-2}$
e) $\dfrac{1 - 2^{-2}}{1 - 4^{-2}}$
f) $\dfrac{x^0 - y^0}{x^0 + y^0}$
g) $x^{-2} + \dfrac{1}{x-2}$
h) $2x^{-1} - \dfrac{1}{x}$
i) $\dfrac{(x^{-1} - y^{-1})^{-1}}{(xy)^{-1}}$
j) $\dfrac{x^{-2} - y^{-1}}{x^{-1} - y^{-1}}$
k) $\dfrac{x^{-1}}{y-1} + \dfrac{x}{y}$
l) $\dfrac{x^{-1}}{y-1} + \dfrac{y}{x}$

5 Write each of the following expressions as a product without fractions in which each variable occurs once.

a) $x^{-2} \cdot x$
b) $\left(\dfrac{x^{2n}}{x^{n-1}}\right)^3$
c) $(x^{3-n} \cdot x^0)^4$
d) $\left(\dfrac{x}{x^2}\right)^{-2}$
e) $\dfrac{x^2 y^{n-1}}{x^3 y^{1-n}}$
f) $(y^{-3} \cdot y^{2n+1})^2$

g) $\left(\dfrac{x^{1-n}}{x^{1+n}}\right)^{-1}$

h) $\left(\dfrac{x^{2n+1}y^{-n}}{x^2 y^{n+1}}\right)^2$

i) $\left(\dfrac{x^{n-1}y^{n+1}}{x^{-1}y}\right)^2$

j) $\left(\dfrac{x^{-1}y^{-2}z^4}{x^0 x^{-3}z^3}\right)^{-3}$

6 Simplify.

a) $\dfrac{2^{4x+2}8^3}{24x2^{10} - 24x + 72^5}$

b) $\dfrac{3^{x-2}27 - 9^{-1}3^x}{3^{2x-4}}$

8 Rational Exponents

In order to define exponents that are positive rational numbers so that Properties i through v of Section 7.1 continue to hold, let us first consider rational exponents that are reciprocals of positive integers.

If $a^{1/q}$ exists, where a is a real number, $a \ne 0$, and q is a positive integer, then it must satisfy the property $(a^m)^n = a^{mn}$. Since $(a^{1/q})^q = a^{q/q} = a$, by this property, $a^{1/q}$ must be defined to represent one of q equal factors of a. The number $a^{1/q}$ if it exists is called the *qth root of a*. Thus, $4^{1/2}$ is one of two equal factors of 4. That is, $4^{1/2}4^{1/2} = 4^{1/2+1/2} = 4$. Since squaring either 2 or -2 gives 4, we could define $4^{1/2}$ to be either 2 or -2. By convention, we agree that $4^{1/2}$ is 2 and we call 2 the *principal square root* of 4. When we wish to designate the negative square root of 4, we do so by $-4^{1/2}$ or -2.

The symbol $-4^{1/2}$ means $-(4^{1/2})$, not $(-4)^{1/2}$. This latter symbol is meaningless in the set of real numbers, since $(-4)^{1/2}(-4)^{1/2} = -4$ contradicts the rule of multiplication of signed numbers, which states that the square of any real number is either zero or positive.

However, odd roots of negative numbers do exist in the set of real numbers. For example, $(-27)^{1/3} = -3$, since $(-3)(-3)(-3) = -27$. Here -3 is called the *cube root* of -27. In general we have the following.

For q a positive integer, $a^{1/q}$ (the qth root of a) exists as a real number if

i a is a positive number, $a^{1/q}$ is also a positive number (*principal qth root*).

ii a is zero and thus $a^{1/q}$ is also zero.

iii a is a negative number and q is an *odd* positive integer. In this case, $a^{1/q}$ is also negative.

Note that if a is negative and q is an *even* positive integer, then, $a^{1/q}$ does not exist. That is, $a^{1/q}$ is not a real number.

EXAMPLE

Compute the values of the following numbers.
a) $16^{1/2}$

b) $(-64)^{1/3}$
c) $-16^{1/4}$
d) $(-4)^{1/2}$

SOLUTION

a) $16^{1/2} = 4$, since $(4)(4) = 16$.
b) $(-64)^{1/3} = -4$, since $(-4)(-4)(-4) = -64$.
c) $-16^{1/4} = -(16)^{1/4} = -2$, since $(2)(2)(2)(2) = 16$.
d) $(-4)^{1/2}$ does not exist in the real number system.

This pattern of the above example can be formalized as follows.

DEFINITION

For a a real number and r a rational number such that $r = p/q$, where p and q are integers with no common factors, we define a^r to be the number $(a^{1/q})^p$ provided that $a^{1/q}$ exists. If $a^{1/q}$ does not exist, we do not define a^r in R.

Let us consider the number $9^{3/2}$. If the property $(a^m)^n = a^{nn}$ is to hold, then $9^{3/2} = (9^{1/2})^3 = 3^3 = 27$. Also, $9^{3/2} = (9^3)^{1/2} = (729)^{1/2} = 27$. Therefore, $9^{3/2} = (9^{1/2})^3 = (9^3)^{1/2}$. This can be generalized as follows:

$$a^{p/q} = (a^{1/q})^p = (a^p)^{1/q}$$

For example, $(-1)^{1/3} = -1$, by definition.
However,

$$(-1)^{1/3} = (-1)^{2/6} = [(-1)^{1/6}]^2 = [(-1)^2]^{1/6} = 1^{1/6} = 1$$

is not valid because $(-1)^{1/6}$ is not defined. Note that $\frac{2}{6}$ is not in its lowest terms, because 2 and 6 have factors in common, namely 2.

For negative rational exponents, we extend the definition to include the exponent $-p/q$, where p/q is a positive rational number and $a \neq 0$. Thus, $a^{-p/q} = 1/a^{p/q}$, $a \neq 0$. For example,

$$8^{-1/3} = \frac{1}{8^{1/3}} = \frac{1}{(2^3)^{1/3}} = \frac{1}{2}$$

EXAMPLE

Compute the value of the following numbers.
a) $27^{4/3}$
b) $32^{-3/5}$
c) $(\frac{4}{9})^{3/2}$

SOLUTION

a) $27^{4/3} = (27^{1/3})^4 = 3^4 = 81$

b) $32^{-3/5} = \dfrac{1}{(32)^{3/5}} = \dfrac{1}{(32^{1/5})^3} = \dfrac{1}{(2)^3} = \dfrac{1}{8}$

c) $(\tfrac{4}{9})^{3/2} = [(\tfrac{4}{9})^{1/2}]^3 = (\tfrac{2}{3})^3 = \tfrac{8}{27}$

We have defined rational exponents so that they satisfy the property $(a^m)^n = a^{mn}$. This definition ensures that they also satisfy the other properties. Thus, the properties of exponents that we originally developed for positive integral exponents, then extended to include all integers, are also valid for rational exponents. Since the set of rational numbers includes, as subsets, the positive integers, zero, and the negative integers, we can summarize the properties of exponents as follows.

8.1 Properties of Rational Exponents

For a and b real numbers, r and s rational numbers, we have (if the individual factors exist)

i $a^r a^s = a^{r+s}$

ii $(a^r)^s = a^{rs}$

iii $(ab)^r = a^r b^r$

iv $\left(\dfrac{a}{b}\right)^r = \dfrac{a^r}{b^r}$ for $b \neq 0$

v $\dfrac{a^r}{a^s} = a^{r-s}$

vi   If $a^r = a^s$, then $r = s$.

It is possible to extend the definition of exponents so as to include all real numbers and still preserve the above properties. We will use this result in Chapter 6 when we graph exponential functions.

EXAMPLES

Use the properties of exponents to simplify the following expressions. Assume that all bases are positive.

1   $\left(\dfrac{8x^6}{y^3}\right)^{2/3}$

SOLUTION

$$\left(\dfrac{8x^6}{y^3}\right)^{2/3} = \dfrac{8^{2/3}(x^6)^{2/3}}{(y^3)^{2/3}} = \dfrac{(8^{1/3})^2 x^4}{y^2}$$

$$= \dfrac{2^2 x^4}{y^2} = \dfrac{4x^4}{y^2}$$

2. $\dfrac{x^{-1/2}x}{x^{-2/3}}$

SOLUTION

$$\dfrac{x^{-1/2}x}{x^{-2/3}} = \dfrac{x^{-1/2+1}}{x^{-2/3}} = \dfrac{x^{1/2}}{x^{-2/3}} = x^{1/2-(-2/3)} = x^{7/6}$$

3. $(x^{-1/2} - y^{-1/2})(x^{-1/2} + y^{-1/2})$

SOLUTION

$$(x^{-1/2} - y^{-1/2})(x^{-1/2} + y^{-1/2}) = (x^{-1/2})^2 - (y^{-1/2})^2$$
$$= x^{-1} - y^{-1} = \dfrac{1}{x} - \dfrac{1}{y}$$

4. $\dfrac{x^{-3}y^2 z^{1/2}}{x^2 y^{-3} z^{-1}}$

SOLUTION

$$\dfrac{x^{-3}y^2 z^{1/2}}{x^2 y^{-3} z^{-1}} = x^{-3-2} \cdot y^{2-(-3)} \cdot z^{1/2-(-1)} = x^{-5}y^5 z^{3/2} = \dfrac{y^5 z^{3/2}}{x^5}$$

5. $\left(\dfrac{x^2 y^{-3}}{z^4}\right)^{-1/2}$

SOLUTION

$$\left(\dfrac{x^2 y^{-3}}{z^4}\right)^{-1/2} = \dfrac{(x^2)^{-1/2}(y^{-3})^{-1/2}}{(z^4)^{-1/2}} = \dfrac{x^{-1} y^{3/2}}{z^{-2}} = \dfrac{z^2 y^{3/2}}{x}$$

6. $\left(\dfrac{(-27)^{-1} x^{-9} y^6}{z^{-12}}\right)^{-1/3}$

SOLUTION

$$\left(\dfrac{(-27)^{-1} x^{-9} y^6}{z^{-12}}\right)^{-1/3} = \dfrac{(-27)^{1/3}(x^{-9})^{-1/3}(y^6)^{-1/3}}{(z^{-12})^{-(1/3)}}$$
$$= \dfrac{(-27)^{1/3} x^3 y^{-2}}{z^4}$$
$$= \dfrac{-3x^3}{y^2 z^4}$$
$$= -\dfrac{3x^3}{y^2 z^4}$$

PROBLEM SET 7

1 Compute the value of each of the following expressions.
   a) $64^{1/2}$
   b) $(-8)^{5/3}$
   c) $(25x^2)^{1/2}$
   d) $32^{3/5}$
   e) $(\frac{16}{25})^{-3/2}$
   f) $(-8)^{4/3}(-32)^{3/5}$
   g) $-125^{1/3}$
   h) $\dfrac{1}{81^{-3/4}}$
   i) $(-0.000027)^{4/3}$
   j) $(-125)^{1/3}$
   k) $(\frac{1}{9})^{-1/2}$
   l) $\dfrac{(81)^{3/4}}{8^{2/3}}$
   m) $(\frac{27}{8})^{2/3}$
   n) $(\frac{27}{64})^{-1/3}$
   o) $(\frac{25}{16})^{-1/2}$
   p) $(\frac{9}{16})^{3/2}$

2 Use the properties of exponents to simplify each of the following expressions. Assume that all bases are positive numbers.
   a) $4^{-1/2} \cdot 9^{1/2}$
   b) $(125)^{2/3} \cdot (27)^{-4/3}$
   c) $(0.0081)^{1/4} \cdot 9^{-1/2}$
   d) $(\frac{16}{81})^{5/4} \cdot (16)^{-3/4}$
   e) $(\frac{8}{125})^{-2/3} \cdot (125)^{1/3}$
   f) $(\frac{16}{25})^{-1/2} \cdot (\frac{25}{16})^{1/2}$
   g) $(32)^{1/5} \cdot (27)^{-1/3}$
   h) $(0.125)^{-1/3} \cdot (0.027)^{1/3}$
   i) $x^{1/3} x^{2/3}$
   j) $x^{1/4} \cdot x^{3/4}$
   k) $x^{1/3}(x^{-2/3} - x^{1/3})$
   l) $(27x^{3/5}x^{-3})^{1/3}$
   m) $x^{-2/3} x^{1/3}$
   n) $(32x^{16})^{1/5}$
   o) $(x+y)^{1/2}[(x+y)^{1/2}(x+y)^{-1/2}]$
   p) $(16x^{-6}x^{2/3})^{-1/2}$
   q) $x^{1/5}(x^{4/5} + x^{-6/5})$
   r) $(4x^{-2}y^4)^{1/2}$
   s) $(x-y)^{1/4}[(x-y)^{-1/4} - (x-y)^{-5/4}]$
   t) $x^{1/2}(x^{1/2} + x^{-1/2})$
   u) $\left(\dfrac{x^{1/2} \cdot y^{3/2}}{z^{1/2}}\right)^4$
   v) $x^{1/3}(x^{2/3} - x^{-1/3})$
   w) $(x^{-1/4}y^{-5/3})^2$
   x) $(9^{3/2}x^{-1/3}y^{-2/3})^3$
   y) $(8x^{-9}y^{3/4})^{-1/3}$
   z) $\dfrac{(27)^{2/3} \cdot x^{-5/3}}{y^{-1/3}}$

3 Simplify each of the following expressions, using only positive exponents.
   a) $\dfrac{5^{2/3} \cdot 3^{-1}}{5^{-1/3}}$
   b) $\dfrac{3^{5/2} \cdot 2^{-1/5}}{3^{-1/2}}$
   c) $\left(\dfrac{x^{-3}y^6}{x^{-4}y^7}\right)^{-5}$
   d) $\left(\dfrac{5^{-3} \cdot 4^6}{5^{-4} \cdot 4^7}\right)^{-2}$
   e) $\left(\dfrac{-5x^{-3}}{12y^2}\right)^{-1}$
   f) $\left(\dfrac{x^2y^3}{16x^{-2}}\right)^{-1/2}$
   g) $\left(\dfrac{81x^{-12}}{y^{16}}\right)^{-1/4}$
   h) $\left(\dfrac{125x^4y^3}{27x^{-2}y^6}\right)^{1/3}$

4 Simplify each of the following expressions, using only positive exponents.
   a) $\dfrac{3}{9^{1/2} + 2^{-1}}$
   b) $\dfrac{5^{-1} - 2^{-1}}{(125)^{1/3} + 3^{-1}}$

c) $\dfrac{x^{-3/4}}{x^{5/4} - x^{-7/4}}$

d) $\dfrac{7 - 5}{7^{1/2} - 5^{1/2}}$

e) $(x^{-2/3} - y^{5/3})^{-1}$

f) $(x^{-2/3} + 3y^{-2/3})(x^{-2/3} - 3y^{-2/3})$

g) $(x^{-1/2} - y^{-2/3})(x^{-1/2} + y^{-2/3})$

h) $(x^{-1} - y^{-1})(x^{-2} + y^{-2})$

i) $\dfrac{1 - 2x^{-2}}{x^{-1/2} - 2y^{-5/2}}$

j) $\dfrac{64x^{-7/3} + x^{-16/3}}{4x^{-1/3} + x^{-4/3}}$

k) $\dfrac{(32)^{4/5}(2^{n+1})^4}{8^3(16)^{n-1}}$

l) $\dfrac{x^{n/2} y^{n/4} z^{4/5}}{z^{1/5}}$

9 Radicals

The symbol $\sqrt{a}$ is another way of writing $a^{1/2}$, the principal square of a. If n is greater than 2, we use the symbol $\sqrt[n]{a}$ as another way of writing $a^{1/n}$, the nth root of a. The symbol $\sqrt[n]{\ }$ is called a *radical*, n is called the *index* (note that for the square root of a the index 2 is omitted), and a is called a *radicand*. Thus $\sqrt{4} = 4^{1/2} = 2$, $\sqrt[3]{8} = 8^{1/3} = 2$, and $\sqrt[5]{-32} = (-32)^{1/5} = -2$. In Section 8 it was pointed out that if a is negative and n is an even positive integer, then $a^{1/n}$ does not exist. For example, $\sqrt{-9}$ and $\sqrt{-16}$ are not real numbers. In general, we have the following principles:

i If $a > 0$, n any positive integer, $\sqrt[n]{a}$ is positive.

ii If $a < 0$, n any positive odd integer, $\sqrt[n]{a}$ is negative.

iii If $a < 0$, n any positive even integer, $\sqrt[n]{a}$ is not defined in the real numbers.

EXAMPLE

Compute the value of each of the following.

a) $\sqrt{25}$
b) $\sqrt[3]{-64}$
c) $\sqrt[4]{16}$
d) $\sqrt[3]{125}$
e) $\sqrt[5]{-32}$

SOLUTION

a) $\sqrt{25} = \sqrt{5^2} = 5$
b) $\sqrt[3]{-64} = \sqrt[3]{(-4)^3} = -4$
c) $\sqrt[4]{16} = \sqrt[4]{2^4} = 2$
d) $\sqrt[3]{125} = \sqrt[3]{5^3} = 5$
e) $\sqrt[5]{-32} = \sqrt[5]{(-2)^5} = -2$

9.1 Properties of Radicals

Since radicals provide alternate symbols for expressions of the form $a^{1/n}$, all properties of exponents also apply to radicals. In particular, we have the following. For m, n and c positive integers and assuming that each of the roots exist, we have

i $\sqrt[n]{a}\sqrt[n]{b} = \sqrt[n]{ab}$ since $a^{1/n}b^{1/n} = (ab)^{1/n}$

ii $\dfrac{\sqrt[n]{a}}{\sqrt[n]{b}} = \sqrt[n]{\dfrac{a}{b}}, \; b \neq 0$ since $\dfrac{a^{1/n}}{b^{1/n}} = \left(\dfrac{a}{b}\right)^{1/n}, \; b \neq 0$

iii $\sqrt[n]{a^m} = (\sqrt[n]{a})^m$ since $(a^m)^{1/n} = (a^{1/n})^m$

iv $\sqrt[m]{\sqrt[n]{a}} = \sqrt[n]{\sqrt[m]{a}} = \sqrt[mn]{a}$ since $(a^{1/n})^{1/m} = (a^{1/m})^{1/n} = a^{1/mn}$

v $\sqrt[cn]{a^{cm}} = \sqrt[n]{a^m}$ since $(a^{cm})^{1/cn} = a^{m/n}$

A negative rational power of a number can be expressed in radical form by considering the denominator of the exponent to be positive. For example,

$$4^{-3/2} = \sqrt{4^{-3}} = \sqrt{\tfrac{1}{64}} = \tfrac{1}{8}$$

and

$$8^{-2/3} = \sqrt[3]{8^{-2}} = \sqrt[3]{\tfrac{1}{64}} = \tfrac{1}{4}$$

If a is a real number, $\sqrt{a^2} = a$ only if a is positive or zero. For example, $\sqrt{2^2} = 2$, since $\sqrt{2^2} = \sqrt{4} = 2$. However, $\sqrt{(-2)^2} \neq -2$, since $\sqrt{(-2)^2} = \sqrt{4} = 2$ (remember the definition of principal roots). In this case, $\sqrt{(-2)^2} = -(-2)$. In general,

$\sqrt{a^2} = a$ if a is positive or zero
$\sqrt{a^2} = -a$ if a is negative

EXAMPLES

Use the properties of radicals to simplify each of the following expressions. Assume that all variables represent positive numbers.

1 $\sqrt{25^3}$

SOLUTION

$\sqrt{25^3} = (\sqrt{25})^3$ (Property iii)
$= 5^3 = 125$

2 $\sqrt[3]{8x^4}$

SOLUTION

$\sqrt[3]{8x^4} = \sqrt[3]{8x^3 \cdot x}$
$= \sqrt[3]{8x^3} \cdot \sqrt[3]{x}$ (Property i)
$= 2x\sqrt[3]{x}$

3 $\dfrac{\sqrt{8x^3}}{\sqrt{2x}}$

SOLUTION

$$\dfrac{\sqrt{8x^3}}{\sqrt{2x}} = \sqrt{\dfrac{8x^3}{2x}} \qquad \text{(Property ii)}$$
$$= \sqrt{4x^2}$$
$$= \sqrt{4}\cdot\sqrt{x^2} \qquad \text{(Property i)}$$
$$= 2x \qquad \text{(since } x > 0\text{)}$$

4 $\sqrt[3]{\sqrt{64}} = \sqrt[6]{64} \qquad \text{(Property iv)}$
$\qquad\qquad = 2$

5 $\sqrt{72}$

SOLUTION

$$\sqrt{72} = \sqrt{36\cdot 2}$$
$$= \sqrt{36}\sqrt{2} \qquad \text{(Property i)}$$
$$= 6\sqrt{2}$$

6 $\sqrt[3]{-125x^8y^{10}}$

SOLUTION

$$\sqrt[3]{-125x^8y^{10}} = \sqrt[3]{-125x^6y^9 x^2 y}$$
$$= \sqrt[3]{-125x^6y^9}\sqrt[3]{x^2y} \qquad \text{(Property i)}$$
$$= -5x^2y^3\sqrt[3]{x^2y}$$

7 $\sqrt[n]{x^{3n}\cdot y^{5n+2}}$

SOLUTION

$$\sqrt[n]{x^{3n}y^{5n+2}} = \sqrt[n]{3^n y^{5n} y^2}$$
$$= \sqrt[n]{3^n y^{5n}}\sqrt[n]{y^2}$$
$$= (3^n y^{5n})^{1/n}\sqrt[n]{y^2}$$
$$= 3^{n/n} y^{5n/n}\sqrt[n]{y^2}$$
$$= 3y^5 \sqrt[n]{y^2}$$

8 $\sqrt[4]{x^3}\cdot\sqrt[3]{x^2}$

SOLUTION

$$\sqrt[4]{x^3}\sqrt[3]{x^2} = x^{3/4}\cdot x^{2/3}$$
$$= x^{3/4+2/3}$$
$$= x^{17/12}$$
$$= x\cdot x^{5/12}$$
$$= x\sqrt[12]{x^5}$$

9.2 Addition and Subtraction of Radicals

Expressions containing radicals can be added and subtracted by combining similar terms. For example, $3\sqrt{2} + 4\sqrt{2} - \sqrt{2} = 6\sqrt{2}$, and $3\sqrt{5} + 7\sqrt{5} - 3\sqrt{5} = 7\sqrt{5}$. *Similar radicals* are those radicals which can be expressed with the same index and radicand. Thus, $\sqrt{50}$ and $\sqrt{72}$ are similar radicals, since $\sqrt{50} = 5\sqrt{2}$ and $\sqrt{72} = 6\sqrt{2}$.

On the other hand, $\sqrt{50}$ and $\sqrt{75}$ are not similar radicals, for $\sqrt{50} = 5\sqrt{2}$, whereas $\sqrt{75} = 5\sqrt{3}$. Similar radicals such as $7\sqrt{3} + 3\sqrt{3}$ and $8\sqrt[4]{x} - 3\sqrt[4]{x}$ can be combined by applying the distribution property of real numbers. Expressions such as $\sqrt{72} + \sqrt{50}$ must be simplified before they can be combined. Thus, $\sqrt{50} + \sqrt{72} = 5\sqrt{2} + 6\sqrt{2} = 11\sqrt{2}$.

EXAMPLES

Simplify by combining similar terms.

1. $4\sqrt{12} + 5\sqrt{8} - \sqrt{50}$

   SOLUTION

   $$\begin{aligned}
   4\sqrt{12} + 5\sqrt{8} - \sqrt{50} &= 4\sqrt{4 \cdot 3} + 5\sqrt{4 \cdot 2} - \sqrt{25 \cdot 2} \\
   &= 4\sqrt{4}\sqrt{3} + 5\sqrt{4}\sqrt{2} - \sqrt{25}\sqrt{2} \\
   &= 4 \cdot 2\sqrt{3} + 5 \cdot 2\sqrt{2} - 5\sqrt{2} \\
   &= 8\sqrt{3} + 10\sqrt{2} - 5\sqrt{2} \\
   &= 8\sqrt{3} + (10 - 5)\sqrt{2} \\
   &= 8\sqrt{3} + 5\sqrt{2}
   \end{aligned}$$

2. $\sqrt{50} + \sqrt[4]{64}$

   SOLUTION

   $$\begin{aligned}
   \sqrt{50} + \sqrt[4]{64} &= \sqrt{2 \cdot 25} + \sqrt[4]{16 \cdot 2^2} \\
   &= 5\sqrt{2} + 2\sqrt[4]{2^2} \\
   &= 5\sqrt{2} + 2\sqrt{2} \\
   &= 7\sqrt{2}
   \end{aligned}$$

3. $\sqrt{\frac{1}{2}} - \sqrt{\frac{1}{8}} + \sqrt{\frac{9}{32}}$

   SOLUTION

   $$\begin{aligned}
   \sqrt{\frac{1}{2}} - \sqrt{\frac{1}{8}} + \sqrt{\frac{9}{32}} &= \frac{1}{\sqrt{2}} - \frac{1}{2\sqrt{2}} + \frac{3}{4\sqrt{2}} \quad \text{(why?)} \\
   &= \frac{1}{\sqrt{2}}\left(1 - \tfrac{1}{2} + \tfrac{3}{4}\right) \\
   &= \frac{1}{\sqrt{2}}\left(\tfrac{5}{4}\right) = \frac{5}{4\sqrt{2}}
   \end{aligned}$$

4    $\sqrt{x^3yz^5} + \sqrt{xy^7z^3} + \sqrt{x^9y^5z}$ x, y, and $z > 0$

SOLUTION

$$\sqrt{x^3yz^5} + \sqrt{xy^7z^3} + \sqrt{x^9y^5z}$$
$$= \sqrt{x^2z^4xyz} + \sqrt{y^6z^2xyz} + \sqrt{x^8y^4xyz}$$
$$= \sqrt{x^2z^4}\sqrt{xyz} + \sqrt{y^6z^2}\sqrt{xyz} + \sqrt{x^8y^4}\sqrt{xyz}$$
$$= xz^2\sqrt{xyz} + y^3z\sqrt{xyz} + x^4y^2\sqrt{xyz}$$
$$= (xz^2 + y^3z + x^4y^2)\sqrt{xyz}$$

5    $\sqrt[3]{2x^4y} + 5\sqrt[3]{16xy} - \sqrt[3]{54x^4y}$ $x > 0, y > 0$

SOLUTION

$$\sqrt[3]{2x^4y} + 5\sqrt[3]{16xy} - \sqrt[3]{54x^4y} = \sqrt[3]{x^3(2xy)} + 5\sqrt[3]{8(2xy)}$$
$$- \sqrt[3]{27x^3(2xy)}$$
$$= x\sqrt[3]{2xy} + 5(2)\sqrt[3]{2xy} - 3x\sqrt[3]{2xy}$$
$$= x\sqrt[3]{2xy} + 10\sqrt[3]{2xy} - 3x\sqrt[3]{2xy}$$
$$= 10\sqrt[3]{2xy} - 2x\sqrt[3]{2xy}$$
$$= (10 - 2x)\sqrt[3]{2xy}$$

9.3   Rationalizing Denominators

We stated that $\sqrt{2}$ is an irrational number. In many problems in mathematics, numbers such as $1/\sqrt{2}$ have irrational denominators. This form is acceptable unless it is desirable to approximate the result by a rational number. Since $\sqrt{2}$ is approximately 1.414, then $1/\sqrt{2}$ is approximately $1/1.414$. However, $1/\sqrt{2}$ can be written as $\sqrt{2}/2$ simply by multiplying the numerator and the denominator of the fraction by $\sqrt{2}$. The form $\sqrt{2}/2$ is easier to approximate than the form $1/\sqrt{2}$, since $1.414/2 = 0.707$, while $1/1.414$ requires a more involved division. Thus,

$$\frac{5}{\sqrt{3}} = \frac{5\sqrt{3}}{\sqrt{3}\cdot\sqrt{3}} = \frac{5\sqrt{3}}{3} \quad \text{and} \quad \frac{4}{\sqrt{5}} = \frac{4\sqrt{5}}{5}$$

The process of eliminating radicals from the denominator of a fraction is called *rationalizing the denominator*. To rationalize a denominator that is an indicated sum of terms involving square roots we use the identity

$$(a + b)(a - b) = a^2 - b^2$$

For example, to rationalize the denominator of $(2 + \sqrt{3})/(\sqrt{5} - \sqrt{3})$, we note that $(\sqrt{5} - \sqrt{3})(\sqrt{5} + \sqrt{3}) = (\sqrt{5})^2 - (\sqrt{3})^2 = 5 - 3 = 2$.

Thus,

$$\frac{2+\sqrt{3}}{\sqrt{5}-\sqrt{3}} = \frac{(2+\sqrt{3})(\sqrt{5}+\sqrt{3})}{(\sqrt{5}-\sqrt{3})(\sqrt{5}+\sqrt{3})}$$
$$= \frac{2\sqrt{5}+2\sqrt{3}+\sqrt{3}\sqrt{5}+3}{2}$$

EXAMPLES

Rationalize the denominators of the following fractions.

1. $\dfrac{5}{\sqrt{3}-1}$

SOLUTION

$$\frac{5}{\sqrt{3}-1} = \frac{5(\sqrt{3}+1)}{(\sqrt{3}-1)(\sqrt{3}+1)}$$
$$= \frac{5\sqrt{3}+5}{(\sqrt{3})^2 - 1^2} = \frac{5\sqrt{3}+5}{3-1}$$
$$= \frac{5\sqrt{3}+5}{2}$$

2. $\dfrac{5}{\sqrt[3]{16}}$

SOLUTION

$$\frac{5}{\sqrt[3]{16}} = \frac{5}{\sqrt[3]{4^2}}$$
$$= \frac{5\sqrt[3]{4}}{\sqrt[3]{4^2}\sqrt[3]{4}}$$
$$= \frac{5\sqrt[3]{4}}{\sqrt[3]{4^3}}$$
$$= \frac{5\sqrt[3]{4}}{4}$$

3. $\dfrac{\sqrt{5}+\sqrt{2}}{\sqrt{5}-\sqrt{2}}$

SOLUTION

$$\frac{\sqrt{5}+\sqrt{2}}{\sqrt{5}-\sqrt{2}} = \frac{(\sqrt{5}+\sqrt{2})(\sqrt{5}+\sqrt{2})}{(\sqrt{5}-\sqrt{2})(\sqrt{5}+\sqrt{2})}$$

SECTION 9 RADICALS

$$= \frac{(\sqrt{5})^2 + 2\sqrt{2}\sqrt{5} + (\sqrt{2})^2}{(\sqrt{5})^2 - (\sqrt{2})^2}$$

$$= \frac{5 + 2\sqrt{10} + 2}{5 - 2}$$

$$= \frac{7 + 2\sqrt{10}}{3}$$

4 $\quad \dfrac{3}{\sqrt{x} - 1} \quad x > 0$

SOLUTION

$$\frac{3}{\sqrt{x} - 1} = \frac{3(\sqrt{x} + 1)}{(\sqrt{x} - 1)(\sqrt{x} + 1)}$$

$$= \frac{3\sqrt{x} + 3}{(\sqrt{x})^2 - (1)^2}$$

$$= \frac{3\sqrt{x} + 3}{x - 1}$$

5 $\quad \dfrac{\sqrt{x} + \sqrt{y}}{\sqrt{x} - \sqrt{y}} \quad x, y > 0$

SOLUTION

$$\frac{\sqrt{x} + \sqrt{y}}{\sqrt{x} - \sqrt{y}} = \frac{(\sqrt{x} + \sqrt{y})(\sqrt{x} + \sqrt{y})}{(\sqrt{x} - \sqrt{y})(\sqrt{x} + \sqrt{y})}$$

$$= \frac{(\sqrt{x})^2 + 2\sqrt{x}\sqrt{y} + (\sqrt{y})^2}{(\sqrt{x})^2 - (\sqrt{y})^2}$$

$$= \frac{x + 2\sqrt{xy} + y}{x - y}$$

PROBLEM SET 8

1 Write each of the following expressions in radical form. b, c, x, and y are positive.

a) $3^{1/2}$
b) $x^{-3/2} + y^{-3/2}$
c) $(16)^{3/4}$
d) $2x^{1/3}$
e) $(x + y)^{-3/2}$
f) $(243)^{1/5}$
g) $(2x)^{1/3}$
h) $-x^{3/4}, x > 0$
i) $\left(\dfrac{-21}{64}\right)^{1/3}$
j) $(25b^2 c^4)^{1/2}$

2 Write each of the following expressions in exponential form. x and y are positive.

a) $\sqrt[3]{8}$
b) $\sqrt[4]{16}$
c) $\sqrt[3]{7}$
d) $\sqrt[6]{5}$

e) $-\sqrt[3]{xy}$
g) $\sqrt{x} - \sqrt{y}$
i) $\sqrt[3]{100x^4y^2}$
f) $-\sqrt[6]{x^{18}}$
h) $\dfrac{x}{\sqrt[3]{y}}$

3 Use the properties of radicals to simplify the following expressions. (Assume the variables represent positive numbers.)

a) $\sqrt{9}$
b) $-\sqrt{64}$
c) $\sqrt{300}$
d) $\sqrt[3]{54}$
e) $\sqrt[4]{162}$
f) $\sqrt{3} \cdot \sqrt{12}$
g) $\sqrt{\sqrt[3]{192}}$
h) $\dfrac{\sqrt{6x}\sqrt{5x}}{\sqrt{10}}$
i) $\dfrac{3x}{\sqrt{x}}$
j) $\dfrac{\sqrt[3]{-2x^2}\sqrt[3]{-4x}}{2x}$
k) $\sqrt[4]{\dfrac{625x^8}{y^4}}$
l) $\sqrt[5]{\dfrac{32x^5}{243y^{10}}}$
m) $\sqrt{\dfrac{81x^6y^6}{49z^2w^2}}$
n) $\sqrt[4]{\dfrac{x^2y^3z}{x^3yz^3}}$
o) $\sqrt[3]{625x^{10}y^5}$
p) $\sqrt[4m]{x^{2m}y^{10m}}$
q) $\sqrt{8} \cdot \sqrt{18}$
r) $\sqrt[4]{8} \cdot \sqrt[4]{32}$
s) $\sqrt[5]{4} \cdot \sqrt[5]{8}$
t) $\sqrt{6xy^3} \cdot \sqrt{2x^2y}$
u) $\sqrt[3]{xy^2} \cdot \sqrt[3]{x^2y}$
v) $\sqrt[4]{x^3y^5} \cdot \sqrt[6]{xy^4}$
w) $\sqrt[5]{xy^2} \cdot \sqrt[3]{x^2y^5}$
x) $\sqrt[5]{x^2y^4} \cdot \sqrt[3]{x^4y} \cdot \sqrt[3]{x^3y^4}$

4 Find the products of the following expressions. (Assume the variables represent positive numbers.)

a) $(\sqrt{5} + \sqrt{2})(\sqrt{5} - \sqrt{2})$
b) $(\sqrt{6} - \sqrt{2})(\sqrt{6} + \sqrt{2})$
c) $(\sqrt{x} + y)(\sqrt{x} - y)$
d) $(\sqrt{x} + \sqrt{y})(\sqrt{x} - \sqrt{y})$
e) $(\sqrt{x} + \sqrt{y})(x - \sqrt{xy} + y)$
f) $(\sqrt{x} - \sqrt{y})(x + \sqrt{xy} + y)$

5 Simplify by combining similar terms, assuming that all variables are positive.

a) $4\sqrt{3} - 5\sqrt{12} + 2\sqrt{75}$
b) $\sqrt[3]{2} + \sqrt[3]{16} - \sqrt[3]{54}$
c) $5\sqrt{2} - \sqrt[4]{64} + 2\sqrt{32}$
d) $\sqrt{x^3} + \sqrt{25x^3} + \sqrt{9x}$
e) $\sqrt[3]{16} + \sqrt[3]{2} + \sqrt[3]{128}$
f) $\sqrt[3]{81} + \sqrt[3]{256} - \sqrt[3]{24} - \sqrt[3]{108}$
g) $\dfrac{\sqrt{25x}}{x} + \dfrac{\sqrt{6x^3}}{x^2} - \dfrac{8}{\sqrt{x}}$
h) $\sqrt[3]{2xy} - 3\sqrt[3]{16xy} - \sqrt[3]{2x^4y}$
i) $\sqrt[3]{8x^3} - \sqrt[3]{125x^2} - \sqrt[3]{x^2} + 6\sqrt{64x^2}$
j) $\sqrt{x^3} - 2x\sqrt[4]{x^2} + 3x\sqrt[6]{x^3} - \sqrt[8]{x^{12}}$

6 Rationalize the denominators of each of the following expressions. Assume that all variables represent positive numbers.

a) $\dfrac{2}{\sqrt{3}}$

b) $\dfrac{\sqrt{2}}{\sqrt{5}}$

c) $\dfrac{8}{\sqrt[3]{49}}$

d) $\dfrac{x-y}{\sqrt[3]{x^2}}$

e) $\dfrac{1}{1+\sqrt{5}}$

f) $\dfrac{3}{2+\sqrt{3}}$

g) $\dfrac{\sqrt{2}+\sqrt{3}}{\sqrt{2}-\sqrt{3}}$

h) $\dfrac{y}{\sqrt{x}-\sqrt{y}}$

i) $\dfrac{x}{x+\sqrt{y}}$

j) $\dfrac{1}{\sqrt{2}+\sqrt{3}+\sqrt{5}}$

k) $\dfrac{\sqrt{3}-3\sqrt{5}}{4\sqrt{3}+3\sqrt{5}}$

l) $\dfrac{1}{1+\sqrt{3}}$

m) $\dfrac{2\sqrt{3}+\sqrt{2}}{\sqrt{6}+2\sqrt{2}}$

n) $\dfrac{2\sqrt{2}}{1-\sqrt{2}+\sqrt{3}}$

REVIEW PROBLEM SET

1 Use the following sets for each of the parts.
   N, the set of natural numbers
   I, the set of integers
   F, the set of quotient numbers (rational numbers which are not integers)
   Q, the set of rational numbers
   L, the set of irrational numbers
   R, the set of real numbers
   ∅, the empty set

Each description corresponds to one or more than one of the above sets. Identify these sets.
a) A set that contains $4 \cdot 9$ but not $-4 \cdot 9$.
b) A set that contains both $\tfrac{2}{3}$ and π.
c) A set that contains $-\tfrac{3}{4}$ but not π.
d) The union of the natural numbers and the integers.
e) The intersection of the quotient numbers and the rational numbers.
f) The union of the irrational numbers and the real numbers.
g) A subset of the quotient numbers which contains $\tfrac{3}{2}$ but not 3.
h) A subset of the real numbers which does not contain 1 but which is not the empty set.
i) A subset of the irrational numbers which does not contain π.
j) A set which is disjoint from the natural numbers and which is not a subset of the integers.

2. Let $A = \{a, b, c\}$, $B = \{c, d, e, f\}$, and $C = \{a, c, d, g\}$. Find each of the following.
   a) $A \cap B$
   b) $B \cup A$
   c) $B \cap C$
   d) $B \cup C$
   e) $A \cap C$
   f) $C \cup \emptyset$
   g) $C \cap \emptyset$
   h) $A \cap \emptyset$
   i) $C \cup A$
   j) $B \cup \emptyset$

3. If N is the set of natural numbers, I is the set of integers, and Q is the set of rational numbers, use Venn diagrams to indicate which of the following are true and which are false.
   a) $N \subset Q$
   b) $N \subset (Q \cup I)$
   c) $I \cap Q = N$
   d) $N \cap Q = N \cap I$
   e) $N \cup Q = I \cup N$

4. Determine which of the following has a terminating decimal representation. If it has a terminating decimal representation, find it.
   a) $\frac{2}{7}$
   b) $\frac{7}{40}$
   c) $\frac{5}{14}$
   d) $\frac{3}{11}$
   e) $\frac{17}{25}$
   f) $\frac{7}{9}$
   g) $\frac{6}{13}$
   h) $\frac{11}{22}$

5. Perform the indicated operations and simplify if possible.
   a) $(\frac{1}{3})^3 (\frac{1}{3})^5$
   b) $\frac{(\frac{1}{5})^5}{(\frac{1}{5})^2}$
   c) $x^7 \cdot x^6 \cdot x^8$
   d) $b^{2n+1} b^3$
   e) $(ab^n)(ab)^n$
   f) $x^{n(n+2)} \div x^n$
   g) $2^n \cdot 4^{n+1} \cdot 8^{n+1}$
   h) $(4^n)^{2n-1} \cdot (16)^{n+2} \div (32)^{2n+1}$

6. Determine the degree and the coefficients of each of the following polynomials.
   a) $4x^2 - 39x + 100$
   b) $x^4 - 3x + 5$
   c) $\frac{81}{35}x^3 - 2x^2 + 25$
   d) $x^3 - \frac{10}{7}x + 13$
   e) $4x^2 + 36x + 14$
   f) $4x^2 + 15x + 3$
   g) $x^4 - 3$
   h) $3x^5 - 3x^4 + 7$

7. Perform the following operations.
   a) $(x^3 - 3x^2y + 2xy^2 - y^3) + (6x^2 - 3xy^2 + y^3)$
   b) $(7x^2 + 5x - 3) + (13x^2 - 15x + 7)$
   c) $(x^8 - 5x^7 + 13) + (2x^8 - x^7 + 6) + (-3x^8 - x^3 + 44)$
   d) $(2x^2 + 1) + (5x^2 + x - 13)$
   e) $(x^2 - 5) + (3 - 2x^2) + (15 + 4x^2)$

8. Perform the following operations.
   a) $(3x^2 + 5x - 3) - (6x^2 - 3x + 2)$
   b) $(7x^2 + 3x + 17) - (13x^2 - 15x + 6)$
   c) $(2x^3 - 5x^2 + 9x + 1) - (x^3 - 3x^2 + 7)$
   d) $(5x^4 - 3x^2 + 14) - (3x^4 - 2x^2 + 5)$

9 Determine the following products.
  a) $(4x + 3)(2x + 7)$
  b) $(5x - 2)(3 + 4x)$
  c) $(-2x)(3x^2 - 7)$
  d) $(2x - 1)(4x^2 + 2x + 1)$
  e) $(13x - 7)(3x + 4)$
  f) $(1 - 3x)(2x + 3)$
  g) $(3x + 5)(7 + 3x)$
  h) $(2x - y)(2x^2 - 5xy + y^2)$
  i) $(39x^2 + 4)(1 - 2x^2)$
  j) $(x + 4)(3x - 13)$
  k) $(x + 2y + 3z)(x - 2y - 3z)$
  l) $(x^2 + 3x - 2)(x^2 + 3x + 2)$
  m) $(5x - 7y + 3)(11x + 5)$
  n) $(2x + 3y + 1)(2x - 3y - 1)$
  o) $(2x - y)^2$
  p) $(3x + 5y)^2$
  q) $(2x - y + z)^2$
  r) $(a - 2b + y)^2$
  s) $(2x - 3y - 5z)^2$
  t) $(x + 2y)^3$
  u) $(2x - 3y)^3$
  v) $(2x - y)^4$
  w) $(x + 2y)^4$
  x) $(x + 2y)^5$

10 Factor each of the following expressions completely.
  a) $a^2b^2c^2 + 5abc + 4$
  b) $x^2 - 4x + 4 - 4y^2 + 4y - 1$
  c) $(ab + 2)^2 - 4$
  d) $x^3y^3 + 27z^3$
  e) $(x + y)^3 - (x - y)^3$
  f) $(x^2 - 1)^2 - 11(x^2 - 1) + 24$
  g) $x^3y - 5x^2y + 4xy$
  h) $x^2 - y^2 + 4(x - y)$
  i) $x^2 - y^2 - (ax - ay)$
  j) $144x^4y^8 - 1 - 18z - 81z^2$
  k) $16x^4 - 625y^4$
  l) $x^4 + x^2 - 42$
  m) $x^2y^2 - x^2 - y^2 + 1$
  n) $125(x - y)^3 - z^3$
  o) $\dfrac{x^3}{8} + \dfrac{y^3}{27}$
  p) $125 + 343x^3$
  q) $2x^5 - 32x$
  r) $x^4 + 4x^3y + 4x^2y^2$
  s) $4x^4 + 1$
  t) $4x^4 + 3x^2 + 1$

11 Reduce each of the following fractions to lowest terms.
  a) $\dfrac{6xy}{9yz}$
  b) $\dfrac{36xy^3z^9}{54x^8y^2z^3}$
  c) $\dfrac{46x(a + b)^3}{69y(a + b)}$
  d) $\dfrac{96a^2bx^2y^3}{84a^2b^2x^2y^2}$
  e) $\dfrac{x^2 - 2xy + y^2}{x^2 - y^2}$
  f) $\dfrac{x^2 - 7x + 12}{x^2 + 3x - 18}$
  g) $\dfrac{(3x - 2y)(2a - b)}{(2x - 3y)(b - 2a)}$
  h) $\dfrac{(a - b)(x + y)}{(b - a)(x - y)}$

12 Perform each of the following operations.
  a) $\dfrac{2}{3x} - \dfrac{1}{4x}$
  b) $\dfrac{a}{x^3y} + \dfrac{b}{x^2y^2}$
  c) $\dfrac{5}{xy} - \dfrac{6}{yz} + \dfrac{7}{xz}$
  d) $\dfrac{3}{2xy} - \dfrac{4}{3yz} + \dfrac{5}{4x^2}$
  e) $\dfrac{1}{x + 3} + \dfrac{1}{x - 2}$
  f) $\dfrac{1}{x - y} + \dfrac{1}{x + y}$

g) $\dfrac{5x}{7x+2y} - \dfrac{6y}{3x+4y}$ h) $\dfrac{1}{x^2-y^2} + \dfrac{1}{x^2+2xy+y^2}$

i) $\dfrac{2}{x-2} - \dfrac{3}{x+3} + \dfrac{4}{x-4}$ j) $\dfrac{1}{(x-y)^2} + \dfrac{x}{x-y} + 1$

13 Perform each of the following operations, and simplify your answer.

a) $\dfrac{a^4 x^2 y^3}{b^2 a z^5} \cdot \dfrac{ab^2 x^2 z^2}{ay^3}$ b) $\dfrac{36 x^2 y}{27 x^3 y^2}$

c) $\left(1 + \dfrac{x}{y}\right) \cdot \left(\dfrac{y-x}{x}\right)$ d) $\left(x + \dfrac{y}{z}\right) \cdot \left(\dfrac{1}{y} - \dfrac{x}{z}\right)$

e) $\left(\dfrac{3xy^2}{10xy^4}\right)^2 \cdot \left(\dfrac{5xy^2}{2xy^2 y^3}\right)^3$ f) $\dfrac{2x-4}{9} \cdot \dfrac{6}{5x-10}$

g) $\dfrac{x^2+x}{y} \cdot \dfrac{y^2}{x+1}$ h) $\dfrac{x^2-16}{x^2-4x} \cdot \dfrac{x-4}{x+4}$

i) $\dfrac{(x+y)^2}{a+b} \cdot \dfrac{(a+b)^2}{x+y}$ j) $\dfrac{x^2+5x+6}{x^2+x-2} \cdot \dfrac{x^2+3x-4}{x^2+7x+12}$

k) $\dfrac{(x+y)^2}{x^2+y} \cdot \dfrac{x^3+y^3}{(x+y)^3}$ l) $\dfrac{x^2-9}{x^2-4} \div \dfrac{x+3}{x-2}$

m) $\dfrac{x+7}{x+2} \div \dfrac{x^2-49}{x^2-4}$ n) $\dfrac{a-b}{(x-y)^2} \div \dfrac{a^2-2ab+b^2}{x^2-y^2}$

o) $\left(1 - \dfrac{a+b}{a^2+b^2}\right) \div \dfrac{(a-b)^2}{a^4-b^4}$

p) $\dfrac{x}{x-y} \cdot \dfrac{x+y}{y} \cdot \dfrac{y^2+x^2}{y^3-x^3} \div \dfrac{x^3+y^3}{x^4+x^2 y^2+y^4}$

14 Simplify each of the following complex fractions.

a) $\dfrac{\tfrac{2}{3}+\tfrac{1}{6}}{1-\tfrac{3}{4}}$ b) $\dfrac{\tfrac{x}{y}+1}{\tfrac{y}{x}-\tfrac{x}{y}}$

c) $\dfrac{\tfrac{y}{y^2-1} - \tfrac{1}{y+1}}{\tfrac{y}{y-1} + \tfrac{1}{y+1}}$ d) $\dfrac{\tfrac{1}{x} - \tfrac{1}{y}}{\dfrac{x^2-y^2}{xy}}$

e) $\dfrac{\tfrac{1}{2x-3} - \tfrac{1}{2x+3}}{4 - \tfrac{9}{x^2}}$ f) $\dfrac{2}{3 - \tfrac{4}{5-\tfrac{6}{7}}}$

g) $\dfrac{1}{1 - \tfrac{1}{x-1}}$ h) $1 - \dfrac{1}{1 - \tfrac{1}{1-y}}$

15 Express each of the following expressions using only positive exponents and simplify your answer when possible.

a) 3^{-4} b) $\left(\tfrac{1}{3}\right)^{-4}$

c) $\dfrac{2^7}{(-2)^5}$ d) $(-32x)^0$

e) $\dfrac{5^{-4}}{5^{-2}}$
f) $5^0 \cdot 5^{-2}$
g) $(3^{-2} - 3^{-1})^{-1}$
h) $(3x^{-2}y^3)^{-3}$
i) $3^4 \cdot 3^{-7} \cdot 9^0$
j) $(5x)^{-3}(-5^{-1}y^{-3})^{-4}$
k) $-(5^{-1} - x^{-1})^{-1}$
l) $\dfrac{wy^{-2} + yw^{-2}}{y^{-1} + w^{-1}}$

16 Simplify the following expressions.
a) $8^{1/3}$
b) $(81)^{1/4}$
c) $(x^8)^{1/2}$
d) $(16x^4)^{1/2}$
e) $(-8x^6)^{1/3}$
f) $(-27x^9)^{2/3}$
g) $(-32x^5)^{2/5}$
h) $(4x^4y^8)^{3/2}$
i) $(16x^{16}y^{12})^{3/4}$
j) $(8x^6y^9)^{2/3}$
k) $\left(\dfrac{16x^4}{y^8}\right)^{-3/4}$
l) $(-27x^{-27})^{-2/3}$
m) $\left(-\dfrac{x^5y^{10}}{32z^{15}}\right)^{-3/5}$
n) $\left(\dfrac{27x^{-27}}{8^{-1} \cdot y^9}\right)^{-1/3}$

17 Change each of the following expressions to simplest radical form.
a) $\sqrt{18}$
b) $\sqrt[3]{81}$
c) $\sqrt[3]{15x^4y^5}$
d) $\sqrt[3]{-16x^{16}}$
e) $\sqrt[4]{32x^5y^{10}}$
f) $\sqrt[3]{-125x^5}$
g) $\sqrt[3]{54x^4y^5z^7}$
h) $\sqrt[4]{32x^5y^{10}z^{15}}$
i) $\sqrt{\dfrac{125xy^3}{4x^2y^4}}$
j) $\sqrt[3]{\dfrac{16x^4y^7}{27x^3y^6}}$
k) $\sqrt[4]{\dfrac{405x^7}{16y^8}}$
l) $\sqrt[n]{\dfrac{x^{2n+1}}{x^{-n+2}}}$
m) $\sqrt[3]{\dfrac{x^3\sqrt{x^5}}{x^4}}$
n) $\sqrt[5]{x^{-3}\sqrt[3]{x^4}\sqrt{x^{-5}}}$

18 Simplify each of the following expressions.
a) $7\sqrt{2} - 3\sqrt{2} + 4\sqrt{2}$
b) $\sqrt{5} - 6\sqrt{5} + 2\sqrt{5}$
c) $4\sqrt{x} + 5\sqrt{x} - 5\sqrt{x}$
d) $3\sqrt[3]{5} - 2\sqrt[3]{5} + 5\sqrt[3]{5}$
e) $\sqrt{63} + 2\sqrt{112} - \sqrt{252}$
f) $\sqrt[3]{16} - \sqrt[3]{54} + \sqrt[3]{250}$
g) $4\sqrt[4]{4} + 6\sqrt[6]{8} + 8\sqrt[8]{16}$
h) $\sqrt{xy} \cdot \sqrt{xz} \cdot \sqrt{yz}$
i) $\sqrt[3]{18x} \cdot \sqrt[3]{15x}$
j) $(\sqrt{6} + \sqrt{5}) \cdot (\sqrt{6} - \sqrt{5})$
k) $(\sqrt{6} + 2) \cdot (\sqrt{6} - 2)$
l) $(2\sqrt{6} - 3\sqrt{2})^2$
m) $(2 - \sqrt{2} + \sqrt{5}) \cdot (2 - \sqrt{2} - \sqrt{5})$

19 Rationalize each of the following denominators.
a) $\dfrac{4}{\sqrt{3}}$
b) $\dfrac{10}{\sqrt[3]{5x}}$
c) $\dfrac{6}{5\sqrt{3x}}$
d) $\dfrac{6x}{\sqrt{162xy}}$

e) $\dfrac{6x}{\sqrt{27xy}}$ f) $\dfrac{6x^2y}{\sqrt[3]{9x^2y}}$

g) $\dfrac{10}{5-\sqrt{15}}$ h) $\dfrac{6}{\sqrt{2}+3\sqrt{5}}$

i) $\dfrac{x-\sqrt{y}}{x-2\sqrt{y}}$ j) $\dfrac{\sqrt{5}-x}{\sqrt{5}+x}$

k) $\dfrac{\sqrt{8}}{3+\sqrt{2}}$ l) $\dfrac{\sqrt{7}-\sqrt{6}}{\sqrt{5}-\sqrt{6}}$

m) $\dfrac{2\sqrt{13}-\sqrt{2}}{\sqrt{13}+\sqrt{2}}$ n) $\dfrac{2}{1+\sqrt{2}-\sqrt{3}}$

o) $\dfrac{7}{1+\sqrt[3]{5}}$

CHAPTER 2

Equations, Order Relations, and the Cartesian Plane

CHAPTER 2

Equations, Order Relations, and Cartesian Plane

1 Introduction

The chapter begins with the study of equality of two algebraic expressions. Next, order relations and their properties are considered in order to investigate solutions of inequalities. Other topics covered in this chapter include absolute value equations and inequalities, the Cartesian coordinate system, and the distance formula.

2 First-Degree Equations

An *equation* is a mathematical statement that expresses the relation of equality between two expressions representing real numbers.

For example, $15 - 2 = 13$ is an equation. Likewise, $3x + 6 = 14$ is an equation.

Equations such as $2x - 1 = 5$, $3(x + 2) = 7 - 2(x + 1)$, and $\frac{1}{3}x + 3 = 5 - x$ are called *first-degree equations* or *linear equations*. In general, a *first-degree equation* is an equation in which the variable or unknown has degree 1. Any value of the variable x that makes an equation a true statement is called a *solution* of the equation. For example, $x = 2$ is a solution of $x^2 = 4$, as is $x = -2$, since $(2)^2 = 4$ and $(-2)^2 = 4$. To solve an equation means to find *all* values of the variables for which the equation becomes a true statement. Hence, $x = 2$ or $x = -2$ expresses the solution of $x^2 = 4$. Also, in the equation $3x + 6 = 14$, $x = \frac{8}{3}$ is the solution to the equation, since $3(\frac{8}{3}) + 6 = 8 + 6 = 14$.

The solutions of an equation form a set, and this set is called the *solution set* of the equation. The solution set of $x^2 = 4$ is $\{-2, 2\}$ and the solution set of the equation $3x + 6 = 14$ is designated by $\{\frac{8}{3}\}$. An equation may have one, more than one, or no solutions. If there are no solutions, then its solution set is designated by the empty set $\emptyset$. In this section we shall be concerned primarily with linear equations in one variable that have one solution.

Two equations are *equivalent* if and only if they have the same solution sets. For example, the equations

$$5x + 2 = 7$$
$$5x = 5$$

and

$$x = 1$$

are equivalent equations, since they each have the same solution set $\{1\}$. If 1 is substituted for x in each of these equations, the equation is satisfied. No other value of x will satisfy any of the equations.

To solve first-degree equations, we make use of the following properties.

1 ADDITION PROPERTY

The addition of the same number to both sides of an equation produces an equivalent equation.

2 MULTIPLICATION PROPERTY

The multiplication of both sides of an equation by the same nonzero number produces an equivalent equation.

The process of solving an equation consists of using the above properties to convert the given equation to an equivalent equation in which the solution is apparent. This process is illustrated in the following examples.

EXAMPLES

Solve the following first-degree equations.

1. $3x - 1 = 11$

    SOLUTION

    $$\begin{aligned} 3x - 1 &= 11 \\ 3x - 1 + 1 &= 11 + 1 \quad &\text{(addition property)} \\ 3x &= 12 \\ \tfrac{1}{3}(3x) &= \tfrac{1}{3}(12) \quad &\text{(multiplication property)} \\ x &= 4 \end{aligned}$$

    Hence, the solution set is $\{4\}$.

    Notice in this example that the process of solution produced three equivalent equations, $3x - 1 = 11$, $3x = 12$, and $x = 4$.

2. $3(x - 1) = 5 - x$

    SOLUTION

    $$\begin{aligned} 3(x - 1) &= 5 - x \\ 3x - 3 + 3 &= 5 - x + 3 \quad &\text{(addition property)} \end{aligned}$$

$$3x = 8 - x$$
$$3x + x = 8 - x + x \quad \text{(addition property)}$$

so that,

$$4x = 8$$
$$\tfrac{1}{4}(4x) = \tfrac{1}{4}(8) \quad \text{(multiplication property)}$$

Therefore,

$$x = 2$$

The solution set is $\{2\}$.

3 $\dfrac{x}{4} + \dfrac{2x}{3} = 5$

SOLUTION

$$\dfrac{x}{4} + \dfrac{2x}{3} = 5$$
$$12\left(\dfrac{x}{4} + \dfrac{2x}{3}\right) = 12(5) \quad \text{(multiplication property)}$$
$$3x + 8x = 60$$
$$11x = 60$$

That is,

$$x = \tfrac{60}{11}$$

The solution set is $\{\tfrac{60}{11}\}$.

4 $\dfrac{3x}{2} - a = x + 3a$, a is a constant, that is, a represents a fixed real number.

SOLUTION

$$\dfrac{3x}{2} - a = x + 3a$$
$$3x - 2a = 2x + 6a \quad \text{(multiplication property)}$$
$$3x - 2x = 6a + 2a \quad \text{(addition property)}$$

Therefore, $x = 8a$. The solution set is $\{8a\}$.

5 $a(x - b) = cx + d$, where a, b, c, and d are constants and $a \neq c$.

SOLUTION

$$a(x - b) = cx + d$$
$$ax - ab = cx + d$$
$$ax = cx + d + ab \quad \text{(addition property)}$$
$$ax - cx = d + ab \quad \text{(addition property)}$$
$$(a - c)x = d + ab$$
$$\frac{1}{a-c}(a - c)x = \frac{1}{a-c}(d + ab) \quad \text{(multiplication property)}$$

Hence,

$$x = \frac{d + ab}{a - c}, \quad a \neq c$$

The solution set is $\left\{\dfrac{d + ab}{a - c}\right\}$, where $a \neq c$.

It should be noted that many applications in algebra result in equations that involve rational expressions. In solving such equations, or any other first-degree equation, we shall find that checking the proposed solutions is often essential. For example, consider the following equations with rational expressions.

EXAMPLES

Solve each of the following equations and check your solutions.

1 $\dfrac{1}{x} + \dfrac{1}{2} = \dfrac{5}{6x} + \dfrac{1}{3}$

SOLUTION. The equation would be simpler if we eliminate the denominators. To do this, multiply both sides of the equation by the least common denominator (L.C.D) of all denominators. Since the L.C.D. here is $6x$, we have, by the multiplication property,

$$6x\left(\frac{1}{x} + \frac{1}{2}\right) = 6x\left(\frac{5}{6x} + \frac{1}{3}\right)$$

so that

$$6x\left(\frac{1}{x}\right) + 6x\left(\frac{1}{2}\right) = 6x\left(\frac{5}{6x}\right) + 6x\left(\frac{1}{3}\right)$$

That is,

$$6 + 3x = 5 + 2x$$
$$3x - 2x = 5 - 6$$

Therefore,

$$x = -1.$$

Check: If $x = -1$, $\dfrac{1}{-1} + \dfrac{1}{2} \stackrel{?}{=} \dfrac{5}{6(-1)} + \dfrac{1}{3}$

$$-1 + \dfrac{1}{2} \stackrel{?}{=} -\dfrac{5}{6} + \dfrac{1}{3}$$

$$-\dfrac{1}{2} \stackrel{?}{=} -\dfrac{5}{6} + \dfrac{2}{6}$$

$$-\dfrac{1}{2} = -\dfrac{3}{6}$$

The solution set is $\{-1\}$.

2 $\quad \dfrac{2}{x-3} - \dfrac{3}{x+3} = \dfrac{12}{x^2-9}$

SOLUTION. Since the L.C.D. is $x^2 - 9$, we multiply both sides of the equation by $x^2 - 9$, so that

$$(x^2 - 9)\left(\dfrac{2}{x-3} - \dfrac{3}{x+3}\right) = (x^2 - 9)\left(\dfrac{12}{x^2-9}\right)$$

or

$$(x-3)(x+3)\left(\dfrac{2}{x-3} - \dfrac{3}{x+3}\right) = (x^2-9)\left(\dfrac{12}{x^2-9}\right)$$

$$(x-3)(x+3)\left(\dfrac{2}{x-3}\right) - (x-3)(x+3)\left(\dfrac{3}{x+3}\right)$$

$$= (x^2-9)\left(\dfrac{12}{x^2-9}\right)$$

That is,

$$2(x+3) - 3(x-3) = 12$$
$$2x + 6 - 3x + 9 = 12$$
$$-x + 15 = 12$$
$$-x = -3$$
$$x = 3$$

Check: If $x = 3$

$$\dfrac{2}{3-3} - \dfrac{3}{3+3} \stackrel{?}{=} \dfrac{12}{9-9}$$

$$\dfrac{2}{0} - \dfrac{3}{6} \stackrel{?}{=} \dfrac{12}{0}$$

Since division by zero is not defined, we must reject $x = 3$ as a solution. The solution set is $\emptyset$.

Notice that in this example, $x = 3$ was obtained as a potential solution and a check showed that it was not a solution. We say in this case that $x = 3$ is an *extraneous solution*.

Let us now consider some examples of applications that result in first-degree equations or equations with rational expressions.

EXAMPLES

1 Find a number such that when two times the number is increased by 5, the result is 11.

SOLUTION. Let x be the symbol for the unknown number. Next, identify the given statement of equality and translate it to a mathematical expression: "*When two times the number is increased by 5, the result is 11*," which translates to the equation $2x + 5 = 11$. Solving this equation, we have

$$2x + 5 = 11$$
$$2x = 6$$

so that

$$x = 3$$

Check

$$2(3) + 5 \stackrel{?}{=} 11$$
$$6 + 5 = 11$$
$$11 = 11$$

The solution is correct.

2 A water tank can be filled by an intake pipe in 4 hours and can be emptied by a drain pipe in 5 hours. How long would it take to fill the tank with both pipes open?

SOLUTION. Let t be the number of hours required to fill the tank with both pipes open. The rate of flow for the intake pipe is one-fourth of a tank per hour, and the rate of flow for the drain pipe is one-fifth of a tank per hour. Hence,

$$\left(\frac{1}{4}\right)t = \frac{t}{4} = \text{the fraction of a tank that the intake pipe can fill in } t \text{ hours}$$

$$\left(\frac{1}{5}\right)t = \frac{t}{5} = \text{the fraction of a tank that the drain pipe can empty in } t \text{ hours}$$

Since both pipes are open and one tank is to be filled, we have the equation

$$\frac{t}{4} - \frac{t}{5} = 1$$

or, equivalently

$$5t - 4t = 20$$

so that $t = 20$ hours.

Check

$$\frac{t}{4} = \frac{20}{4} = 5 \text{ tank loads}$$

$$\frac{t}{5} = \frac{20}{5} = 4 \text{ tank loads}$$

and

$$5 - 4 = 1 \text{ tank filled}$$

Hence, $t = 20$ is the solution.

3 The sum of the ages of Joe and Jammie is 35. Joe is 5 years older than twice Jammie's age. How old are they?

SOLUTION. Let x years be Jammie's age; then $2x + 5$ years in Joe's age, and $2x + 5 + x$ years is the sum of their ages. Therefore,

$$2x + 5 + x = 35$$
$$3x = 30$$
$$x = 10$$

Therefore, Jammie's age is 10 years and Joe's age is 25 years.

PROBLEM SET 1

1 Solve the following first-degree equations.

a) $x - 10 = 15$

b) $5(x + 8) = x + 40$

c) $7(4 + 8x) - 3 = 6x + 75$

d) $\frac{5x}{4} - 1 = \frac{3x}{4} + \frac{1}{2}$

e) $\frac{15x + 12}{3} - 2 = 3x + 2$

f) $x + \frac{16}{3} = \frac{3x}{2} + \frac{25}{6}$

g) $\frac{2}{3} - 5x = -3x - \frac{2}{15}$ h) $\frac{23}{100} - x = \frac{19}{20} - 3x$

i) $\frac{4x}{9} - \frac{3}{5} = \frac{5x}{6} - \frac{13}{10}$ j) $\frac{33}{10} + \frac{2x+9}{5} = -\frac{3x-5}{4}$

k) $\frac{3x}{2} - \frac{3x}{5} = \frac{3}{5}$ l) $\frac{4-2x}{3} = \frac{21}{12} - \frac{5x-3}{4}$

2 Solve the following equations. Identify extraneous roots.

a) $\frac{1}{x} + \frac{2}{x} = 3 - \frac{3}{x}$ b) $\frac{2}{3x} + \frac{1}{6x} = \frac{1}{4}$

c) $\frac{8}{x-3} = \frac{12}{x+3}$ d) $\frac{x}{x-1} - \frac{3}{x+1} = 1$

e) $\frac{9}{5x-3} = \frac{5}{3x+7}$ f) $\frac{x}{x+1} + 2 = \frac{3x}{x+2}$

g) $\frac{3(x-1)}{x+7} = -\frac{3}{5}$ h) $\frac{1}{4x} - \frac{2}{x(x-1)} = \frac{1}{2(1-x)}$

i) $\frac{x-1}{x+1} + x = 2 + x$ j) $\frac{3}{1-x} + \frac{1}{1+x} = \frac{x+2}{1-x^2}$

k) $2 + \frac{2-x}{x+2} = \frac{3x+7}{x+5} - 2$ l) $\frac{3}{x+4} - \frac{2}{x+2} = \frac{1}{x-2}$

m) $\frac{8x+10}{5} + \frac{6x+1}{4} = 2x + 3$

n) $\frac{x+3}{4} + 2 = \frac{3}{2} - \frac{2x-1}{3}$ o) $\frac{1-5x}{1+5x} - \frac{5x}{1+5x} = \frac{5}{2}$

p) $\frac{x+1}{x-4} - \frac{x}{x-2} = \frac{3}{x-6}$ q) $\frac{2}{x^2-4} = \frac{4}{2x^2-5x+12}$

r) $\frac{x+1}{x-1} + \frac{x^2-3}{1-x^2} = \frac{4}{1+x}$

3 In each of the following problems, solve for the indicated variable. Treat the remaining letters as constants.

a) $f = mx + b$, for x b) $u = at + b^2$, for t

c) $s = gt^2 + ut$, for u d) $l = a + (m-1)d$, for m

e) $\frac{3}{x} - \frac{4}{b} = \frac{5}{3c}$, for x f) $x = \frac{ax^n - a}{x-1}$, for a

g) $\frac{ax}{7} - \frac{bc}{3} = \frac{2x}{2b^2} + d$, for x h) $7x^2 - 3y + 5c = 12$, for y

i) $\frac{-5}{a} + 3 = 16 + \frac{9}{a}$, for a j) $\frac{x-a}{x} - \frac{x}{x-a} = \frac{a}{x}$, for x

k) $2 = \frac{n(n+a)}{2}$, for a

Solve the following word problems and check your solution.

4 Three more than twice a certain number is 57. Find the number.

5 Find two numbers whose sum is 18 if one number is 8 larger than the other.

6 A is twice as old as B, but in eight years he will be $\frac{6}{5}$ as old as B. How old is A now?

7 Tom and John start a bicycle race. If Tom starts at 15 miles per hour and 1 hour later John starts at 23 miles per hour, how long will it take John to overtake Tom?

8 Jimmy invested $2,000 in a bank: part at 4 percent interest, and the rest at 5 percent. If the total income from both investments is $90, how much was invested at each rate?

9 A had marks of 72 and 84 on his last two tests and he wishes to raise his average to 87. What must he get on a third test to accomplish this?

10 Find the radius of a circle that has a 30-inch circumference.

11 Suppose there is a pouch containing a mixture of 16 coins consisting of quarters, dimes, and pennies. Find how many coins there are of each type if the total value is $1.99 and there are two more dimes than quarters.

12 The sum of ages of A and B is 14 years. A is 2 years older than twice B's age. How old are they?

13 James has invested $3,000 at x percent and $2,000 at $(x + 1)$ percent. If this annual interest on both investments was $220, at what rates did he invest his money?

3   Geometry of Line

The set of real numbers discussed in Chapter 1, Section 3, can be represented geometrically as the set of all points on a straight line. This geometric representation is possible because of a correspondence which associates each real number with exactly one point on the line, and, conversely, associates each point on the line with exactly one real number. This type of correspondence is called a *one-to-one correspondence*. For example, the two sets $A = \{a, b, c, d\}$ and $B = \{1, 2, 3, 4\}$ can be placed in one-to-one correspondence as follows:

$$
\begin{array}{cc}
A & B \\
a \longleftrightarrow 1 \\
b \longleftrightarrow 2 \\
c \longleftrightarrow 3 \\
d \longleftrightarrow 4
\end{array}
\quad \text{or} \quad
\begin{array}{cc}
A & B \\
a \longleftrightarrow 2 \\
b \longleftrightarrow 1 \\
d \longleftrightarrow 4 \\
c \longleftrightarrow 3
\end{array}
$$

Note that there are other ways of establishing a one-to-one correspondence between sets A and B. On the other hand, the two sets

$C = \{a, b\}$ and $D = \{1, 2, 3\}$ cannot be placed in one-to-one correspondence since, after associating the two elements a and b with elements from the set $\{1, 2, 3\}$, there is always one element remaining. That is,

$$\begin{array}{cc} C & D \\ a \longleftrightarrow & 1 \\ b \longleftrightarrow & 2 \\ \longleftarrow ? \longrightarrow & 3 \end{array}$$

In general, two sets, A and B, can be put in *one-to-one correspondence* if it is possible to associate each member of set A with exactly one member of set B, and, conversely if it is possible to associate each member of set B with exactly one member of set A.

Thus, the concept of one-to-one correspondence is not difficult to understand when the two sets are finite. Quite simply, finite sets can be placed in one-to-one correspondence when they have the same number of elements. The concept of one-to-one correspondence is not so easy to understand, however, if the sets have infinitely many elements. For example, the set of positive integers $\{1, 2, 3, 4, \ldots\}$ can be placed in one-to-one correspondence with the even positive integers $\{2, 4, 6, 8, \ldots\}$ by using the following scheme:

$$\begin{array}{cc} 1 & \longleftrightarrow 2 \\ 2 & \longleftrightarrow 4 \\ 3 & \longleftrightarrow 6 \\ \cdots & \cdots \\ n & \longleftrightarrow 2n \\ \vdots & \vdots \end{array}$$

3.1 Number Line

Let us now examine a one-to-one correspondence between the real numbers and points on a line. The resulting "numbered line" is called a *real line* or *real axis*.

An arbitrary point on the line is selected to represent 0 and another arbitrary point to the right of 0 is selected to represent 1. The point associated with 0 is called the *origin*, and the line segment determined by the point 0 and the point 1 is called the *scale unit* (Figure 1). By repeating

Figure 1

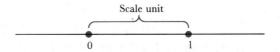

the scale unit, moving from left to right and starting at 0, we can associate the set of positive integers {1, 2, 3, 4, ...} with points on the line. Moving from right to left, starting with 0, we can associate the set of negative integers {−1, −2, −3, ...} with points on the line (Figure 2). The remaining real numbers can be "located" or "plotted" on the real line by using decimal representations or by using the geometry of the number line.

Figure 2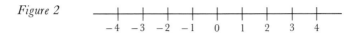

EXAMPLES

1 Locate 2.3 and 2.38 on the real line.

SOLUTION. We can locate 2.3 by subdividing the portion of the number line between 2 and 3 into 10 equal parts; then, starting at 2, we move 3 parts to the right to 2.3 (Figure 3). 2.38 can be located by subdividing

Figure 3

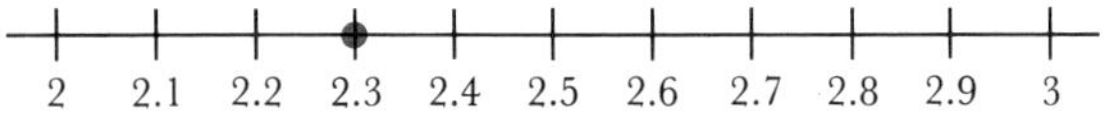

the segment between 2.3 and 2.4 into 10 equal parts; then, starting at 2.3, we move 8 parts to the right (Figure 4).

Figure 4

2 Locate $\frac{17}{8}$ on the real line.

SOLUTION. $\frac{17}{8} = 2.125$ is located on the real line in Figure 5.

Figure 5

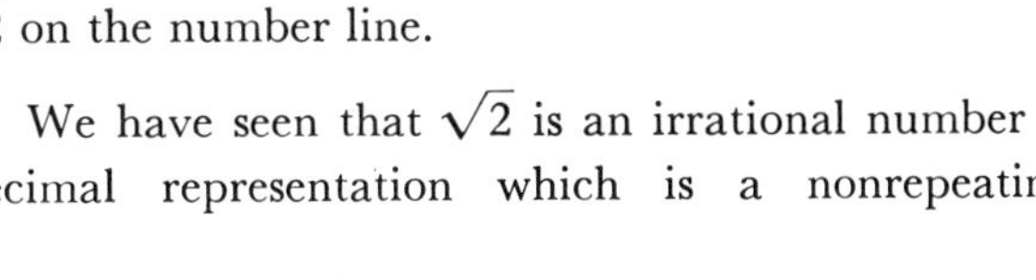

3 Locate $\sqrt{2}$ on the number line.

SOLUTION. We have seen that $\sqrt{2}$ is an irrational number so that $\sqrt{2}$ has a decimal representation which is a nonrepeating decimal

($\sqrt{2} = 1.41421\ldots$). Hence, if we were to attempt to locate $\sqrt{2}$ by using decimal representation, we would become involved in an unending process in which we would "approach" but never actually locate the point. We can, however, locate $\sqrt{2}$ using the geometry of the number line as illustrated in Figure 6.

Figure 6

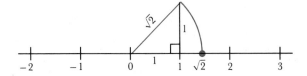

3.2 Order

The real numbers are located on the real line as follows: The numbers located to the right of zero are the *positive real numbers*, and they are arranged in order of increasing magnitude starting from zero and moving to the right. The numbers located to the left of zero are the *negative real numbers*, and they are arranged in decreasing order starting from zero and moving to the left. Accordingly, we have the following properties:

TRICHOTOMY

If a is a real number, then one and only one of the following situations can hold:

a is positive   (a is to the right of 0)
a is negative   (a is to the left of 0)
a is zero

Figure 7

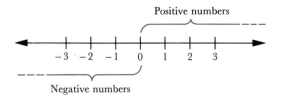

POSITIVE NUMBER AXIOM. *The set of positive numbers is closed under the operations of addition and multiplication. That is, the sum of two positive numbers is a positive number, and the product of two positive numbers is also a positive number.*

Geometrically, any number whose corresponding point on the number line lies to the right of the corresponding point of a second number is said to be *greater than* (denoted by ">") the second number. The second number is also said to be *less than* (denoted by "<") the first number.

For example,

$$7 > 3 \quad \text{or} \quad 3 < 7$$

and

$$1 > -5 \quad \text{or} \quad -5 < 1$$

since the point that corresponds to the number 7 lies to the right of the point that corresponds to the number 3, and the point that corresponds to the number -5 lies to the left of the point that corresponds to the number 1 (Figure 8).

Figure 8
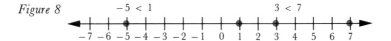

In general, if $a < b$ or $b > a$, then in a geometric sense the point that corresponds to a lies to the left of the point that corresponds to b (Figure 9). This ordering is formalized in the following definition.

Figure 9

DEFINITION

For any two real numbers a and b, *a is less than b* $(a < b)$ or, equivalently, *b is greater than a* $(b > a)$ if $b - a$ is a positive number.

For example, $3 < 7$ (or $7 > 3$), since the difference $7 - 3$ is 4, a positive number. Also, $-5 < -3$ (or $-3 > -5$) since $-3 - (-5) = 2$, which is positive. We can also indicate that a number is positive or negative by use of the inequality relation. If the number a is positive, then $a > 0$ (Figure 10), that is, a lies to the right of 0. Similarly, if the

Figure 10
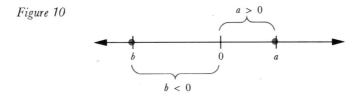

number b is negative (Figure 10), then $b < 0$; that is, b lies to the left of zero.

PROPERTIES OF ORDER RELATIONS. Although the geometry of the number line can be used to *illustrate* the properties of order relations, it is the definition of order and the positive number axiom which enable us to *prove* those properties, that will be used later to solve inequalities.

If a lies to the left of b ($a < b$) and b lies to the left of c ($b < c$), then a must lie to the left of c; that is, $a < c$ (Figure 11). For example, since

Figure 11

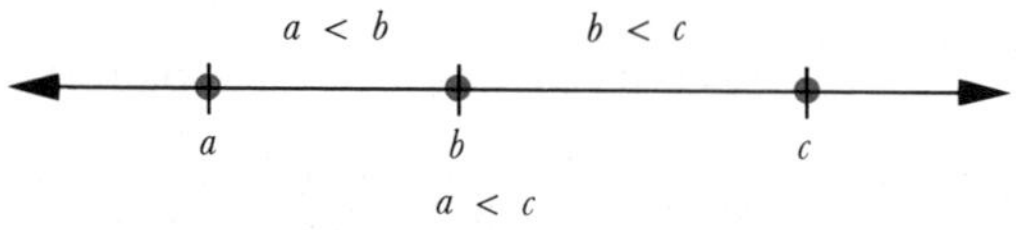

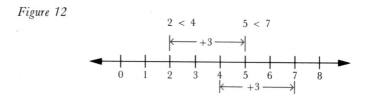

$3 < 5$ and $5 < 7$, then $3 < 7$. This property can be stated as follows:

PROPERTY 1   TRANSITIVE LAW

If a, b, and c are real numbers such that $a < b$ and $b < c$, then $a < c$.

PROOF.   Using the definition above, $a < b$ means that $b - a > 0$. Let $p = b - a$, so that $p > 0$. $b < c$ means that $c - b > 0$. Let $q = c - b$, so that $q > 0$. Then $(b - a) + (c - b) = p + q$, or $c - a = p + q$. Applying the positive number axiom, $p + q > 0$; that is, $c - a$ is a positive number. Hence, by the definition above, $a < c$.

Let us now consider the effect of adding the same number to both sides of an inequality. For instance, suppose that 3 is added to both sides of the inequality $2 < 4$. We have $2 + 3 < 4 + 3$, or, equivalently, $5 < 7$. The result of adding is illustrated in Figure 12. By adding 3 to both sides,

Figure 12

we have moved or translated both points in the same direction and the same distance along the number line. The relation between the sums will be the same as that between the original two numbers. This example can be generalized as follows:

PROPERTY 2   ADDITION LAW OF INEQUALITIES

If $a < b$ and c is any number, then $a + c < b + c$. In other words, if the same number is added to both sides of an equality, the order is preserved.

PROOF. $a < b$ implies that $b - a > 0$. Let $b - a = p$, so that $p > 0$. Since $c - c = 0$, we have $b - a + c - c = p + 0$, or $(b + c) -$

$(a + c) = p$, where p is positive. Hence, by the definition above, $a + c < b + c$.

Since subtraction is defined in terms of addition, we can deduce the fact that "if the same number is subtracted from both sides of an inequality the order is preserved." Thus, if a and b are real numbers, and $a < b$, then for any real number c,

$$a + (-c) < b + (-c) \quad \text{(Property 2)}$$

However, by definition of subtraction

$$a + (-c) = a - c \quad \text{and} \quad b + (-c) = b - c$$

so that if $a < b$, then $a - c < b - c$, that is, *subtracting the same number from both sides of a given inequality does not change the order.*

EXAMPLES

1 Given $x < y$, what can be said about the relation between $x + 1$ and $y + 1$?

SOLUTION

$$x < y$$
$$x + 1 < y + 1 \quad \text{(Property 2)}$$

2 Show that if $x + 2 < 5$, then $x < 3$.

SOLUTION

$$x + 2 < 5$$
$$x + 2 - 2 < 5 - 2 \quad \text{(why?)}$$
$$x < 3$$

We shall use specific examples to illustrate the multiplication properties of order geometrically. Consider $-3 < 4$ and multiply both sides by 2 (Figure 13). We see from the figure that multiplying both sides of the inequality $-3 < 4$ by 2 produces a new inequality $(-3)2 < (4)2$. The original order relation was preserved under the operation of multiplying by a positive number.

Figure 13

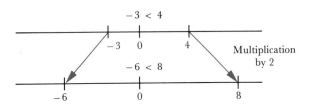

On the other hand, if we multiply both sides of the inequality $-3 < 4$ by -2, we have $(-3)(-2) > 4(-2)$, as shown in Figure 14. Here the

Figure 14

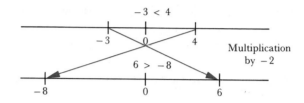

multiplication by -2 reverses the order relation. This can be generalized as follows:

PROPERTY 3 MULTIPLICATION LAWS OF INEQUALITY

a) If $a < b$ and $c > 0$, then $ac < bc$.
b) If $a < b$ and $c < 0$, then $ac > bc$.

In other words, if an equality is multiplied on each side by the same positive number, the order is preserved; whereas, if each side is multiplied by the same negative number, the order is reversed.

PROOF

a) $a < b$ means that $b - a > 0$. Let $p = b - a$, so that $p > 0$. $c(b - a) = cp$. Since $c > 0$ and $p > 0$, then $cp > 0$, by the positive number axiom. However, $bc - ac = cp$, which is positive, so that $ac < bc$ by the definition on page 91.

b) $a < b$ means that $b - a > 0$. Let $p = b - a$, so that $p > 0$. Since $c < 0$, $-c > 0$. (The negative of a negative number is positive.) Hence, $p(-c) > 0$ by the positive number axiom, so that $p(-c) = (b - a)(-c) = ac - bc > 0$. Thus, by the definition on page 91, $ac > bc$.

EXAMPLES

1 Which of the following statements are true and which are false. Justify your answer.
a) Since $-3 < 5$, then $(-4)(-3) < (-4)(5)$.
b) Since $3 < 8$, then $2(3) < 2(8)$.
c) If $0 < a < 1$, then $a^2 < a$.

SOLUTION

a) Since $-4 < 0$, and $-3 < 5$ we have
$(-4)(-3) > (-4)(5)$ [Property 3(b)].
Therefore, the statement is false.

b) Since $2 > 0$, and $3 < 8$ we have $2(3) < 2(8)$ [Property 3(a)]. Therefore, the statement is true.

c) If $0 < a < 1$, then

$$a \cdot a < a \cdot 1 \quad \text{(why?)} \quad \text{or} \quad a^2 < a$$

Therefore, the statement is true.

2. Show that if $x \neq 0$, then $x^2 > 0$.

PROOF. If $x \neq 0$, then $x > 0$ or $x < 0$. If $x > 0$, then $x \cdot x > x \cdot 0$ [Property 3(a)] or $x^2 > 0$. On the other hand, if $x < 0$, then $x \cdot x > x \cdot 0$ [Property 3(b)] or $x^2 > 0$. Hence, for $x \neq 0$, $x^2 > 0$.

3. Prove that if x and y are real numbers such that $x < y$, then $-x > -y$.

PROOF. Let $c = -1$. Then, by Property 3(b), $(-1)x > (-1)y$ or, equivalently, $-x > -y$.

Since division by a nonzero real number can be expressed as a multiplication, we can conclude from Property 3 that division on both sides of an inequality by a positive number does not change the order of the inequality, whereas division on both sides of an inequality by a negative number reverses the order; that is,

a) If $a < b$ and $c > 0$, then $a/c < b/c$.

b) If $a < b$ and $c < 0$, then $a/c > b/c$.

For example, since $3 < 4$ and if $c = 5$, then $\tfrac{3}{5} < \tfrac{4}{5}$; however, if $c = -5$, then $-\tfrac{3}{5} > -\tfrac{4}{5}$.

In summary, we have the following properties of order relations:

1. If $a < b$ and c any number, then $a + c < b + c$.

2. If $a < b$ and $c > 0$, then $ac < bc$.

3. If $a < b$ and $c < 0$, then $ac > bc$.

We have already considered similar properties for the equality relation (Section 2); that is,

1. If $a = b$ and c is any number, then $a + c = b + c$ (addition property).

2. If $a = b$ and c is any number, then $ac = bc$ (multiplication property).

We can combine the two sets of properties into one set of properties that includes both the relation of equality and inequality. (We interpret the symbol "$\leq$" to mean *less than or equal to*. That is, if $a \leq b$, then either $a < b$ or $a = b$.) Hence, we have the following result:

1. If $a \leq b$ and c is any number, then $a + c \leq b + c$.
2. If $a \leq b$ and $c > 0$, then $ac \leq bc$.
3. If $a \leq b$ and $c < 0$, then $ac \geq bc$.

EXAMPLES

1. Suppose that a and b are real numbers. Prove that $a^2 + b^2 \geq 2ab$.

   PROOF

   $$(a - b)^2 \geq 0 \quad \text{(why?)}$$

   so that

   $$a^2 - 2ab + b^2 \geq 0$$

   or

   $$a^2 + b^2 \geq 2ab$$

2. Suppose that $a > 0$. Prove that $a + 1/a \geq 2$.

   PROOF

   $$(a - 1)^2 \geq 0 \quad \text{(why?)}$$

   so that

   $$a^2 - 2a + 1 \geq 0$$

   or

   $$a^2 + 1 \geq 2a$$

   Since $a > 0$, we have

   $$a + \frac{1}{a} \geq 2$$

3.3 Notation for Order Relations

There is some standard shorthand notation which is used in connection with the order relation to describe subsets of the number line.

1. *Betweenness.* $a < b < c$ means that $a < b$ and simultaneously $b < c$ (Figure 15). For example, $-1 < 4 < 5$ is correct, since $-1 < 4$ and, simultaneously, $4 < 5$. But $2 < a < 1$ is an incorrect usage of this notation; $2 < a < 1$ suggests that $2 < a$ and, *simultaneously*, $a < 1$, from which we conclude, by the transitive property, that $2 < 1$, which is false. In other words, $a < b < c$ means that b is *between* a and c on the real line and $a < c$.

Figure 15

2. $a \not< b$ means that a is not less than b. Because of trichotomy, this implies that $a = b$ or $a > b$. For example, if $3 \not< x$, then $3 = x$ or $3 > x$ (Figure 16).

Figure 16

3. $a \leq b$ means that either $a = b$ or $a < b$. For example, $x \leq 2$ suggests that $x < 2$ or $x = 2$ (Figure 17).

Figure 17

In view of statements 1, 2, and 3, the meanings of $a \not> b$, $a \geq b$, $a \not\leq b$, and $a \not\geq b$ should be obvious. Notice that different notations may be used to represent the same relation. For example, $a \leq b$ means the same as $a \not> b$ and $a \geq b$ means the same as $a \not< b$.

Now we can combine concepts from three areas—set theory, order relations, and the geometric interpretation of order relations—to introduce some convenient notation.

4. *Bounded intervals.* We will assume here that a and b are real numbers such that $a < b$. The *open interval from a to b*, denoted by (a, b), is defined as

i $(a, b) = \{x \mid a < x < b\}$ (Figure 18)

Figure 18

Notice that $a \notin (a, b)$ and $b \notin (a, b)$.

The *closed interval from a to b*, denoted by $[a, b]$, is defined as

ii $[a, b] = \{x \mid a \leq x \leq b\}$ (Figure 19)

Figure 19

The closed interval includes the endpoints, whereas the open interval does not include the endpoints. For example, $[0, 1]$ is the set of all real numbers between 0 and 1 including 0 and 1, whereas $(0, 1)$ is the set of all real numbers between 0 and 1 excluding 0 and 1 (Figure 20).

Figure 20

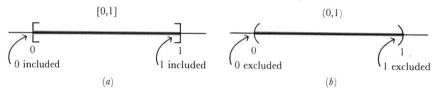

An interval from a to b including one endpoint but excluding the other endpoint is said to be *half-open* (or *half-closed*). This can happen in one of two ways.

i $[a, b) = \{x \mid a \leq x < b\}$ (Figure 21a)

Figure 21a

ii $(a, b] = \{x \mid a < x \leq b\}$ (Figure 21b)

Figure 21b

5   *Unbounded intervals.* We use the symbols ∞ and $-\infty$ (∞ and $-\infty$ are *not* real numbers) to describe unbounded intervals as follows. If a is a real number, then

i $[a, \infty) = \{x \mid x \geq a\}$ Figure 22a

Figure 22a

ii $(-\infty, a] = \{x \mid x \leq a\}$ (Figure 22b)

Figure 22b

iii $(a, \infty) = \{x \mid x > a\}$ (Figure 22c)

Figure 22c

iv $(-\infty, a) = \{x \mid x < a\}$ (Figure 22d)

Figure 22d

For example,

$$(-\infty, 3) \cap (0, \infty) = (0, 3) \quad \text{(Figure 23)}$$

Figure 23

Finally, we use interval notation $(-\infty, \infty)$ to denote the set of all real numbers R.

EXAMPLES

1. If $2 \leq x \leq 4$, the set of values of x determines the interval in Figure 24 and this set can be denoted by $[2, 4]$.

Figure 24

2. $[3, 3] = \{3\}$ (why?)

3. $(2, 4) \subset [2, 4]$, since $2 \in [2, 4]$ but $2 \notin (2, 4)$.   (Figure 25)

Figure 25

In Examples 4 through 8, perform the indicated set operations.

4 $(-\infty, 2) \cap (-3, \infty) = (-3, 2)$ (Figure 26)

Figure 26

5 $[2, 5) \cup (-3, 0]$ is illustrated in Figure 27.

Figure 27

6 $(-\infty, 3) \cap [3, 4) = \emptyset$ (Figure 28)

Figure 28

7 $[2, 3] \cap (2, 3) = (2, 3)$ (Figure 29)

Figure 29

8 $\{x \mid x > 1 \text{ or } x < -2\} = \{x \mid x > 1\} \cup \{x \mid x < -2\}$
$= (1, \infty) \cup (-\infty, -2)$ (Figure 30)

Figure 30

The *complement* A^c of a set A of real numbers is the set of all real numbers that are not contained in A. Thus, if A is a set of real numbers, $A^c = \{x \mid x \notin A\}$. For example, if A is the interval set $[2, \infty)$, then the complement of the set A is $(-\infty, 2)$ (Figure 31). If Q is the set of rational numbers, then the complement of Q, Q^c, relative to the set of numbers R, is the set of irrational numbers.

Figure 31

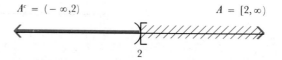

EXAMPLE

Find the complement of each of the following sets relative to the set of real numbers R.

a) R b) $(3, \infty)$ c) $(8, 9)$

SOLUTION

a) $R^c = \emptyset$
b) $(3, \infty)^c = (-\infty, 3]$ (Figure 32)

Figure 32

c) $(8, 9)^c = (-\infty, 8] \cup [9, \infty)$ (Figure 33)

Figure 33

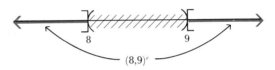

3.4 Linear Inequalities in One Unknown

Linear inequalities in one unknown can be solved in much the same manner as linear equations. We replace an inequality by an equivalent inequality which has a solution that is obvious. As with equations, two *inequalities* are *equivalent* if their solution sets are the same. The properties of order relations enable us to convert a given inequality to an equivalent one. A few examples will help to clarify the method for solving linear inequalities in one unknown.

EXAMPLES

Solve each of the following inequalities and represent the solution in set notation. Finally, show the solution set on the real line.

1    $3x - 2 < 7$

SOLUTION

$\{x \mid 3x - 2 < 7\} = \{x \mid 3x - 2 + 2 < 7 + 2\}$ (Property 2)
$= \{x \mid 3x < 9\}$
$= \{x \mid (\tfrac{1}{3})(3x) < (\tfrac{1}{3})(9)\}$ (Property 3)
$= \{x \mid x < 3\}$
$= (-\infty, 3)$ (Figure 34)

Figure 34

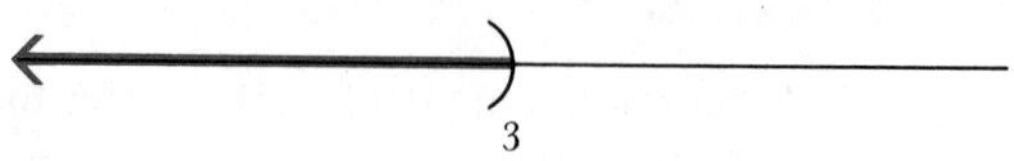

2 $x + 2 < 7x - 1$

SOLUTION

$$\begin{aligned}
\{x \mid x + 2 < 7x - 1\} &= \{x \mid x - 7x + 2 < 7x - 1 - 7x\} \\
&\qquad\qquad\text{(Property 2)} \\
&= \{x \mid -6x + 2 < -1\} \\
&= \{x \mid -6x + 2 - 2 < -1 - 2\} \\
&\qquad\qquad\text{(Property 2)} \\
&= \{x \mid -6x < -3\} \\
&= \{x \mid (-\tfrac{1}{6})(-6x) > (-\tfrac{1}{6})(-3)\} \\
&\qquad\qquad\text{(Property 3b)} \\
&= \{x \mid x > \tfrac{1}{2}\} \\
&= (\tfrac{1}{2}, \infty) \qquad\qquad\text{(Figure 35)}
\end{aligned}$$

Figure 35

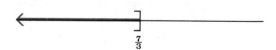

3 $3x + 5 \leq 12$

SOLUTION

$$\begin{aligned}
\{x \mid 3x + 5 \leq 12\} &= \{x \mid 3x + 5 - 5 \leq 12 - 5\} \quad \text{(Property 2)} \\
&= \{x \mid 3x \leq 7\} \\
&= \{x \mid (\tfrac{1}{3})(3x) \leq (\tfrac{1}{3})(7)\} \qquad \text{(Property 3a)} \\
&= \{x \mid x \leq \tfrac{7}{3}\} \\
&= (-\infty, \tfrac{7}{3}] \qquad\qquad\qquad\text{(Figure 36)}
\end{aligned}$$

Figure 36

PROBLEM SET 2

1. Sketch a number line and locate the following rational numbers:

$$0, \quad \frac{+2}{1}, \quad \frac{+3}{1}, \quad \frac{+3}{2}, \quad \frac{+2}{3}, \quad \frac{-5}{2}, \quad \frac{-1}{1}, \quad \frac{-2}{2}, \quad \frac{-3}{2}$$

2. Locate each of the following real numbers on the number line.
   a) -2
   b) 1.8
   c) $\sqrt{7}$
   d) -3.22

e) $\frac{17}{2}$ f) $-\frac{1}{2}$
g) $2\sqrt{2}$ h) 3.77
i) 6.99 j) 6.999

3 Show that $a < b$ in each case by showing that $b - a$ is a positive number.
   a) $4 < 8$
   b) $-3 < 5$
   c) $-2 < -1$
   d) $0 < 5$
   e) $-3 < 4$
   f) $-2 < 0$

4 Illustrate each of the following inequalities on the real line.
   a) $2 < 3$ and $2 + 3 < 3 + 3$
   b) $1 < 2$ and $2 < 3$, then $1 < 3$
   c) $-1 > -2$ and $-2 > -4$, then $-1 > -4$
   d) $3 > 2$ and $3 - 1 > 2 - 1$
   e) $3 < 4$ and $3(2) < 4(2)$
   f) $-2 < -1$ and $(-2)(-1) > (-1)(-1)$

5 Apply the multiplication property to the following inequalities by multiplying each side by the given number.
   a) $2 < 3$, 2
   b) $2 < 5$, 3
   c) $-1 < 0$, -2
   d) $-3 < -2$, -1
   e) $x < y$, -3
   f) $x < y$, 3

6 Indicate the order property that validates each of the following statements.
   a) Since $-3 < 2$, then $-3 - x < 2 - x$.
   b) If $x < y$, then $-2x > -2y$.
   c) If $-\frac{1}{2} < x$ and $x < y$, then $y > -\frac{1}{2}$.
   d) If $x > 0$, then $-2/x < -1/x$.
   e) If a is a real number, then either $a = 3$ or $a < 3$ or $a > 3$.

7 Determine whether or not the following statements are true.
   a) $-[2 - (4 - 5)] < 0$
   b) If $a < b$, then $1/a < 1/b$ for $a, b > 0$.
   c) If $0 < a < 1$, then $a^3 < a$.
   d) If $a < b < c$, then $(a + b + c)/3 < c$.
   e) If $a < b$, then $-1/a < -1/b$ for $a > 0$ and $b > 0$.

8 Prove that if $0 < a < b$, then $a^2 < b^2$.

9 If $a^2 < b^2$, does it follow that $a < b$? Use numerical examples to support your answer.

10 Prove that if $a < b$ and $c < d$, then $a + c < b + d$.

11 Express each of the given sets as intervals and illustrate each on a number line.
    a) $\{x \mid x \leq 1\}$
    b) $\{x \mid x \geq 1\}$
    c) $\{x \mid x < 2\} \cup \{x \mid x \geq -2\}$
    d) $\{x \mid x \leq 1\} \cap \{x \mid x \geq -1\}$
    e) $\{x \mid x \geq 0\} \cap \{x \mid x \leq 2\}$

12 Show that the Trichotomy Principle of real numbers can also be expressed as: For any two real numbers x and y, either $x < y$, $x = y$, or $x > y$. (*Hint:* Let $a = x - y$.)

13 Show that if $a < b$ and $c < d$, where $a, b, c, d > 0$ then $a \cdot c < b \cdot d$.

14 Show that if $a < b$, then $a < (a + b)/2 < b$.

15 Show that if $0 < x < 1$, then $x^2 < x$ and $x^3 < x^2$.

16 Prove and illustrate each of the following statements.
   a) If $a < 0$ and $b < 0$, then $ab > 0$.
   b) If $a > 0$ and $b < 0$, then $ab < 0$.

17 Use interval notation and set operations to represent each of the following sets.

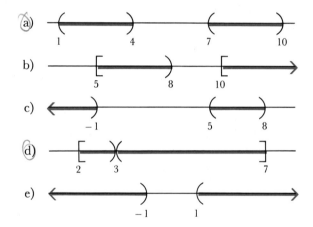

18 Give two examples of each of the following identities.
   a) $(A^c)^c = A$
   b) $(A \cup B)^c = A^c \cap B^c$
   c) $(A \cap B)^c = A^c \cup B^c$

19 Find the complement of each of the following sets relative to the set of real numbers R.
   a) $A = \{x \mid x < 3\}$
   b) $B = \{x \mid 3 < x < 4\}$
   c) $C = \{x \mid x^2 \neq 9\}$
   d) $D = \{x \mid 5x > 3x\}$

20 Find the solution set of the following inequalites and represent the solution in interval form. Also show the solution set on the real line.
   a) $5x < 5$
   b) $5x + 6 < 13$
   c) $2x - 7 > 8$
   d) $3x - 5 > 7$
   e) $\dfrac{3x - 5}{9} > 0$
   f) $\dfrac{x}{3} + 2 < \dfrac{x}{4} - 2x$
   g) $3x + 1 < 0$
   h) $3(x + 2) \leq \frac{2}{5}(x + 1)$
   i) $5 + x < -x + 3$
   j) $x + 6 \leq 4 - 3x$
   k) $4x + 3 \geq 12$
   l) $\frac{2}{5}(3x - 2) - \frac{1}{10}(6x + 7) \leq 0$

m) $\frac{2}{3}(2x - 1) - \frac{2}{5}(x + 2) \leq 4$ n) $\frac{4 - 5x}{2} < -15$

o) $\frac{4x - 1}{3} \geq 5$ p) $\frac{3x - 7}{2} \leq 13$

q) $\frac{3x - 2}{5} - 4 < 0$ r) $\frac{1}{3}(2x + 3) \leq \frac{3x}{4}$

s) $\frac{2x + 1}{5} - 2 \leq \frac{3x + 1}{2}$ t) $\frac{3x - 7}{6} - 13 \geq 1 - \frac{x}{2}$

4 Absolute Value Equations and Inequalities

If a and b are real numbers such that $a \leq b$, then the "distance" between a and b is considered to be the nonnegative number $b - a$ (Figure 1).

Figure 1

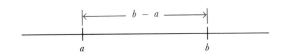

For example, the distance between -1 and 3 is given by $3 - (-1) = 4$, and the distance between -5 and -2 is equal to $(-2) - (-5) = 3$ (Figure 2).

Figure 2

Suppose that we are interested in finding the distance between 0 and any real number x. For convenience, the notation $|x|$, which is read the *absolute value of x*, will be used to represent the distance between x and 0. Thus, if $x > 0$, we have

$$|x| = x - 0 = x \quad \text{(Figure 3)}$$

Figure 3

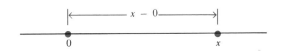

If $x = 0$,

$$|x| = 0 - 0 = 0$$

If $x < 0$,

$$|x| = 0 - x = -x \quad \text{(Figure 4)}$$

Note that for $x < 0$, $-x > 0$.

Figure 4

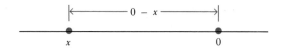

This concept can be formalized as follows:

DEFINITION

For x a real number, the *absolute value of* x, denoted by $|x|$, is defined by

$$|x| = \begin{cases} x & \text{if } x \geq 0 \\ -x & \text{if } x < 0 \end{cases}$$

Hence, from the definition, $|3| = 3$, because $3 > 0$, $|0| = 0$; and $|-4| = -(-4) = 4$ because $-4 < 0$.

EXAMPLE

If $x = 4$ and $y = -7$, compute each of the following expressions.
a) $|x + 2y|$
b) $|x| + |2y|$
c) $|xy|$
d) $|x| \, |y|$
e) $|x - y|$
f) $|x| - |y|$
g) $\left|\dfrac{x}{y}\right|$
h) $\dfrac{|x|}{|y|}$
i) $|x|^2$
j) $|y|^2$

SOLUTION

a) $|x + 2y| = |4 + 2(-7)| = |4 - 14| = |-10| = 10$
b) $|x| + |2y| = |4| + |2(-7)| = |4| + |-14| = 4 + 14 = 18$
c) $|xy| = |4(-7)| = |-28| = 28$
d) $|x| \, |y| = |4| \, |-7| = (4)(7) = 28$
e) $|x - y| = |4 - (-7)| = |11| = 11$
f) $|x| - |y| = |4| - |-7| = 4 - 7 = -3$
g) $\left|\dfrac{x}{y}\right| = \left|\dfrac{4}{-7}\right| = \dfrac{4}{7}$
h) $\dfrac{|x|}{|y|} = \dfrac{|4|}{|-7|} = \dfrac{4}{7}$
i) $|x|^2 = |4|^2 = 4^2 = 16$
j) $|y|^2 = |-7|^2 = 7^2 = 49$

4.1 Properties of Absolute Value Equalities

We have seen from the definition on page 106 that the absolute value of a number is either a positive number or zero. That is, if a is a real number, then $|a| \geq 0$.

If a and b are real numbers and $a \leq b$, then the distance between a and b is $b - a$, as illustrated in Figure 5. Using the absolute value

Figure 5

notation, we can also represent the distance between the points a and b by $|a - b|$ without regard to which is the smaller number. For example, $|5 - (-1)| = |6| = 6$ and $|-1 - 5| = |-6| = 6$, so that the distance between 5 and -1 is 6 and between -1 and 5 is 6 (Figure 6).

Figure 6

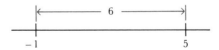

We have the following properties of absolute value equalities.

PROPERTY 1

$|-a| = |a|$, for any real number a.

The property can be interpreted geometrically as follows. $|-a| = |a|$ means that the distance between 0 and a is the same as the distance between 0 and $-a$ (Figure 7). For example, the distance from 0 to 5

Figure 7

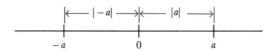

is 5 units, and the distance from 0 to -5 is also 5 units. Using absolute value notation, we have $|-5| = |5|$ (Figure 8).

Figure 8

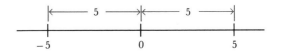

PROPERTY 2

$|a|^2 = a^2$, for all real numbers a.

PROOF. For $a \geq 0$, $|a| = a$, so that

$$|a|^2 = (a)^2 = a^2$$

For $a < 0$, $|a| = -a$, so that

$$|a|^2 = (-a)^2 = a^2$$

Hence, for all possible values of a, $|a|^2 = a^2$.

PROPERTY 3

$|ab| = |a| \, |b|$, for real numbers a and b.

PROOF. See Problem 2a of Problem Set 3.

PROPERTY 4

$|a/b| = |a|/|b|$, for real numbers a and b and $b \neq 0$.

PROOF. See Problem 2b of Problem Set 3.

EXAMPLES

1 Find the value of $|a|/a$, for $a \neq 0$.

SOLUTION. For $a > 0$, $|a| = a$, so that

$$\frac{|a|}{a} = \frac{a}{a} = 1$$

For $a < 0$, $|a| = -a$, so that

$$\frac{|a|}{a} = \frac{-a}{a} = -1$$

Hence,

$$\frac{|a|}{a} = \begin{cases} 1 & \text{if } a > 0 \\ -1 & \text{if } a < 0 \end{cases}$$

2 Prove that $|a - b| = |b - a|$ for all real numbers a and b.

PROOF. By Property 1, we have

$$|a - b| = |-(a - b)|$$

SECTION 4 ABSOLUTE VALUE EQUATIONS AND INEQUALITIES 109

Since $-(a - b) = b - a$, we have by substitution

$$|a - b| = |b - a|$$

We shall now use the definition of absolute value to solve absolute value equations. The procedure for solving such equations is illustrated in the following examples.

EXAMPLES

Solve each of the following absolute value equations.

1 $|x| = 3$

SOLUTION. By definition, we know that $|3| = 3$ and $|-3| = 3$. Therefore, $x = 3$ or $x = -3$; that is, the solution set is $\{-3, 3\}$. The solution can be illustrated on the real line (Figure 9).

Figure 9

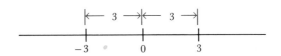

2 $|3x - 4| = 5$.

SOLUTION. Since $|-5| = |5| = 5$, $3x - 4 = 5$ or $3x - 4 = -5$ (Figure 10), so that $x = 3$ or $x = -\frac{1}{3}$; that is, the solution set is $\{-\frac{1}{3}, 3\}$.

Figure 10

3 $|x + 2| = |x - 7|$.

SOLUTION. We can solve the equation by considering the various cases suggested by the definition on page 106.

i) $|x + 2| = x + 2$ if $x + 2 \geq 0$, that is, if $x \geq -2$; and $|x + 2| = -x - 2$ if $x + 2 < 0$, that is, if $x < -2$ (Figure 11a).

Figure 11a

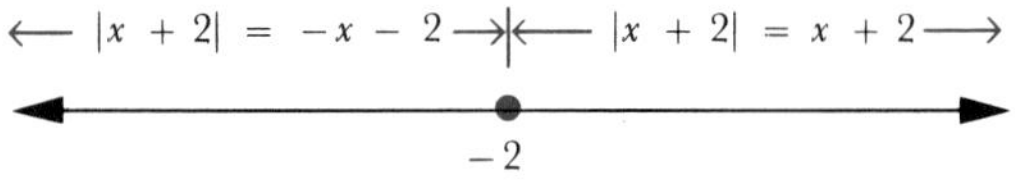

ii) $|x - 7| = x - 7$ if $x - 7 \geq 0$, that is, if $x \geq 7$; and $|x - 7| = -x + 7$ if $x - 7 < 0$, that is, if $x < 7$ (Figure 11b).

110 CHAPTER 2 EQUATIONS, ORDER RELATIONS, AND CARTESIAN PLANE

Figure 11b

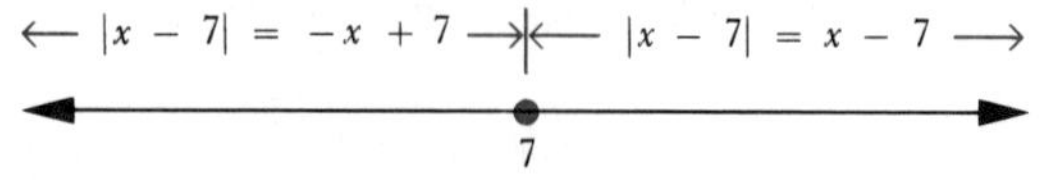

$\leftarrow |x - 7| = -x + 7 \rightarrow\!\!\leftarrow |x - 7| = x - 7 \rightarrow$

$\bullet$
7

Combining the above cases, we have the following possibilities:

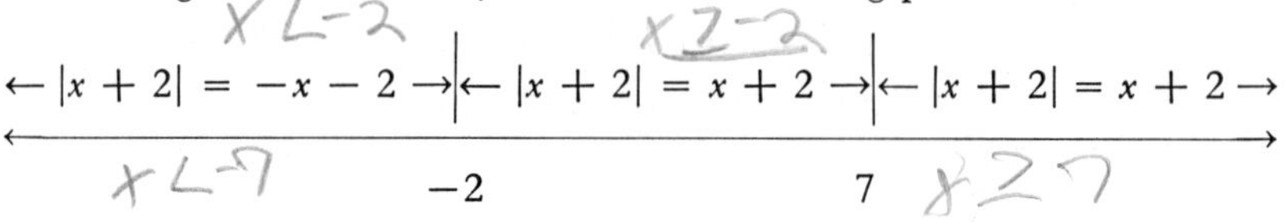

$\leftarrow |x + 2| = -x - 2 \rightarrow\!\!\leftarrow |x + 2| = x + 2 \rightarrow\!\!\leftarrow |x + 2| = x + 2 \rightarrow$

$x < -2$ \qquad -2 \qquad\qquad 7 \qquad $x > 2$

(annotations: $x < -2$, $x > -2$, $x < 7$, $x > 7$)

and \qquad\qquad and \qquad\qquad and

$|x - 7| = -x + 7$ \qquad $|x - 7| = -x + 7$ \qquad $|x - 7| = x - 7$

so that \qquad\qquad so that \qquad\qquad so that

$-x - 2 = -x + 7$ \qquad $x + 2 = -x + 7$ \qquad $x + 2 = x - 7$

which is impossible. \qquad $2x = 5$ \qquad\qquad which is impossible.

$x = \tfrac{5}{2}$

Hence, the solution set is $\{\tfrac{5}{2}\}$.

4.2 Properties of Absolute Value Inequalities

We know that $|x|$ represents the distance between 0 and x as shown in Figure 12, where x is illustrated as a positive number. Now we can use this geometric interpretation, together with the results above, to get a clear understanding of absolute value inequalities of the forms $|x| < a$ or $|x| > a$, where a is a positive number.

Figure 12

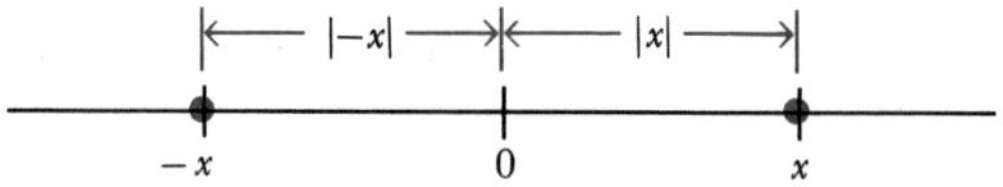

$\leftarrow |-x| \rightarrow\!\!\leftarrow |x| \rightarrow$

$-x \qquad 0 \qquad x$

THEOREM 1

If $|x| < a$, then $-a < x < a$, where $a > 0$; that is,

$$\{x \mid |x| < a\} = \{x \mid -a < x < a\} = (-a, a)$$

GEOMETRIC INTERPRETATION. Quite simply, $|x| < a$ means that the distance between 0 and x is less than a units, or, equivalently, x is within a units of 0 (Figure 13). Using interval notation, $x \in (-a, a)$.

Figure 13

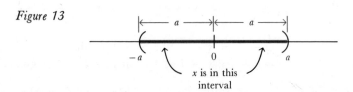

Using inequalities, this means that

$$-a < x < a$$

PROOF OF THEOREM. If $|x| < a$, then $-a < -|x|$. (Why?) By the definition on page 106,

$$|x| = x \quad \text{or} \quad |x| = -x$$

Hence,

$$-a < -|x| \leq x \leq |x| < a \quad \text{(see Problem 8a)}$$

so that, by the transitive property of inequalities,

$$-a < x < a$$

EXAMPLES

Solve the following inequalities.

1 $|x| < 3$.

SOLUTION. By Theorem 1,

$$\{x | |x| < 3\} = \{x \mid -3 < x < 3\} = (-3, 3) \quad \text{(Figure 14)}$$

Figure 14

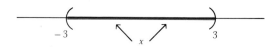

2 $|3x - 2| < 8$.

SOLUTION. By Theorem 1,

$$\begin{aligned}\{x | |3x - 2| < 8\} &= \{x \mid -8 < 3x - 2 < 8\} \\ &= \{x \mid -6 < 3x < 10\} \\ &= \{x \mid -2 < x < \tfrac{10}{3}\} \\ &= (-2, \tfrac{10}{3}) \quad \text{(Figure 15)}\end{aligned}$$

Figure 15

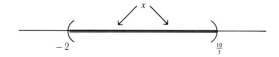

THEOREM 2

If $|x| > a$, then $x < -a$ or $x > a$, where $a > 0$; that is,

$$\begin{aligned}\{x | |x| > a\} &= \{x \mid x < -a\} \cup \{x \mid x > a\} \\ &= (-\infty, -a) \cup (a, \infty)\end{aligned}$$

GEOMETRIC INTERPRETATION. $|x| > a$ means that the distance between 0 and x is more than a units, or, equivalently, x is more than a units from 0 (Figure 16).

Figure 16

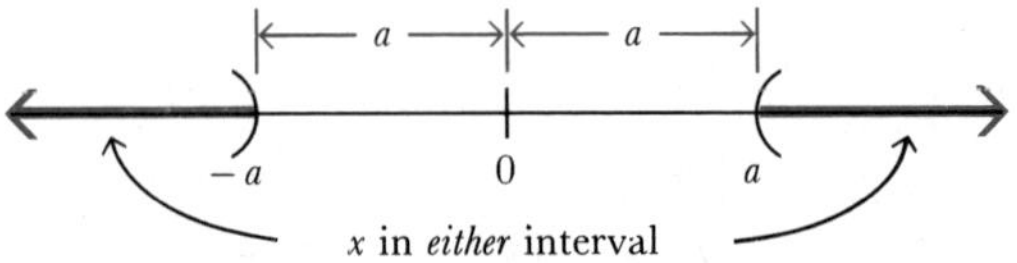

x in *either* interval is more than a units from 0

PROOF OF THEOREM. By the definition on page 106, either $|x| = x$ or $|x| = -x$. Hence,

$$|x| = x > a \quad \text{or} \quad |x| = -x > a$$

That is,

$$x > a \quad \text{or} \quad -x > a$$

But $-x > a$ implies $x < -a$ (why?), so that

$$x > a \quad \text{or} \quad x < -a$$

That is,

$$\{x \mid |x| > a\} = \{x \mid x < -a\} \cup \{x \mid x > a\}$$
$$= (-\infty, -a) \cup (a, \infty)$$

EXAMPLES

Solve the following inequalities.

1. $|x| > 7$.

SOLUTION. By Theorem 2,

$$\{x \mid |x| > 7\} = \{x \mid x < -7\} \cup \{x \mid x > 7\} = (-\infty, -7) \cup (7, \infty)$$
(Figure 17)

Figure 17

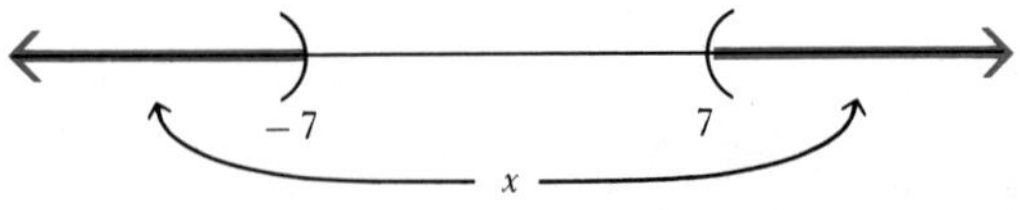

2. $|2x - 3| \geq 5$.

SOLUTION. $|2x - 3| \geq 5$ suggests that

$$2x - 3 \leq -5 \quad \text{or} \quad 2x - 3 \geq 5 \qquad \text{(Figure 18)}$$

Figure 18

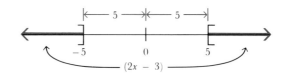

so that

$$2x \leq -2 \quad \text{or} \quad 2x \geq 8$$

That is,

$$x \leq -1 \quad \text{or} \quad x \geq 4 \quad \text{(Figure 19)}$$

Figure 19

Hence,

$$\{x \mid |2x - 3| \geq 5\} = \{x \mid x \leq -1\} \cup \{x \mid x \geq 4\}$$
$$= (-\infty, -1] \cup [4, \infty)$$

3 $|2x + 7| \geq 11$.

SOLUTION. $|2x + 7| \geq 11$ suggests that

$$2x + 7 \leq -11 \quad \text{or} \quad 2x + 7 \geq 11 \quad \text{(Figure 20)}$$

Figure 20

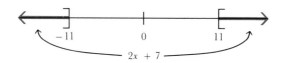

so that

$$2x \leq -18 \quad \text{or} \quad 2x \geq 4$$

That is,

$$x \leq -9 \quad \text{or} \quad x \geq 2$$

Hence,

$$\{x \mid |2x + 7| \geq 11\} = \{x \mid x \leq -9\} \cup \{x \mid x \geq 2\}$$
$$= (-\infty, -9] \cup [2, \infty) \quad \text{(Figure 21)}$$

Figure 21

THEOREM 3 TRIANGLE INEQUALITY

If a and b are real numbers, then $|a + b| \leq |a| + |b|$.

PROOF. We have that $-|a| \leq a \leq |a|$ and $-|b| \leq b \leq |b|$ (see Problem 8a). Adding inequalities (see Problem 10 of Problem Set 2), we get

$$-(|a| + |b|) \leq a + b \leq (|a| + |b|)$$

Now, if $a + b \geq 0$,

$$|a + b| = a + b \leq |a| + |b|$$

whereas if $a + b < 0$,

$$|a + b| = -(a + b)$$

so that we have, after multiplying each side of $-(|a| + |b|) \leq a + b$ by -1,

$$|a + b| = -(a + b) \leq |a| + |b|$$

In either case,

$$|a + b| \leq |a| + |b|$$

EXAMPLES

1   Show that if $|x| < 3$ and $|y| < 1$, then $|x + y| < 4$.

SOLUTION. By Theorem 3,

$$|x + y| \leq |x| + |y| < 3 + 1$$

so that

$$|x + y| < 4$$

2   If x, y, and z are real numbers, then $|x - y| \leq |x - z| + |y - z|$.

PROOF. Let $a = x - z$ and $b = z - y$ in Theorem 3 to get

$$|(x - z) + (z - y)| \leq |x - z| + |z - y|$$

That is,

$$|x - y| \leq |x - z| + |z - y|$$

from which it follows that

$$|x - y| \leq |x - z| + |y - z|$$

(See Example 2, Section 4.1, page 108.)

3 Use the triangle inequality to find a real number c such that $|x^3 + 3x^2 - 2x + 5| \leq c$ for all values of x such that $|x| \leq 2$.

SOLUTION. By repeated use of the triangle inequality, we have

$$\begin{aligned}
|x^3 + 3x^2 - 2x + 5| &\leq |x^3 + 3x^2 - 2x| + |5| \\
&\leq |x^3 + 3x^2| + |-2x| + |5| \\
&\leq |x^3| + |3x^2| + |-2x| + |5| \\
&= |x^3| + 3x^2 + 2|x| + 5 \\
&\leq 8 + 3(4) + 2(2) + 5 \\
&= 29 \qquad \text{so that } c = 29
\end{aligned}$$

4 Express $\{x \mid -2 < x < 4\}$ as an absolute value inequality.

SOLUTION. Interval $(-2, 4)$ is of length 6 and has midpoint 1 (Figure 22). Now, we saw in Section 4.1 that $|x - 1|$ represents the distance between x and 1; hence,

$$|x - 1| < 3$$

Figure 22

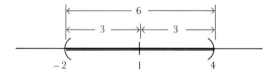

That is,

$$\{x \mid -2 < x < 4\} = \{x \mid |x - 1| < 3\}$$

PROBLEM SET 3

1 If $x = 5$ and $y = -7$, compute each of the following numbers.
  a) $|x| + |y|$ b) $|x + y|$
  c) $|x - y|$ d) $|x| - |y|$
  e) $|xy|$ f) $|x| |y|$

g) $|y|^2$

h) $\left|\dfrac{x}{y}\right|$

i) $|3x| + |-4y|$

j) $3|x| - 4|y|$

2 Prove each of the following properties.

a) $|ab| = |a||b|$

b) $\left|\dfrac{a}{b}\right| = \dfrac{|a|}{|b|}$

3 Solve the following absolute value equalities.

a) $|x| = 5$
b) $|x| = 2$
c) $|x - 2| = 3$
d) $|x| = -2$
e) $|x + 5| = 6$
f) $|2x + 1| = 3$
g) $|x| = |3 - 2x|$
h) $|-3x| = 15$
i) $|x - 5| = |-3x + 7|$
j) $2|x + 1| = 3|1 - x|$

4 Match each set in the left-hand column with an equal set from the right-hand column.

a) The set of numbers whose distance from 5 is less than 4.

A) $\{x \mid |x - 1| > |a - 1|\}$

b) The set of numbers whose distance from -2 is greater than 5.

B) $\{x \mid |x + 1| < |x - 1|\}$

c) The set of numbers whose distance from -2 is less than 4 and greater than 1.

C) $\{x \mid |x - 5| < 4\}$

d) The set of numbers whose distance from 1 is greater than the distance from a to 1.

D) $\{x \mid |x + 2| > 5\}$

e) The set of numbers whose distance from -1 is less than their distance from 1.

E) $\{x \mid 1 < |x + 2| < 4\}$

5 Solve the following absolute value inequalities and represent the solutions on the real line.

a) $|x| < 3$
b) $|x| > 3$
c) $|x - 1| \leq 3$
d) $|2x - 1| < 5$
e) $|3x - 1| > 1$
f) $|4 - 7x| \leq 3$
g) $|2x + 1| \geq 5$
h) $|5x - 2| > 0$
i) $|x + 5| < |x + 1|$
j) $|4x - 2| \leq 1$
k) $\left|5 - \dfrac{1}{x}\right| < 1$

6 a) Compute $|x/|x||$ if x is a nonzero real number.
b) For what values of x, if any, is $|x^3| = x^3$?
c) Under what conditions does $|x + y| = |x| + |y|$?
d) Under what conditions does $|x| = |y|$?

7 What restrictions, if any, must be put on x in order for each of the following statements to be true?

a) $|5x| = 5x$ b) $|x - 2| = 2 - x$
c) $|x + 2| = |x| + |2|$ d) $\left|\dfrac{1}{x}\right| = |x|$
e) $|3x| > 0$

8 Give two examples and prove each of the following assertions.
    a) $-|x| \leq x \leq |x|$ b) $|x| + |y| = ||x| + |y||$
    c) $|x - y| \leq |x| + |y|$ d) $||x| - |y|| \leq |x - y|$
    e) If $|x - a| < \tfrac{1}{10}$ and $|a - y| < \tfrac{1}{10}$ show that $|x - y| < \tfrac{1}{5}$.
    (*Hint:* Use the triangle inequality for parts c, d, and e.)

9 For what values of x, if any, is each of the following statements true?
    a) $|x - 2| \leq |x| + 2$ b) $|x - 5| \leq |x - 3|$
    c) $|x| > |x - 1|$ d) $|x| < -3$
    e) $|x| > -3$

10 Write each of the following inequalities as an absolute value inequality.
    a) $-3 < x < 3$ b) $1.99 < x < 2.0$
    c) $-3.1 < x < -3$

5 Cartesian Coordinate System and the Distance Formula

We saw in Section 3 that a real line provides us with a geometric representation of real numbers as points on a line. We used this geometric representation to investigate the order of the real numbers and the notion of the distance between points on a line. In this section we will investigate a method of representing "ordered pairs" of real numbers as points in a plane. We will use this geometric representation to determine a way to find the distance between two points in a plane.

5.1 Ordered Pairs

Consider the words "*on*" and "*no.*" If we were to consider a word merely as a set of letters, then there would be no distinction between these two words. However, since we consider a word as a set of letters with a prescribed order, *on* and *no* are different because, even though they are composed of the same letters, they have a different order.

A similar situation holds for sets. For example, the elements of the set do not have to be listed in any particular order. The set consisting of the two objects a and b could be written either as $\{a, b\}$ or as $\{b, a\}$; in other words, $\{a, b\} = \{b, a\}$. By contrast, (a, b) is an *ordered pair* consisting of the *first element a* and the *second element b*.

Two ordered pairs are considered to be equal when they have equal first members and equal second members. For example, $(1, 2) \neq (2, 1)$, even though each pair contains the same entries. Likewise, $(4, 3) \neq (4, 4)$ and $(7, 8) \neq (-7, 8)$, whereas $(9, x) = (y, 8)$ if and only if $x = 8$ and $y = 9$.

If $A = \{1, 2\}$ and $B = \{c, d\}$, the set of all possible ordered pairs formed by selecting the first member of the ordered pair from A and the second member of the ordered pair from B is denoted by $A \times B$ and is given by $A \times B = \{(1, c), (1, d), (2, c), (2, d)\}$. A simple method of grouping the elements in order to form the ordered pairs of two sets is to arrange the elements in a rectangular pattern as shown in the following table:

Set B

×	c	d
1	(1, c)	(1, d)
2	(2, c)	(2, d)

Set A

$A \times B$

On the other hand, $B \times A = \{(c, 1), (d, 1), (c, 2), (d, 2)\}$.

In general if S and T are two sets, the notation $S \times T$ (which reads "S cross T") denotes the set of all ordered pairs of the form (x, y) with x an element of set S and y an element of set T. Symbolically,

$$S \times T = \{(x, y) \mid x \in S \text{ and } y \in T\}$$

The set $S \times T$ is called the *Cartesian product* of the sets S and T.

EXAMPLES

1 Given sets $A = \{2, 3\}$ and set $B = \{x, y, z\}$, form the set $A \times B$.

   SOLUTION

   $$A \times B = \{(2, x), (2, y), (2, z), (3, x), (3, y), (3, z)\}$$

2 Let $A = \{2, 3\}$ and $B = \{2, 7, 5\}$.
   a) Determine $A \times A$ and $B \times B$.
   b) Compare $A \times B$ and $B \times A$.

   SOLUTION

   a) $A \times A = \{(2, 2), (2, 3), (3, 2), (3, 3)\}$

and

$$B \times B = \{(2, 2), (2, 7), (2, 5), (7, 2), (7, 7), (7, 5), (5, 2),$$
$$(5, 7), (5, 5)\}$$

b) $A \times B = \{(2, 2), (2, 7), (2, 5)(3, 2), (3, 7), (3, 5)\}$

whereas

$$B \times A = \{(2, 2), (2, 3), (7, 2), (7, 3), (5, 2), (5, 3)\}$$

which shows that $A \times B$ and $B \times A$ are not equal in this case. It should be noted here that if neither A nor B is empty, then $A \times B = B \times A$ if and only if $A = B$.

3 Form the following set of ordered pairs.

$$A = \{(x, y) \mid y = 2x, x \in \{-1, 0, 1\}\}$$

SOLUTION. For $x = -1$, $\quad y = 2(-1) = -2$
For $x = 0$, $\quad y = 2(0) = 0$
For $x = 1$, $\quad y = 2(1) = 2$

Hence,

$$A = \{(-1, -2), (0, 0), (1, 2)\}$$

5.2 The Cartesian Coordinate System

The set of real numbers R can be represented geometrically on a number line. We can extend this concept to represent $R \times R$, the set of all ordered pairs of real numbers, as the set of points in a *plane*, using a two-dimensional indexing system called the *Cartesian coordinate system*. This system is constructed as follows.

First, two perpendicular lines L_1 and L_2 are constructed (Figure 1). The point of intersection of the two lines is called the *origin*. Next, L_1 and L_2 are scaled as real lines by using the origin as the 0 point (zero point) for each of the two lines. The portion of L_2 above the origin is the positive direction and the portion below the origin is the negative direction of the number line; the portion of L_1 to the right of the origin is the positive direction and the portion to the left is the negative direction (Figure 2).

Figure 1

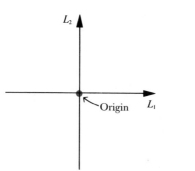

Figure 2

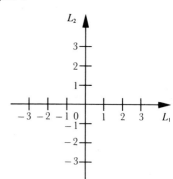

The resulting two real lines are called the coordinate axes. The coordinate axes, L_1 and L_2, are often referred to as the *horizontal axis or the x axis* and the *vertical axis or the y axis*, respectively. Given an ordered pair of real numbers (x, y) (the first member of the pair, x, is called the *abscissa;* the second member of the pair, y, is called the *ordinate;* x and y are called the *coordinates*), we can use the coordinate system to represent (x, y) as a point on the plane as follows:

The abscissa x is located on the x axis. Then a line is drawn perpendicular to this axis at point x; the ordinate y is located on the y axis at point y and a line is drawn perpendicular to the axis at point y. The intersection of the two lines we have just constructed is the point in the plane used to represent the ordered pair (x, y) (Figure 3).

Figure 3

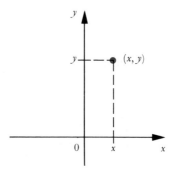

Thus, for each ordered pair in $R \times R$ we can associate a point in the plane. Conversely, for each point in the plane, we can associate an ordered pair in $R \times R$. Hence, there is a one-to-one correspondence between $R \times R$ and the points in the plane.

For example, the ordered pair $(1, 1)$ is located by moving 1 unit to the right of 0, on the x axis, then 1 unit up from the x axis. Similarly, $(7, 5)$ is located by moving 7 units to the right of 0 and 5 units up;

$(-\frac{1}{2}, 0)$ is located by moving $\frac{1}{2}$ unit to the left of 0 and no units up from the x axis (Figure 4).

Figure 4

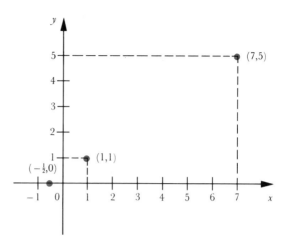

The set of all points in the plane whose coordinates correspond to the ordered pairs of a given set is called the *graph* of the set of ordered pairs. Locating these points in the plane by using the Cartesian coordinate system is called *graphing* the set of ordered pairs.

The coordinate axes divide the plane into four disjoint regions called quadrants. Thus, *quadrant I* includes all points (x, y) such that $x > 0$ and simultaneously $y > 0$. *Quadrant II* includes all points (x, y) such that $x < 0$ and simultaneously $y > 0$. *Quadrant III* includes all points (x, y) such that $x < 0$ and simultaneously $y < 0$. *Quadrant IV* includes all points (x, y) such that $x > 0$ and simultaneously $y < 0$ (Figure 5).

Figure 5

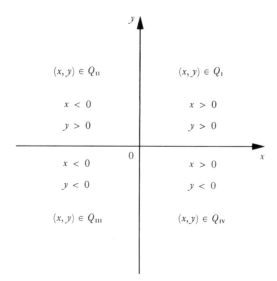

EXAMPLES

1. What are the coordinates of a point 5 units to the left of the y axis and 3 units above the x axis?

   SOLUTION. The abscissa is -5 and the ordinate is 3; therefore, the coordinates of the point are given by $(-5, 3)$ (Figure 6).

2. Plot each of the points $(1, -2)$, $(3, 4)$, $(-2, -3)$, and $(2, 0)$, and indicate which quadrant, if any, contains the points.

   SOLUTION. These points are plotted in Figure 7. $(1, -2)$ lies in quadrant

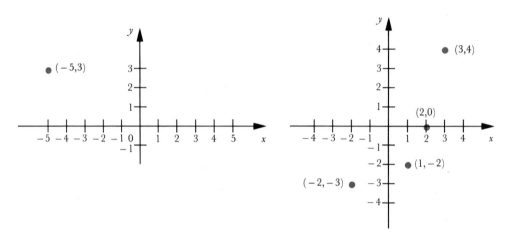

Figure 6

Figure 7

IV. $(3, 4)$ lies in quadrant I. $(-2, -3)$ lies in quadrant III. $(2, 0)$ does not lie in any quadrant.

3. If $A = \{1, 2, 3\}$ and $B = \{2, 1\}$, graph the members of $A \times B$ in the plane.

   SOLUTION. $A \times B = \{(1, 2), (1, 1), (2, 2), (2, 1), (3, 2), (3, 1)\}$, which is shown in Figure 8.

4. Graph all points with abscissas greater than 1 and ordinates less than or equal to -2; that is, indicate the region on the plane containing all points corresponding to $\{(x, y) \mid x > 1 \text{ and } y \leq -2\}$.

   SOLUTION. The points whose abscissas are greater than 1 and whose ordinates are less than or equal to -2 comprise the shaded region in Figure 9.

5. Graph $S = \{(x, y) \mid x = 0\}$.

   SOLUTION. The set of all points with an abscissa of 0 is the same as the set of all points on the y axis; hence, *the graph of S is the y axis* (Figure 10).

Figure 8

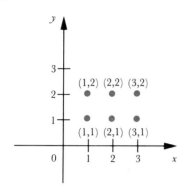

Figure 9

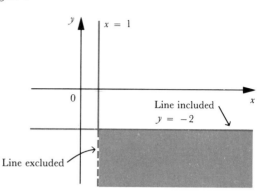

6   Graph the set $A = \{(x, y) \mid y = 2x + 1, x \in \{0, 1, 2, 3\}\}$.

SOLUTION.   The set A can be enumerated as follows:

$$\begin{aligned} \text{For } x = 0, \quad & y = 2 \cdot 0 + 1 = 1 \\ \text{For } x = 1, \quad & y = 2 \cdot 1 + 1 = 3 \\ \text{For } x = 2, \quad & y = 2 \cdot 2 + 1 = 5 \\ \text{For } x = 3, \quad & y = 2 \cdot 3 + 1 = 7 \end{aligned}$$

Hence, $A = \{(0, 1), (1, 3), (2, 5), (3, 7)\}$, which is graphed in Figure 11. Notice that we do not connect the four points with a line, for in-between points of the line are not part of the graph.

Figure 10

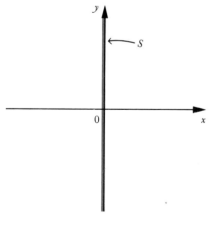

Figure 11

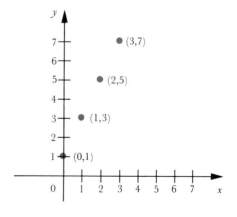

7   Graph $\{(x, y) \mid 0 < x < 2\} \cap \{(x, y) \mid 2 < y < 3\}$.

SOLUTION.   The shaded region described by the dashed lines (Figure 12) is the graph of $\{(x, y) \mid 0 < x < 2\} \cap \{(x, y) \mid 2 < y < 3\}$. Notice the use of dashed lines, which indicates that points on the boundary of the rectangle are excluded.

Figure 12

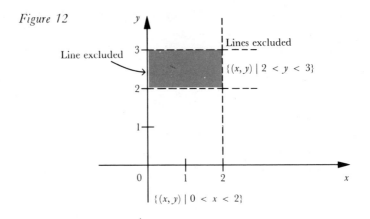

5.3 Distance Between Points

Suppose that a Cartesian coordinate system is established using the same scale for the x axis and the y axis. We could then find the distance between any two points, say P_1 and P_2, by connecting the two points with a line segment, and then "measuring" or marking off the number of scale units along this segment. The total number of such units would be the length of the segment or the distance between the two points (Figure 13). This process is purely geometric, and the accuracy of the result is dependent upon physical measurement. The coordinates of P_1 and P_2 have played no part in this measurement.

Figure 13

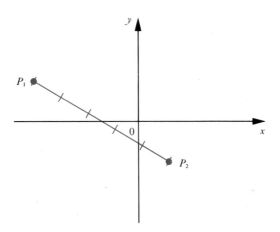

The question arises: Is it possible to use the coordinates of the two given points to get an *exact* value for the distance between them? This question will be answered in the affirmative by deriving the *distance formula*. Given any two points P_1 and P_2 with coordinates (x_1, y_1) and (x_2, y_2), respectively, a formula for the distance d between P_1 and P_2 in terms of coordinates x_1, y_1, and x_2, and y_2 can be derived by considering three cases.

i If the two points lie on the same vertical line, that is, $x_1 = x_2$, then $|y_1 - y_2| = d$ (Figure 14).

ii If the two points lie on the same horizontal line, that is, $y_1 = y_2$, then $|x_1 - x_2| = d$ (Figure 15).

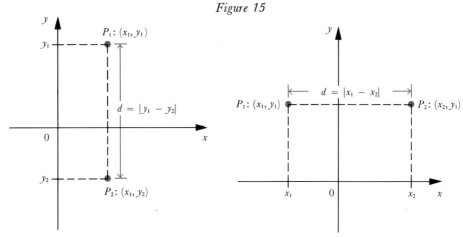

Figure 14

Figure 15

iii If the two points lie on a line which is neither horizontal nor vertical, then a right triangle (one with a 90-degree angle) is determined (Figure 16).

Figure 16

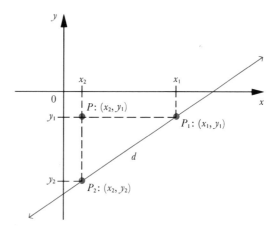

Now we use the Pythagorean theorem, so that

$$(\text{length of segment } PP_1)^2 + (\text{length of segment } PP_2)^2 = d^2$$

That is,

$$|x_1 - x_2|^2 + |y_1 - y_2|^2 = d^2$$

so that

$$(x_1 - x_2)^2 + (y_1 - y_2)^2 = d^2 \quad \text{(Why?)}$$

Hence, the distance formula is

$$d = \sqrt{(x_1 - x_2)^2 + (y_1 - y_2)^2}$$

Notice that this formula is also applicable in the special cases where P_1 and P_2 are on the same vertical line or the same horizontal line. (Why?) Since $(a - b)^2 = (b - a)^2$, the "order" of subtracting the abscissas or the ordinates is irrelevant when using the formula.

EXAMPLES

1 Plot the points $(2, -4)$ and $(-2, -1)$, and then find the distance d between them.

SOLUTION. The points $P_1 : (2, -4)$ and $P_2 : (-2, -1)$ are plotted in Figure 17, and the distance d is given by

$$\begin{aligned} d &= \sqrt{[2 - (-2)]^2 + [-4 - (-1)]^2} \\ &= \sqrt{4^2 + (-3)^2} \\ &= \sqrt{16 + 9} = \sqrt{25} = 5 \end{aligned}$$

2 Plot the points $(-1, -2)$ and $(3, -4)$, and then find the distance d between them.

SOLUTION. The points $P_1 : (-1, -2)$ and $P_2 : (3, -4)$ are plotted in Figure 18, and the distance d is given by

$$\begin{aligned} d &= \sqrt{[3 - (-1)]^2 + [-4 - (-2)]^2} = \sqrt{4^2 + (-2)^2} \\ &= \sqrt{16 + 4} = \sqrt{20} = 2\sqrt{5} \end{aligned}$$

3 Find the distance between $(4, -7)$ and $(-2, 1)$.

SOLUTION

$$\begin{aligned} d &= \sqrt{[4 - (-2)]^2 + (-7 - 1)^2} \\ &= \sqrt{6^2 + (-8)^2} \\ &= \sqrt{36 + 64} = \sqrt{100} = 10 \end{aligned}$$

Figure 17

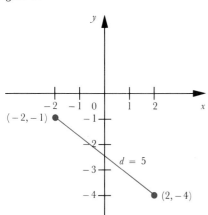

Figure 18

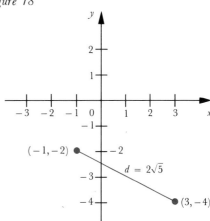

4 Derive a formula for the distance between the origin and any point (x, y) in quadrant II.

SOLUTION. The distance d between $(0, 0)$ and (x, y) (Figure 19) is given by

$$d = \sqrt{(x - 0)^2 + (y - 0)^2}$$

so that

$$d = \sqrt{x^2 + y^2}$$

Figure 19

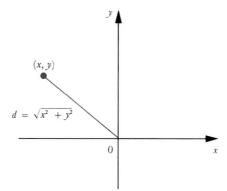

PROBLEM SET 4

1 Determine which of the following ordered pairs are equal to $(4, 5)$.
  a) $(4, 7)$
  b) $(4, 3 + 2)$
  c) $(3 + 1, 1 + 4)$
  d) $(2 + 2, 7 - 2)$
  e) $(4 + 0, 8 - 3)$
  f) $(5 + 0, 3 - 6)$

2 Which of the following statements are true? Which are false? Explain.
   a) $(x, y) = (y, x)$
   b) $\{x, y\} = \{y, x\}$
   c) $(x, y) = (x, y)$
   d) $\{(y, x)\} = \{\{y, x\}\}$
   e) If $(x, y) = (y, x)$, then $x = y$.

3 Find the missing number so that the ordered pair (x, y) will satisfy the equation $x - 3y = 5$.

$$(0, \ ), (\ , 0), (1, \ ), (\ , 5), (3, \ )$$

4 Let $A = \{1, 2, 3, 4\}$ and $B = \{1, 2\}$. Find $A \times B$ and $B \times A$ by arranging the elements of A and B in a rectangular pattern in the form of a table.

5 $S = \{2, 4, 6\}$ and $T = \{a, b\}$. List the following sets.
   a) $S \times T$
   b) $S \times S$
   c) $T \times T$
   d) $T \times S$

6 If $A \times B = \{(2, 3), (1, 7), (2, 5), (2, 7), (1, 3), (1, 5)\}$, then what are the members of set A? What are the members of set B?

7 Form each of the following sets of ordered pairs.
   a) $A = \{(x, y) \mid y = 3x + 1, x \in \{-1, 0, 1, 2\}\}$
   b) $B = \{(x, y) \mid y = -2x + 3, x \in \{-2, -1, 0, 1\}\}$
   c) $C = \{(x, y) \mid y = 2x, x \in \{1, 2, 3, 4, 5\}\}$
   d) $D = \{(x, y) \mid y = 2x, x \in \{-1, 0, 1, 2, 3, 4, 5, 6\}\}$

8 Given $A = \{a, d, c\}$ and $B = \{x, y\}$:
   a) List all the ordered pairs of $A \times B$.
   b) List all the ordered pairs of $B \times A$.

9 Let $A = \{1, 2, 3, 4, 5, 6, 7, 8\}$ and $B = \{1, 2, 3, 4, 5, 6, 7, 8, 9\}$. How many elements of $A \times B$ are of the form (x, x)?

10 a) If $A \times B = B \times A$, does $A = B$? Give an example to illustrate this situation.
    b) If $A = B$, does $A \times B = B \times A$? Explain.

11 Locate each of the following points and indicate what quadrant, if any, contains the points.
    a) $(3, 3)$
    b) $(-2, 4)$
    c) $(\pi, \sqrt{2})$
    d) $(0, 7)$
    e) $(-3, 0)$
    f) $(-1, -5)$
    g) $(-4, 2)$
    h) $(2, -4)$
    i) $(\frac{1}{2}, -\frac{3}{2})$

12 Given $A = \{-2, 0, 1\}$ and $B = \{-1, 3, 4\}$, form $A \times B$. Graph $A \times B$.

13 Give the coordinates of four points which are exactly 1 unit from the origin.

14 a) Give the coordinates of any five points of the x axis.
    b) What is common to the coordinates of all points of the x axis?

15 a) Draw the line segment between points (1, 3) and (−2, 5). Do all the points on the line segment lie in either quadrants I or II?
    b) Find the intersection of the line segment and the vertical axis.

16 Graph each of the following sets.
    a) $\{(x, y) \mid x > 1 \text{ and } y < 2\}$
    b) $\{(x, y) \mid x > 2\} \cap \{(x, y) \mid x < 5\}$
    c) $\{(x, y) \mid y < 0\} \cap \{(x, y) \mid x > 0\}$
    d) $\{(x, y) \mid y < 0\}$
    e) $\{(x, y) \mid x < 0\} \cap \{(x, y) \mid y < 0\}$
    f) $\{(x, y) \mid -1 \leq x < 0\} \cap \{(x, y) \mid y \geq 0\}$
    g) $\{(x, y) \mid -2 \leq x \leq 2 \text{ and } 0 \leq y \leq 2\}$
    h) $\{(x, y) \mid x \geq 2\} \cap \{(x, y) \mid y \geq 1\}$
    i) $\{(x, y) \mid y = 2x + 1, x \in \{-1, 0, 1\}\}$
    j) $\{(x, y) \mid y = -x + 1, x \in \{-2, 3, 4\}\}$

17 Plot the following pairs of points in a Cartesian coordinate system and find the distance between each.
    a) $(-\tfrac{1}{2}, 1)$, $(2, 3)$ b) $(-3, -4)$, $(-5, -7)$
    c) $(5, 0)$, $(-7, 3)$ d) $(6, 8)$, $(6, 7)$
    e) $(1, 1)$, $(-3, 2)$ f) $(-3, -4)$, $(8, 0)$

18 First plot the following pairs of points in a Cartesian coordinate system; then use the distance formula to determine if the triangle whose vertices are (3, 1), (4, 3), and (6, 2) is an isosceles triangle (a triangle with at least two equal sides).

19 Given points $P_1: (-3, -2)$, $P_2: (1, 2)$, and $P_3: (3, 4)$:
    a) Plot the points.
    b) Find the lengths of segments P_1P_2, P_2P_3, and P_1P_3, where P_1P_2 means the line segment determined by P_1 and P_2; and so on.
    c) Three points are said to be *collinear* if they all lie on the same straight line. Are P_1, P_2, and P_3 collinear? Explain.

20 Given points $P_1 = (a, b)$, and $P_2 = (c, d)$, and $P_3 = [(a + c)/2, (b + d)/2]$:
    a) Find the lengths of segments P_1P_3 and P_2P_3 in terms of a, b, c, and d by using the distance formula.
    b) How do these two lengths compare?
    c) What can you conclude about the geometric position of P_3 with respect to P_1 and P_2?

d) By using the formula above, find the coordinates of the midpoints of the following pairs of points. Also, use the distance formula to check the midpoints.

    i $(5, 6), (-7, 8)$
    ii $(-4, 7), (-3, 0)$
    iii $(3, 8), (8, 3)$

21 Find the length of the sides of the triangle whose vertices are given.
    a) $(5, -3), (-2, 4),$ and $(2, 5)$
    b) $(-6, -3), (2, 1),$ and $(-2, -5)$
    c) $(10, 1), (3, 1),$ and $(5, 9)$
    d) $(0, 6), (9, -6),$ and $(-3, 0)$

REVIEW PROBLEM SET

1 Solve each of the following equations for x.
    a) $x - 7(4 + x) = 5x - 6(3 - 4x)$
    b) $2(x - 1) + 3(x - 2) + 4(x - 3) = 0$
    c) $3(x + 5)(x - 3) + x = 3(x + 4)(x - 2) - 5$
    d) $(x + 1)(x - 2) = x^2 + 3x + 6$
    e) $5(1.2 - 3.5x) = 6(3.5 - 1.2x)$
    f) $2a + 3(x + b) = 5a + 6$
    g) $2(x - a) = 3(2x + a + 1)$
    h) $ax + b = cx + d$
    i) $3xa + 2b = 6c - 3xb$
    j) $(x - a)(x - b) = (x - c)(x - d)$
    k) $\dfrac{1}{x - 6} = \dfrac{2}{x + 2}$
    l) $\dfrac{2x}{x - 3} - \dfrac{5}{x + 1} = 2$
    m) $\dfrac{x}{2x - 3} - \dfrac{2}{3x + 4} = \dfrac{1}{2}$
    n) $\dfrac{1}{x + 3} + \dfrac{1}{x - 3} = \dfrac{6}{x^2 - 9}$
    o) $\dfrac{12}{x^2 - 25} = \dfrac{1}{x + 5} + \dfrac{2}{x - 5}$

2 Locate each of the following real numbers on the real line.
    a) -2.4 b) 1.72
    c) $\sqrt{7}$ d) $-\frac{2}{3}$
    e) $\frac{13}{4}$ f) $\frac{5}{3}$
    g) 3.99 h) $3\sqrt{2}$

3 Suppose that a and b are real numbers, with $a < b$. Indicate which of the following are true and which are false.
    a) $a < 3b$ b) $a - 4 < b - 4$
    c) $2a > -(-2)b$ d) $a + 2 < b + 3$
    e) $a + c < b + c, \quad c \in R$ f) $1/a < 1/b$
    g) $a + 3 < b + 4$ h) $|a| < b$

4 Let x, y, and z be real numbers. Which of the following are true?
   a) If $0 < x < \frac{1}{2}$, then $-\frac{1}{3} < x < \frac{1}{3}$.
   b) If $-\frac{1}{2} < x < \frac{1}{2}$, then $-1 < x < 1$.
   c) If $x > y$, then $x - z > y - z$.
   d) If $x > y$ and $z > 0$, then $x/z > y/z$.
   e) If $x > y$, with $x > 0$, $y > 0$, then $x^2 > y^2$, and $1/x < 1/y$.
   f) If $x^3 > y^3$, and $x > 0$, $y > 0$, then $x < y$.

5 Solve each of the following first-degree inequalities. Illustrate your solution on the real line.
   a) $3x + 5 < 2x + 1$
   b) $3x - 5 > 2x + 7$
   c) $3x - 2 < 5 - x$
   d) $1 - 2x > 5 + x$
   e) $1 - 3x > 1 + 2x$
   f) $3x + 4 < 8x + 7$
   g) $3(x - 4) < 2(4 - x)$
   h) $5x + 2 > 3x + 1$
   i) $3(x + 2) > 2(x + 3)$
   j) $5x + 2(3x - 1) < -1 - 5(x - 3)$

6 Solve each of the following absolute value equations for x. Illustrate your solution on the real line.
   a) $|2x| = 5$
   b) $|7x| = 6$
   c) $|3x + 4| = 12$
   d) $|1 - 2x| = 2$
   e) $|7x - 3| = 5$
   f) $|3 - x| = |-3|$
   g) $|5x - 8| = 3$
   h) $|3(x - 4)| = |2(4 - x)|$

7 Solve each of the following absolute value inequalities. Illustrate your solution on the real line.
   a) $|x| < 7$
   b) $|x| > \frac{5}{2}$
   c) $|x - 5| < \frac{3}{2}$
   d) $|x - 2| \leq |2 - x|$
   e) $|3x + 1| > 8$
   f) $|2x - 5| < 9$
   g) $|x - 3| > 0$
   h) $|x - 2| \geq 5$
   i) $|x - 2| > 3$
   j) $|2 - 3x| < 1$

8 In an election of the president of a faculty senate there were two candidates. The successful candidate received 5 more votes than the defeated candidate. If 243 votes were cast all together, how many did each candidate receive?

9 How many pounds of water must be evaporated from 50 pounds of a 3 percent salt solution so that the remaining portion will be a 5 percent solution?

10 Two airplanes left airports which are 600 miles apart and flew toward each other. One plane flew 20 miles per hour faster than the other. If they passed each other at the end of 1 hour and 12 minutes, what were their speeds?

11 A garage builder found out that it would cost $1\frac{1}{2}$ times as much to build a garage with brick as it would with aluminum siding, whereas lumber

would be $\frac{3}{4}$ the cost of aluminum siding. If it cost $1,200 to build the garage with wood siding, how much would it cost to build it with brick?

12 Plot each of the following points and indicate which quadrant, if any, contains the points.
   a) $(-1, 3)$ b) $(3.1, -1.7)$
   c) $(-3, 2)$ d) $(-\sqrt{3}, -2)$
   e) $(-3, 9)$ f) $(-2, -4)$
   g) $(3, 0)$ h) $(-3, 7)$

13 Let $A = \{3, -3, 4\}$ and $B = \{-1, 2, 4\}$. List all the ordered pairs of the following.
   a) $A \times A$ b) $B \times B$
   c) $A \times B$ d) $B \times A$

14 Let $A = \{2, 3, 5, 7\}$ and $B = \{2, 3, 5, 7, 8\}$.
   a) Find $A \times B$ and $B \times A$.
   b) How many elements (x, y) of $A \times B$ satisfy the condition that x is less than y?

15 Use the distance formula to show that each triangle whose vertices are the following points is an isosceles triangle, that is, a triangle with at least two sides of equal length.
   a) $(2, 1)$, $(9, 3)$, and $(4, -6)$
   b) $(0, 2)$, $(-1, 4)$, and $(-3, 3)$
   c) $(5, -2)$, $(6, 5)$, and $(2, 2)$

CHAPTER 3

Relations and Functions

CHAPTER 3

Relations and Functions

1 Introduction

One of the most useful and universal concepts in mathematics is that of a *function;* hence, we shall devote this chapter to the study of functions, paying particular attention to certain special functions which are of fundamental importance. This chapter will begin with the study of special sets called *relations;* functions evolve as certain types of relations.

2 Relations

The statements "Brian is married to Doreen," "Joan is a sister of Kathy's," "New York is larger than Michigan," and "5 is less than 8" involve what is commonly understood to be a relationship. Expressions of the type "is married to," "is a sister of," "is larger than," and "is less than" are also classified as relations. Accordingly, a relation suggests a correspondence or an association between the elements of two sets. For example, Table 1 states a relation between the numbers in the x column

Table 1	x	y
	1	2
	2	4
	3	6
	4	8
	5	10
	...	...
	n	$2n$

and the numbers in the y column. Here the correspondence between the numbers in the x column and the numbers in the y column is given by the formula $y = 2x$, where x represents a positive integer.

In the above example there are three main elements to notice: a first set, a second set, and a correspondence between the members of the two sets. In order to describe a relation so that the corresponding members of the two sets are clearly identified the ordered pair notation can be used. For example, the relation in Table 1 can be described as $\{(x, y) \mid y = 2x, x \text{ is a positive integer}\}$. In general, we define a relation as follows.

DEFINITION

A *relation* is a set of ordered pairs. The *domain* of a relation is the set of all first members of the ordered pairs, and the *range* of a relation is the set of all second members of the ordered pairs.

Since we will consider only those relations that are formed from real numbers, we can use the Cartesian coordinate system to represent relations as points on a plane. This kind of representation is called the *graph* of the relation; the graph provides us with a "geometric picture" of the relation.

EXAMPLES

1. For the relations R_0, R_1, and R_2, below, let U be $\{1, 2, 3\}$. Enumerate the members of the relation, indicate the domain and range, and graph the relation.
   a) $R_0 = \{(1, 2), (2, 3), (3, 1)\}$
   b) $R_1 = \{(x, y) \mid y = x, x \in U \text{ and } y \in U\}$
   c) $R_2 = \{(x, y) \mid y > x, x \in U \text{ and } y \in U\}$

   SOLUTION

   a) $R_0 = \{(1, 2), (2, 3), (3, 1)\}$, so that the domain is the set $\{1, 2, 3\}$ and the range is the set $\{1, 2, 3\}$. The graph consists of three points (Figure 1).

   Figure 1

   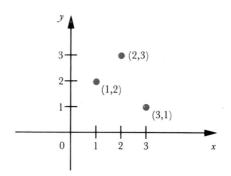

   b) $R_1 = \{(1, 1), (2, 2), (3, 3)\}$, so that the domain and range are the same set, that is, $\{1, 2, 3\}$. The graph is composed of three points (Figure 2).
   c) $R_2 = \{(x, y) \mid y > x\} = \{(1, 2), (1, 3), (2, 3)\}$, so that the domain of R_2 is $\{1, 2\}$ and the range of R_2 is $\{2, 3\}$. The graph has three points (Figure 3).

2. Graph the relation $\{(x, y) \mid y = 3x, x \in R\}$.

   SOLUTION. Since we cannot list all members of this relation, we plot

Figure 2

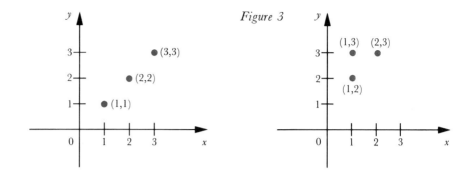

Figure 3

enough points to determine the pattern of the graph (Figure 4a). Here we use a table to list those members of the relation that were plotted.

Figure 4a

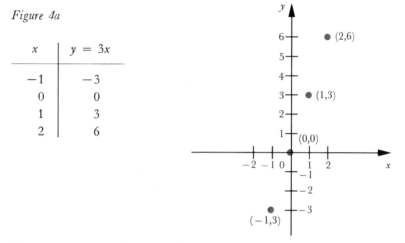

If we were to continue to plot members of this relation, the pattern would continue in that the points would form a linear pattern. In fact, the graph would actually turn out to be a line (Figure 4b).

Figure 4b

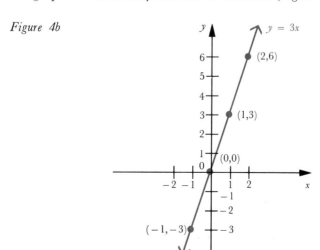

138 CHAPTER 3 RELATIONS AND FUNCTIONS

In general, the *graph of any relationship that is defined by first-degree equations such as* $y = 3x$, $y = -8x$, *and* $y = 2x + 7$ *is a straight line.* This topic will be justified in Chapter 4.

EXAMPLES

1 Graph the relation $\{(x, y) \mid y < 2x - 1, x \in R\}$.

SOLUTION. The graph of $y < 2x - 1$ consists of all points (x, y) that lie below the points on the graph of $y = 2x - 1$. Since $y = 2x - 1$ is a first-degree equation, the graph is a straight line (Figure 5a). Thus, the graph of $y < 2x - 1$ is the shaded region below the graph of the line $y = 2x - 1$ (Figure 5b).

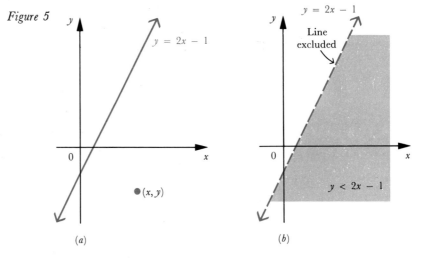

Figure 5

2 Graph the relation $\{(x, y) \mid y \geq 2x, x \in R\}$.

SOLUTION. The graph of $y \geq 2x$ consists of all points (x, y) that lie above the point (x, y_1), where (x, y_1) is a point on the graph of $y = 2x$ (Figure 6a). Thus, the graph of $y \geq 2x$ is the shaded region above, and including, the graph of the line $y = 2x$ (Figure 6b).

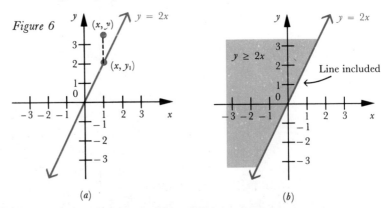

Figure 6

3 Graph the relation $\{(x, y) \mid y \leq -3x + 2, x \in R\}$.

SOLUTION. The graph of $y < -3x + 2$ consists of all points (x, y) that lie below the point (x, y_1), where (x, y_1) is a point on the graph of the line $y = -3x + 2$ (Figure 7a). Hence, the graph of $y \leq -3x + 2$ is the shaded region below, and including, the graph of $y = -3x + 2$ (Figure 7b).

Figure 7

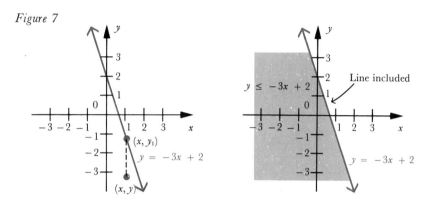

Quite often set notation is not used to describe a relation. When this is the case, the set of ordered pairs is implied. For example, $x < y$ is an abbreviated way of writing the relation $\{(x, y) \mid x < y, x \in R\}$.

Moreover, the domain and/or range of a relation is not always given explicitly. When this is the case, we determine the domain and/or range by inspection. Finally, if the universal set is not given, we will assume the universal set to be $R \times R$, where R is the set of real numbers.

Note that if (x, y) is a member of a relation, we can consider the real numbers x and y from two viewpoints. On the one hand, x is a member of the domain and y is the corresponding member of the range of the relation. On the other hand, x represents the abscissa and y represents the ordinate of a point on a plane. On the graph, then, the members of the

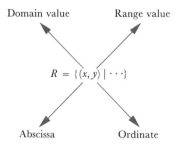

domain are the abscissas and the members of the range are the ordinates. Any restriction on the domain is a restriction on the horizontal position of the graph, and any restriction on the range is a restriction on the vertical position of the graph.

EXAMPLES

1. Graph $y = x^2$ and indicate the domain and range.

   SOLUTION. $y = x^2$ is an abbreviated way of writing the relation $\{(x, y) \mid y = x^2\}$ with universal set $R \times R$. Since any real number can be squared and since the square of a real number is always nonnegative the domain is R and the range is the set $\{y \mid y \geq 0\}$. This means that the graph "lies" in the region above and on the x axis (Figure 8a).

   Figure 8a

   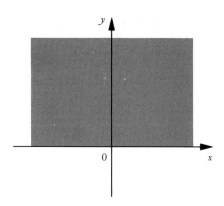

   Finally, we can use a table to determine the pattern of the graph, which we complete "by inspection" (Figure 8b).

   Figure 8b

x	y
0	0
1	1
-1	1
2	4
-2	4

   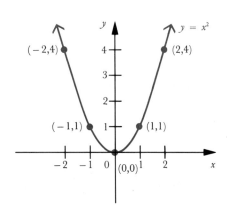

2. Let $R \times R$ be the universal set, where R is the set of real numbers. Indicate the domain and range of the relation $R_1 = \{(x, y) \mid x = \sqrt{4 - y^2}\}$. Also sketch the graph of R_1.

   SOLUTION. We seek values of y for which x is real, that is, for which $4 - y^2 \geq 0$, or $4 \geq y^2$, so that $|y| \leq 2$. Thus, the range of R_1 is $\{y \mid |y| \leq 2\}$. Since $0 \leq \sqrt{4 - y^2} \leq 2$ for these values of y, then $0 \leq x \leq 2$, so that the domain of R_1 is $\{x \mid 0 \leq x \leq 2\}$. Finally, we can use a table to graph the relation R_1 (Figure 9).

Figure 9

x	y
0	−2
$\sqrt{3}$	−1
2	0
$\sqrt{3}$	1
0	2

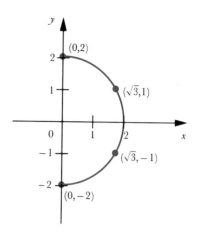

3 Determine the domain and range of the relation defined by $y = 1/x$. Also sketch the graph of R.

SOLUTION. The domain of the relation $y = 1/x$ consists (by convention) of all values of x for which $1/x$ is defined. Thus, the domain of the relation $y = 1/x$ is $\{x \mid x \neq 0\}$. Solving the equation $y = 1/x$ for x, we obtain $x = 1/y$. The range of the relation $y = 1/x$ consists of all values of y for which $1/y$ is defined, or $\{y \mid y \neq 0\}$. Using the table we can graph the relation (Figure 10).

Figure 10

x	y
−2	$-\frac{1}{2}$
−1	−1
$-\frac{1}{2}$	−2
$\frac{1}{2}$	2
1	1
2	$\frac{1}{2}$

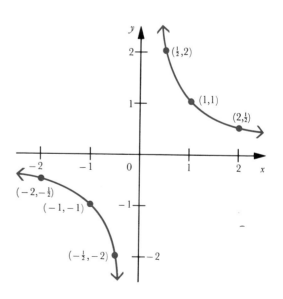

PROBLEM SET 1

1 Which of the following sets are relations?
   a) $\{(0, 1), (2, 5), (6, 3)\}$
   b) $\{(a, b), (b, a), (a, c)\}$
   c) $\{2, 4, 6, 8\}$
   d) $\{(-1, 1), (-1, 3), (-1, 5)\}$

2. Indicate the domain and range of each of the following relations.
   a) $\{(-1, 3), (2, -5), (3, 6)\}$
   b) $\{(-3, 4), (-2, 5), (1, 6)\}$
   c) $\{(-1, 1), (-1, 0), (-1, 3)\}$
   d) $\{(x, y) \mid y^2 = x, x \in \{1, 4, 9\}\}$
   e) $\{(x, y) \mid y = x^3, x \in \{-2, -1, 0, 1, 2\}\}$
   f) $\{(x, y) \mid x - y = 3, x \in \{-1, 0, 1\}\}$

3. Let $U \times U$ be the universal set, where $U = \{-1, 0, 1, 2\}$ for each of the following relations. Indicate the domain and range and graph the relation.
   a) $R_1 = \{(x, y) \mid y = x^2\}$
   b) $R_2 = \{(x, y) \mid y = x^3\}$
   c) $R_3 = \{(x, y) \mid y^2 = x^2\}$
   d) $R_4 = \{(x, y) \mid x^2 + y^2 = 4\}$
   e) $R_5 = \{(x, y) \mid 3x^2 + y^2 = 12\}$

4. Find the domain and range of each of the following relations. List four members of the relation in table form and sketch the graph of the relation.
   a) $y \geq -\frac{1}{3}x$
   b) $3y \leq 2x + 1$
   c) $2x - 3y \leq 0$
   d) $y \geq 2 - x$
   e) $x - 2y \leq 1$
   f) $y \leq -2x + 5$
   g) $y = -1$
   h) $y = 3$
   i) $y = |x - 1|$
   j) $|x| + |y| = 2$
   k) $|x| - |y| = 1$
   l) $|y| = 3$

3 Functions

We have already encountered the concept of function in everyday living. For example, the amount of sales tax charged on a purchase of $5.00 is a function of the sales tax rate; the number of books to be ordered for a course is a function of the number of students in the course; the number of congressional representatives for a particular state is a function of the population of the state.

Intuitively, a function suggests some kind of correspondence. In each of the examples above, there is an established correspondence between numbers—the amount of sales tax corresponds with the cost, the number of books with the number of students, and the number of representatives with the number of people. In general, we have the following definition.

DEFINITION 1

A *function* is a correspondence that assigns to each member in a certain set, called the *domain* of the function, exactly one member in a second set, called the *range* of the function.

For example, let us consider the situation at the beginning of the term. As each student registers for classes, his tuition charge is recorded with the student account number, as illustrated by the following partial table:

Student account number	Tuition charge
895	315.00
475	323.50
182	260.90
743	315.00
234	370.00

Here the set of student account numbers is the domain and the set of tuition charges is the range. The correspondence between the domain members and range members is suggested by the following table:

Domain		Range
895	"corresponds to"	315.00
475	"corresponds to"	323.50
182	"corresponds to"	260.90
743	"corresponds to"	315.00
234	"corresponds to"	370.00

Notice in the above table, that the tuition charge is a function of the student account number. Now we can represent the function as a set of ordered pairs, as follows:

$$\{\ldots, (895, 315.00), (475, 323.50), (182, 260.90), (743, 315.00), (234, 370.00), \ldots\}$$

In general, ordered pair notation is well suited to representing functions. If x is a member of the domain of a function and y is the member of the range corresponding to x, we can represent the correspondence between x and y as the ordered pair (x, y). In fact, we say that the function is the set of all such ordered pairs. Accordingly, we define a function as follows, recalling that a relation is a set of ordered pairs.

DEFINITION 2

A *function* is a relation in which no two different ordered pairs have the same first member. The set of all first members (of the ordered pairs) is called the *domain* of the function. The set of all second members (of the ordered pairs) is called the *range* of the function. For each ordered pair (x, y) of the function we say that x in the domain has y as the *corresponding* member in the range.

From Definition 2 we conclude that all functions are relations; however, not all relations are functions. For example, $\{(1, 1), (2, 2), (3, 7), (3, 5)\}$ is a relation, with domain $\{1, 2, 3\}$ and range $\{1, 2, 7, 5\}$, but it is not a function, because (3, 7) and (3, 5) have the same first members (see Definition 2). By contrast, {(1, 2), (3, 4), (4, 4)} is a relation which is a function with domain {1, 3, 4} and range {2, 4}. Note that in the latter example, there are two pairs which have the same second member; this does not violate the definition of a function.

Hence, if at least two different ordered pairs of a relation have the same first member, the relation is not a function. In other words, if a domain member appears with more than one range member, the relation is not a function. Geometrically, this means that if the graph of a relation has more than one point with the same abscissa, the relation is not a function.

Consider the graphs of the relations in Figure 1. The relation

Figure 1

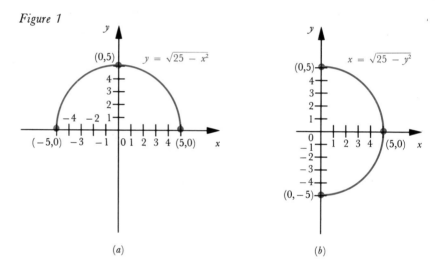

(a)   (b)

$\{(x, y) \mid y = \sqrt{25 - x^2}\}$ in part (a) represents a function, since the graph does not have two different points with the same x coordinates, whereas the relation $\{(x, y) \mid x = \sqrt{25 - y^2}\}$ in part (b) is not a function, since there are two different points with the same x coordinates—for example, the two points (3, 4) and (3, −4).

EXAMPLES

1 In each of the following parts, indicate whether or not the relation is a function. What is the domain and range? Graph the relation.
a) {(1, 2), (2, 3), (3, 4), (4, 4), (5, 6)}
b) $y = 3$
c) $x = 1$
d) $y = 3x$, $x \in \{-1, 0, 2, 3\}$
e) $y = 3x$

SOLUTION

a) {(1, 2), (2, 3), (3, 4), (4, 4), (5, 6)} is a function with domain {1, 2, 3, 4, 5} and range {2, 3, 4, 6} (Figure 2). Note that we do not draw lines connecting these points, since these are the only points that belong to the relation.

Figure 2

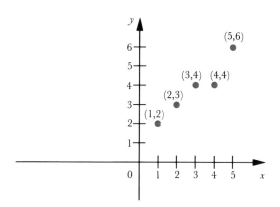

b) $y = 3$ is an abbreviated way of writing the relation $\{(x, y) \mid y = 3\}$. The domain is the set of all real numbers because there is no restriction on x, whereas the range is $\{3\}$. The relation is a function because no two different ordered pairs have the same first members. The graph is the line parallel to the x axis (Figure 3). This function is called a *constant function*, since regardless of which domain element we choose, the corresponding range element is always the same number, 3.

Figure 3

x	y
1	3
$\frac{3}{2}$	3
-2	3
3	3
0	3

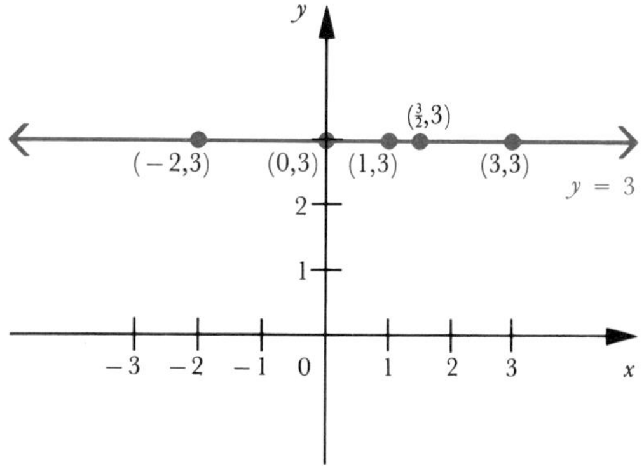

c) $x = 1$ is the relation $\{(x, y) \mid x = 1\}$. This relation is not a function because $(1, 3)$ and $(1, 4)$ are members of the set. The domain of the relation is $\{1\}$, and the range is the set of all real numbers. The graph is the line parallel to the y axis (Figure 4).

d) $y = 3x$, with $x \in \{-1, 0, 2, 3\}$ represents the ordered pairs $\{(-1, -3), (0, 0), (2, 6), (3, 9)\}$. It is a function with domain

Figure 4

x	y
1	0
1	2
1	−1
1	−2

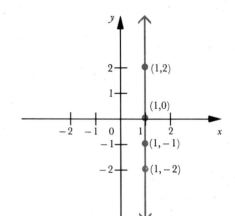

$\{-1, 0, 2, 3\}$ and range $\{-3, 0, 6, 9\}$, and the graph is composed of four points (Figure 5).

Figure 5

x	y
−1	−3
0	0
2	6
3	9

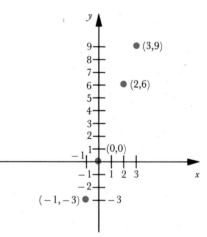

e) $y = 3x$ is the set of ordered pairs $\{(x, y) \mid y = 3x\}$. It is a function and has as both its domain and range the set of all real numbers. The graph is a line since $y = 3x$ is a linear equation (Figure 6).

Figure 6

x	y
0	0
1	3
−1	−3
2	6
3	9

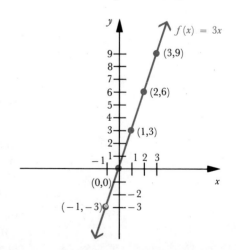

It can be seen geometrically that this set of ordered pairs is a function, since no two points have the same abscissa.

2 Examine the graphs of each of the relations given in Figure 7 to decide whether or not the relation is a function.

Figure 7

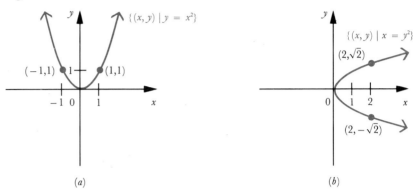

(a)   (b)

SOLUTION. Since the graph in Figure 7a does not have two different points with equal abscissas, no two different ordered pairs of the relation have equal first members, so that the relation $\{(x, y) \mid y = x^2\}$ is a function. It can be seen from the graph in Figure 7b that there are two different points with abscissa 2, so that the relation $\{(x, y) \mid x = y^2\}$ is not a function.

3.1 Function Notation

If x represents members of the domain of $y = x^2$, then y is considered to be a function of x and we indicate this by writing $f(x) = x^2$, which reads "f of x equals x squared" and means "the value of the function f at the number x is the number x^2." Therefore, $f(x) = x^2$ is another way of defining the function $\{(x, y) \mid y = x^2\}$.

In other words, $y = f(x)$ means that (x, y) is a member of the function f. For example, if $4x - 2y = 1$ and $y = f(x)$, then after solving for y in terms of x, we could write the function either as $\{(x, y) \mid y = 2x - \frac{1}{2}\}$ or as $f(x) = 2x - \frac{1}{2}$.

It is important to realize that letters other than f can be used to denote functions. For example, $h(r) = r^2 - 1$ defines the function given by $\{(r, h(r)) \mid h(r) = r^2 - 1\}$; $3r + 5t = 3$, with $t = g(r)$, defines the function given by $\{(r, t) \mid t = (3 - 3r)/5\}$; and $c(d) = \pi d$ defines the function given by $\{(d, c(d)) \mid c(d) = \pi d\}$.

When functional notation is used, sometimes it is helpful to think of the variable which represents the members of the domain as a "blank." For example, $g(t) = t^2$ can be thought of as $g(\ ) = (\ )^2$; hence, if any expression (representing a real number) is to be used to represent a member of the domain, it is easy to see where this same expression is to be substituted in the equation describing the function. For example, using $g(t) = t^2$ again, $g(x + h)$ can be determined by first writing the function as

$$g(\ ) = (\ )^2$$

so that, by substitution,

$$g(x + h) = (x + h)^2 = x^2 + 2xh + h^2$$

Similarly,

$$g(3 - 5x) = (3 - 5x)^2 = 9 - 30x + 25x^2$$

EXAMPLES

1 Let $f(x) = x + 1$. Find $f(2)$, $f(3)$, $f(5)$, $f(a - 6)$, and $f(a) - f(6)$. What is the domain of f?

SOLUTION

$$f(2) = (2) + 1 = 3$$
$$f(3) = (3) + 1 = 4$$
$$f(5) = (5) + 1 = 6$$
$$f(a - 6) = (a - 6) + 1 = a - 5$$
$$f(a) - f(6) = [(a) + 1] - [(6) + 1] = a - 6$$

The domain is the set of real numbers R.

2 Let $f(x) = \sqrt{25 - x^2}$. Find $f(0)$, $f(3)$, $f(4)$, and $f(5)$.

SOLUTION

$$f(0) = \sqrt{25 - (0)^2} = 5$$
$$f(3) = \sqrt{25 - (3)^2} = \sqrt{25 - 9} = \sqrt{16} = 4$$
$$f(4) = \sqrt{25 - (4)^2} = \sqrt{25 - 16} = \sqrt{9} = 3$$
$$f(5) = \sqrt{25 - (5)^2} = \sqrt{25 - 25} = 0$$

3 Given that $f(x) = x^2 - 1$, find the domain and range of f and evaluate each of the following expressions.

a) $f(3) + f(5)$ b) $f(2a + 4)$ c) $2f(a - 2)$

SOLUTION. The domain of the function f is the set of real numbers R and the range of f is the set $\{y \mid y \geq -1\}$ since $x^2 \geq 0$ implies that $y = x^2 - 1 \geq -1$.

a) $f(3) = (3)^2 - 1 = 8$

and

$f(5) = (5)^2 - 1 = 24$

so that

$f(3) + f(5) = 8 + 24 = 32$

b) $f(2a + 4) = (2a + 4)^2 - 1 = 4a^2 + 16a + 15$
c) $2f(a - 2) = 2[(a - 2)^2 - 1] = 2[a^2 - 4a + 3] = 2a^2 - 8a + 6$

4   Given that $f(x) = x^2$ and $g(x) = x + 1$, evaluate $f[g(x)]$ and $g[f(x)]$, and then compare the results.

SOLUTION

$$f[g(x)] = f(x + 1) = (x + 1)^2 = x^2 + 2x + 1$$

and

$$g[f(x)] = g(x^2) = x^2 + 1$$

Since $x^2 + 2x + 1 \neq x^2 + 1$, we can conclude that $f[g(x)] \neq g[f(x)]$.

$f[g(x)]$ and $g[f(x)]$ are called *composite functions*. Composite functions will be discussed again in Chapter 6.

It is often helpful to illustrate a function as a mapping. For example, consider the function f defined by the equation $f(x) = 3x$ with domain $\{1, 2, 3\}$ and range $\{3, 6, 9\}$. In this example, 1 can be considered as mapped to 3, 2 mapped to 6, and 3 mapped to 9 (Figure 8).

Figure 8

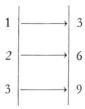

In general, if x is an element of the domain of a function f, then f associates x with $f(x)$, the value of f at x, and we say that f *maps* x *to* $f(x)$ or x *is mapped to* $f(x)$ or, equivalently, $f(x)$ *is the image of* x *under* f. Using arrow notation, we can represent this symbolically by $f: x \longrightarrow f(x)$

or $x \xrightarrow{f} f(x)$. Hence, in the above example

$1 \xrightarrow{f} 3,$
$2 \xrightarrow{f} 6,$

and

$3 \xrightarrow{f} 9$

EXAMPLES

1. Interpret $f(x) = 3x - 1$, $x \in \{-1, 1, 2, 3\}$ as a mapping.

    SOLUTION. In mapping notation, $-1 \xrightarrow{f} -4$, $1 \xrightarrow{f} 2$, $2 \xrightarrow{f} 5$, and $3 \xrightarrow{f} 8$. Here -4 is the image of -1, 2 is the image of 1, 5 is the image of 2, and 8 is the image of 3.

2. Describe the function f defined by the equation $f(x) = -3x + 5$, $x \in \{-1, 0, 1, 2\}$, using the mapping notation, listing the ordered pairs and function notation.

    SOLUTION

Mapping $f: x \longrightarrow -3x + 5$	Ordered pairs (x, y)	$y = f(x)$
$-1 \longrightarrow 8$	$(-1, 8)$	$f(-1) = 8$
$0 \longrightarrow 5$	$(0, 5)$	$f(0) = 5$
$1 \longrightarrow 2$	$(1, 2)$	$f(1) = 2$
$2 \longrightarrow -1$	$(2, -1)$	$f(2) = -1$

3. Given the *identity function* $f(x) = x$, find $f(-2)$, $f(-1)$, $f(0)$, $f(1)$, and $f(2)$. Describe the domain and range of f. Sketch the graph of f.

    SOLUTION. $f(-2) = -2$, $f(-1) = -1$, $f(0) = 0$, $f(1) = 1$, and $f(2) = 2$. The domain is the set of real numbers R and the range is also the set of real numbers R (Figure 9).

    Figure 9

x	y
-2	-2
-1	-1
0	0
1	1
2	2

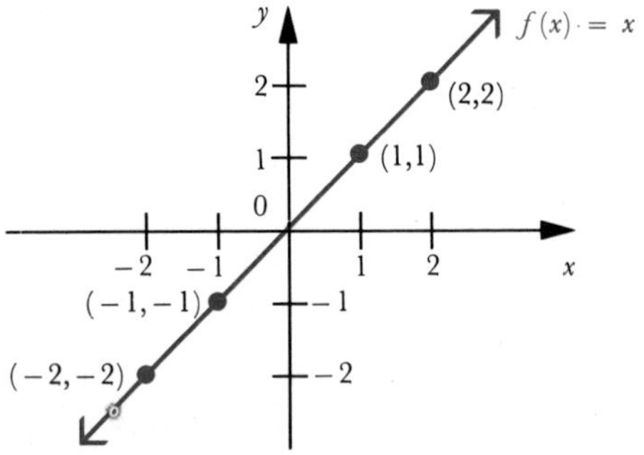

4. Given the *absolute value function* $f(x) = |x|$, find $f(-2)$, $f(-1)$, $f(0)$, $f(1)$, and $f(2)$. Describe the domain and range of f. Sketch the graph of f.

SOLUTION. $f(-2) = 2$, $f(-1) = 1$, $f(0) = 0$, $f(1) = 1$, and $f(2) = 2$. The domain is the set of real numbers R, and the range is the set of all nonnegative real numbers (Figure 10).

Figure 10

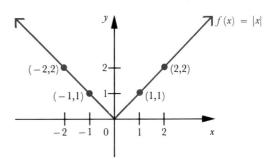

x	y
0	0
1	1
−1	1
2	2
−2	2

5 Given the *greatest integer function* f defined by the equation $f(x) = [\![x]\!]$. $[\![x]\!]$ indicates the "nearest" integer less than or equal to x. Find $f(3\frac{1}{4})$, $f(-2\frac{1}{3})$, $f(-\frac{1}{5})$, $f(\sqrt{3})$, and $f(2)$. Describe the domain and range of f. Sketch the graph of f.

SOLUTION

$f(3\frac{1}{4}) = [\![3\frac{1}{4}]\!] = 3$
$f(-2\frac{1}{3}) = [\![-2\frac{1}{3}]\!] = -3$
$f(-\frac{1}{5}) = [\![-\frac{1}{5}]\!] = -1$
$f(\sqrt{3}) = [\![\sqrt{3}]\!] = 1$
$f(2) = [\![2]\!] = 2$

Figure 11

x	y
$3\frac{1}{4}$	3
$-2\frac{1}{3}$	−3
$-\frac{1}{5}$	−1
$\sqrt{3}$	1
2	2

The domain of f is the set of real numbers R and the range is the set of integers I (Figure 11).

In Example 5 notice that the graph of the greatest integer function $f(x) = [\![x]\!]$ has "breaks" at each of the integers, whereas the graphs of the identity function $f(x) = x$ and the absolute function $f(x) = |x|$, Examples 3 and 4, do not have "breaks" anywhere. We say that $f(x) = [\![x]\!]$ is a *discontinuous* function, whereas $f(x) = x$ and $f(x) = |x|$ are continuous functions.

The term "continuous function" describes those functions which can be drawn without lifting the pen from the paper. That is, one can move "continuously" from one point on the graph to any other point without crossing any gaps. This intuitive description of continuous functions will suffice for the functions considered in this text. A more detailed study of this topic is considered in calculus.

6   A rectangular area of 4,000 square feet is to be fenced on three sides with fencing costing 20 cents per foot and on the fourth side with fencing costing 60 cents per foot. If x denotes the fourth side and C denotes the corresponding cost of the fence in cents, express C as a function of x.

SOLUTION.   Let x feet be the length; then the width is $4000/x$ feet. The cost C of the fence is

$$C = 0.20\left(x + \frac{4{,}000}{x} + \frac{4{,}000}{x}\right) + 0.60x$$

or

$$C = 0.20\left(x + \frac{8{,}000}{x}\right) + 0.60x$$

$$C = 0.80x + \frac{1{,}600}{x}$$

Width = x Area = 4000 ft²

$\dfrac{4000}{x}$ = length

7   Consider a new car dealer who sells a certain model car with anywhere from 0 to 7 extras (such as power steering, radio, air conditioning, special bumper, and so on). If x is the number of extras he puts on these cars, he knows from past experience that he can expect to sell $y = 7x + 21$ cars during the model year. This equation expresses the functional relationship between x, the number of extras, and y, the number of cars

he can expect to sell during the model year. Find the domain and the range of the function. How many cars does he expect to sell if the number of extras is 6?

SOLUTION. The domain of the function is the set $\{0, 1, 2, 3, 4, 5, 6, 7\}$. The range of the function is the set $\{21, 28, 35, 42, 49, 56, 63, 70\}$. If $x = 6$, then $y = 7(6) + 21 = 42 + 21 = 63$ cars.

Two important types of functions that are widely used in many scientific laws will be introduced in this section. We say that *y is directly proportional to x* or *y varies directly as x* if there is a constant number k such that $y = kx$ for every ordered pair (x, y). The number k is called the *constant of proportionality*. Any function defined by the relation $y = kx$ is an example of direct variation. That is, if y is a function of x and y is directly proportional to x, then $y = f(x) = kx$.

For example, if $y = 4$ when $x = 1$ for $y = kx$, then $4 = k1$ or $k = 4$. Thus, $y = f(x) = 4x$. The graph of this function is a straight line (Figure 12).

Figure 12

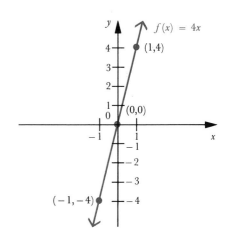

x	y
1	4
−1	−4
0	0

EXAMPLES

1. Express y as a function of x if y is directly proportional to x^2 and if $y = 98$ when $x = 7$.

   SOLUTION. Since y is directly proportional to x^2, there is a constant number k such that $y = kx^2$. The fact that $y = 98$ when $x = 7$ tells us that $98 = 49k$, or $k = 2$, so that the equation is $y = 2x^2$.

2. If y is directly proportional to x and $y = f(x)$, show that $f(ax) = af(x)$.

   SOLUTION. Since y is directly proportional to x, there is a constant k such that $y = kx$, but $y = f(x)$, so that $f(x) = kx$; finally, $f(ax) = k(ax) = a(kx) = af(x)$.

3 The area of a sphere is directly proportional to the square of the radius. If a sphere of radius of 4 inches has an area of 64π square inches, express the area of a sphere as a function of its radius.

SOLUTION. Let A represent the area of the sphere and r represent its radius, in inches. Since A is directly proportional to r^2, there is a constant k such that $A = kr^2$. Since $A = 64\pi$ when $r = 4$, then $64\pi = 16k$, so that $k = \frac{64}{16}\pi = 4\pi$. Hence, $A = 4\pi r^2$.

Another type of variation that we shall consider here is called *inverse variation*, which is stated as follows:

y is inversely proportional to x, or *y varies inversely as x*, if there is a number k such that $y = k/x$ for every ordered pair (x, y), $x \neq 0$. For example, if y is inversely proportional to x and $y = 2$ when $x = 6$, then y can be expressed as a function of x by $y = k/x$. Since $y = 2$ when $x = 6$, then $k = 12$, so that $y = 12/x$. The graph of this function is shown below (Figure 13).

Figure 13

x	y
-2	-6
-1	-12
1	12
2	6

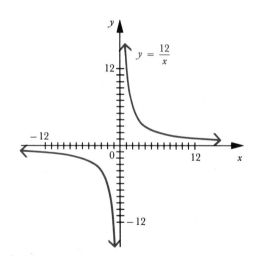

EXAMPLES

1 Express y as a function of x if y is inversely proportional to x^2 and $y = 12$ when $x = 2$.

SOLUTION. Since y is inversely proportional to x^2, there is a number k such that $y = k/x^2$. Given $y = 12$ when $x = 2$, then $12 = k/2^2$, so that $k = 48$. Thus, $y = 48/x^2$.

2 The intensity of a floodlight is inversely proportional to the square of the distance from the floodlight. How far from an object must the floodlight be placed in order for the object to receive three times the intensity it receives at 6 feet?

SOLUTION. Since the intensity I is inversely proportional to the square of the distance x, there is some constant k such that $I = k/x^2$. When $x = 6$ feet, we have $I = k/6^2$ or $I = k/36$. At some unknown distance x the intensity will be three times the intensity at 6 feet, that is, $3I = k/x^2$. Solving these two equations for k, we have $k = 36I$ and $k = 3Ix^2$. Thus, $3Ix^2 = 36I$, so that $x^2 = 12$, or $x = \sqrt{12} = 2\sqrt{3}$. Hence, the required distance is $2\sqrt{3}$ feet.

3   Boyle's law states that the pressure P is inversely proportional to the volume V of an ideal gas. Find the constant of variation if the pressure P is 30 pounds per square inch when the volume V of the gas is 100 cubic inches.

SOLUTION. Since the pressure P is inversely proportional to V, there is a number k such that $P = k/V$ or $k = PV$. At $P = 30$ pounds per square inch, $V = 100$ cubic inches, so that $k = (30)(100) = 3,000$ pounds per inch.

4   If z varies directly as the product of two variables x and y, the relationship between z and the product xy is known as *joint variation*. Thus, z varies jointly as x and y if there is a constant real number k such that $z = kxy$.

The volume V of a right circular cone varies jointly as its altitude h and the square of its base radius r. If $V = 12\pi$ cubic inches when $r = 3$ inches and $h = 4$ inches, find V in terms of r and h.

SOLUTION. Since V varies jointly as r^2 and h, there is a real number k such that $V = kr^2h$. Substituting for r, h, and V, we have

$$12\pi = k(9)(4) \quad \text{or} \quad k = \frac{12\pi}{36} = \frac{\pi}{3}$$

Hence, the required formula is $V = \frac{1}{3}\pi r^2 h$.

3.2 Algebra of Functions

So far we have been concerned with individual functions and with their domains, ranges, and graphs. We now examine ways of combining two functions under the operations of addition, subtraction, multiplication, and division to form new functions. We will find uses of the following material later in this text.

Let f and g be two functions and assume that D_f and D_g are the domains of f and g, respectively; then the sum $f + g$, the difference $f - g$, the product fg, and the quotient f/g are defined as follows:

$$f + g = \{(x, y) \mid y = f(x) + g(x) \text{ and } x \in D_f \cap D_g\}$$
$$f - g = \{(x, y) \mid y = f(x) - g(x) \text{ and } x \in D_f \cap D_g\}$$
$$f \cdot g = \{(x, y) \mid y = f(x) \cdot g(x) \text{ and } x \in D_f \cap D_g\}$$

and

$$\frac{f}{g} = \left\{ (x, y) \mid y = \frac{f(x)}{g(x)}, g(x) \neq 0 \text{ and } x \in D_f \cap D_g \right\}$$

EXAMPLES

1. Let $f = \{(2, -1), (-1, 2), (3, 5), (6, 1)\}$ and $g = \{(2, 3), (3, 1), (5, 3), (8, 7)\}$. Form each of the following functions.

    a) $f + g$ b) $f - g$ c) $f \cdot g$ d) $\dfrac{f}{g}$

    SOLUTION. $D_f = \{2, -1, 3, 6\}$ and $D_g = \{2, 3, 5, 8\}$; then $D_f \cap D_g = \{2, 3\}$. Hence, $\{2, 3\}$ is the domain of $f + g$, $f - g$, $f \cdot g$, and f/g since $g(2) \neq 0$ and $g(3) \neq 0$.

    a) Since $(f + g)(2) = f(2) + g(2) = -1 + 3 = 2$ and $(f + g)(3) = f(3) + g(3) = 5 + 1 = 6$, then $f + g = \{(2, 2), (3, 6)\}$.
    b) Since $(f - g)(2) = f(2) - g(2) = -1 - 3 = -4$ and $(f - g)(3) = f(3) - g(3) = 5 - 1 = 4$, then $f - g = \{(2, -4), (3, 4)\}$.
    c) Since $(f \cdot g)(2) = f(2) \cdot g(2) = -1 \cdot 3 = -3$ and $(f \cdot g)(3) = f(3) \cdot g(3) = 5 \cdot 1 = 5$, then $f \cdot g = \{(2, -3), (3, 5)\}$.
    d) Since $(f/g)(2) = f(2)/g(2) = -1/3$ and $(f/g)(3) = f(3)/g(3) = 5/1$, then $f/g = \{(2, -\frac{1}{3}), (3, 5)\}$.

2. Let $f(x) = 2x^2 + 3$ and $g(x) = 2x$. Form each of the following functions:

    a) $f + g$ b) $f - g$ c) $f \cdot g$ d) $\dfrac{f}{g}$

    SOLUTION. The domain of both f and g is the set of real numbers, so that the domain of $f + g$, $f - g$, and $f \cdot g$ is also the set of real numbers. The domain of f/g is the set of real numbers except $x = 0$, since $g(0) = 0$.
    a) $(f + g)(x) = f(x) + g(x) = 2x^2 + 3 + 2x$
    b) $(f - g)(x) = f(x) - g(x) = 2x^2 + 3 - 2x$
    c) $(f \cdot g)(x) = f(x) \cdot g(x) = (2x^2 + 3)2x = 4x^3 + 6x$
    d) $\left(\dfrac{f}{g}\right)(x) = \dfrac{f(x)}{g(x)} = \dfrac{2x^2 + 3}{2x}$

3. The *difference quotient* of a function f is defined as $[f(x + h) - f(x)]/h$, $h \neq 0$ and is a fundamental building block in calculus. Compute the difference quotient for each of the following functions.

    a) $f(x) = -3x + 4$ b) $f(x) = 4$

    SOLUTION

    a) $f(x + h) = -3(x + h) + 4$, so that

    $$\frac{f(x + h) - f(x)}{h} = \frac{[-3(x + h) + 4] - (-3x + 4)}{h}$$

    $$= \frac{-3x - 3h + 4 + 3x - 4}{h} = -3$$

b) $f(x + h) = 4$, so that

$$\frac{f(x + h) - f(x)}{h} = \frac{4 - 4}{h} = 0$$

PROBLEM SET 2

1 In each of the following parts, identify which of the relations is a function and which is not a function. If it is not a function, state the reason. Also identify the domain and range and graph the relation.
   a) $\{(2, 4), (3, 6), (7, 2), (9, -3)\}$
   b) $\{(-8, 0), (-7, 0), (-6, 2), (-7, 3), (5, 1)\}$
   c) $\{(x, y) \mid y = 2x - 1\}$
   d) $\{(x, y) \mid y \leq 3x + 2\}$
   e) $\{(x, y) \mid |y| = x\}$
   f) $\{(x, y) \mid |y| = |x|\}$
   g) $\{(x, y) \mid y = \sqrt[3]{x}\}$
   h) $y = -3x^2$
   i) $y = -x^3$
   j) $y = \dfrac{1}{\sqrt{x - 2}}$
   k) $\{(x, y) \mid 3y \geq -2x + 5\}$
   l) $x^2 + y^2 = 9$
   m) $y^2 = -x^3$

2 Which of the graphs in Figure 14 represent functions?

Figure 14

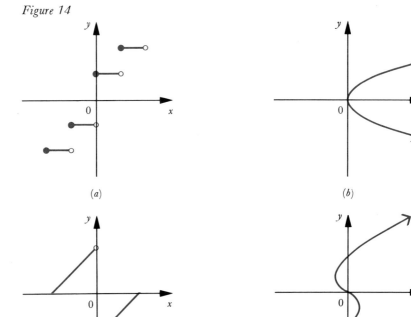

(a) (b) (c) (d)

3. What numbers in the domain of each of the following functions have 9 as an image?
   a) $f(x) = 2x + 1$
   b) $f(x) = -2x$
   c) $f(x) = -2x^2$
   d) $f(x) = 7x$.

4. Find the domain and the range of each of the following functions; find $f(1)$, $f(3)$, and $f(5)$; also sketch the graph of f.
   a) $f(x) = 3x + 1$
   b) $f = \{(1, 2), (3, 4), (5, 7)\}$
   c) $f(x) = \sqrt{x - 1}$
   d) $f(x) = \dfrac{1}{2x - 3}$
   e) $f(x) = \dfrac{1}{x + 2}$
   f) $f(x) = x^2 + 1$
   g) $f(x) = \dfrac{2}{x}$
   h) $f(x) = x^3 - 1$
   i) $f(x) = |x| + x$
   j) $f(x) = 2[\![x]\!]$

5. Describe each of the functions in Problem 4 using the mapping notation, the ordered pairs interpretation, and the function notation.

6. Find $f(0)$, $f(2)$, $f(3)$, $f(-3)$, $f(a)$, $f(a + h)$, and the difference quotient $[f(a + h) - f(a)]/h$ for each of the given functions. What is the domain and the range of f?
   a) $f(x) = 3x - 7$
   b) $f(x) = x^2 - 3x - 4$
   c) $f(x) = 3x^2 + 1$
   d) $f(x) = \dfrac{3x}{4} - 5$
   e) $f(x) = x^3 - 2$

7. Let $f(x) = |x| - x$.
   a) Determine which of the following points lie on the graph of f: $(1, 0)$, $(-1, 0)$, $(-2, 4)$, $(-3, 6)$.
   b) What are the domain and the range of f?
   c) Evaluate $f(3)$, $2f(3)$, $f(2)$, $f(3 - 2)$, and $f(a + 1)$.
   d) Sketch the graph of f.

8. Assume that $f(x) = 2x + 1$ and $g(x) = (x - 1)/2$. Find each of the following expressions.
   a) $f[g(x)]$
   b) $g[f(x)]$
   c) $f[f(x)]$
   d) $g[g(x)]$

9. Repeat Problem 8 for $f(x) = |x|$ and $g(x) = [\![x]\!]$.

10. Compare the graphs of $y = f(x)$ and $y = |f(x)|$ for each of the following functions.
    a) $f(x) = x$
    b) $f(x) = -x$
    c) $f(x) = x^2$
    d) $f(x) = x^3$
    e) $f(x) = -x^3$
    f) $f(x) = 1 - x$

11 Let the radius and height of a right circular cylinder be represented by r and h units, respectively. If h is constant, express the total surface area S as a function of r. If r is 5 inches and h is 9 inches, what is the total surface area?

12 The demand for a certain kind of candy bar is such that the product of the demand (in thousands of cartons) and the price per carton (in cents) is always equal to 250,000, so long as the price is not less than $2.00 or more than $3.50. Express the relationship between the price and the demand by means of an equation, and use it to compute the demand when the price is a) $2.00, b) $2.50, c) $2.90, d) $3.00, and e) $3.50.

13 A gardener wishes to fence a rectangular garden along a straight river that require no fence on one of the four sides. He has enough wire to build a fence 300 feet long. If the side bordering on the river is represented by x, express the area of the garden as a function of x.

14 Form each of the following functions: $f + g, f - g, f \cdot g$, and f/g. Indicate the domain and range of each function.
  a) $f = \{(-3, -6), (-2, -4), (-1, -2), (0, 0), (1, 2), (2, 4)\}$ and
     $g = \{(-4, 6), (-2, 1), (-1, -2), (1, -2), (3, 6)\}$
  b) $f = \{(x, y) \mid y = 2x\}$ and $g = \{(x, y) \mid y = x^2 - 3\}$
  c) $f = \{(x, y) \mid y = 2x - 3\}$ and $g = \{(x, y) \mid y = 4 - x\}$

15 Let y be directly proportional to x, and $y = f(x)$. If $y = 8$ when $x = 4$, find a formula for f and sketch the graph of f. Also find
  a) $f(2) + f(3)$ b) $f(x + 2)$
  c) $\dfrac{f(x + h) - f(x)}{h}$ where $h \neq 0$

16 If y is directly proportional to x^2 and $y = f(x)$, does $f(ax) = af(x)$?

17 If y is directly proportional to x^3, express y as a function of x in each of the following cases:
  a) $y = 4$ when $x = 2$ b) $y = 12$ when $x = -2$
  c) $y = 3$ when $x = 1$ d) $y = -2$ when $x = 3$
  e) $y = 14$ when $x = 11$ f) $y = 10$ when $x = -3$

18 If y is inversely proportional to x^2 and $x = 9$ when $y = 2$, find x when $y = 105$.

19 If y is inversely proportional to $\sqrt[3]{x}$ and $y = 9$ when $x = 8$, find y when $x = 216$.

20 If V varies directly as T and inversely as P, and $V = 40$ when $T = 300$ and $P = 30$, find V when $T = 324$ and $P = 24$.

21 The surfaces of two spheres have the ratio $9:4$. What is the ratio of their
  a) Radii?
  b) Volume ($V = \frac{4}{3}\pi r^3$)?

22 The total surface area S of a cube is directly proportional to the square of the edge x. If a cube whose edge is 3 inches has a surface area of 36 square inches, express the surface area S as a function of x. Then find the surface area of a cube whose edge is 12 inches.

23 Coulomb's law states that the magnitude of the force F that acts on two charges q_1 and q_2 varies directly as the product of the magnitude of q_1 and q_2 and inversely as the square of the distance r between them. If the force on two charges, each of 1 coulomb, that are separated in air by a distance of 1,000 meters is 9,000 newtons, find the force when the two charges are separated by 2,000 meters.

24 Newton's law of gravitational attraction states that the force F with which two particles of mass m_1 and m_2, respectively, attract each other varies directly as the product of the masses and inversely as the square of the distance r between them. If one of the masses is tripled, and the distance between the masses is also tripled, what happens to the force?

4 Even and Odd Functions

Up to this point we have graphed functions by constructing tables, listing in one column values of x for which the function is defined, and in a second column the value $f(x)$ assigned to each x. This procedure involved selecting enough values of x so that the pattern suggested by the corresponding points $(x, f(x))$ would enable us to sketch the complete graph. We can sometimes reduce the number of points needed to graph some functions by employing special features of the emerging pattern such as symmetry. Therefore, before we consider types of functions, we shall first consider the kinds of *symmetry* that graphs of functions may possess.

Figure 1

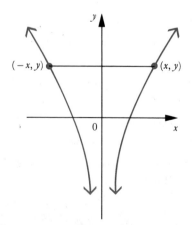

The graph of a function f is *symmetric with respect to the y axis* if whenever the point (x, y) is on the graph of f: the point $(-x, y)$ is also on the graph of f, that is, $f(-x) = f(x)$ (Figure 1).

A function possessing this property is called an *even function*. For example, the graph of $f(x) = x^2$ is symmetric with respect to the y axis, since $f(-x) = (-x)^2 = x^2 = f(x)$ (Figure 2). $f(x) = x^2$ is an even function. Thus, in graphing such a function, we need only to plot points to the right of the y axis and then "copy" the graph onto the left side, as though the y axis were a mirror.

Figure 2

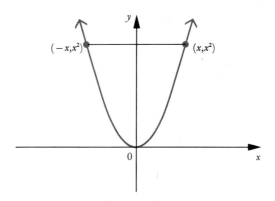

The graph of a function f is *symmetric with respect to the origin*, if whenever the point (x, y) is on the graph of f, then the point $(-x, -y)$ is also on the graph of f, that is, $f(-x) = -f(x)$ (Figure 3). In this case, the function is said to be an *odd function*. For example, the graph of $f(x) = x^3$

Figure 3

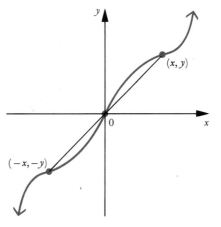

is symmetric with respect to the origin, since $f(-x) = (-x)^3 = -x^3 = -f(x)$ (Figure 4). $f(x) = x^3$ is an odd function.

Figure 4

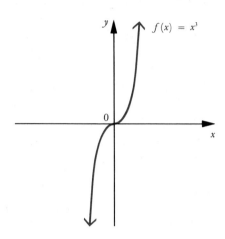

EXAMPLES

Discuss the symmetry of each of the following functions, sketch the graphs using the symmetry, and indicate whether the function is even or odd.

1 $f(x) = -3x^2$

SOLUTION. $f(-x) = -3(-x)^2 = -3x^2 = f(x)$, so that the graph of f is symmetric with respect to the y axis; hence, it is only necessary to graph the function f for nonnegative values of x (Figure 5a). The remainder of the graph is determined by "reflection" across the y axis (Figure 5b). f is an even function.

Figure 5

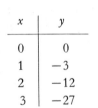

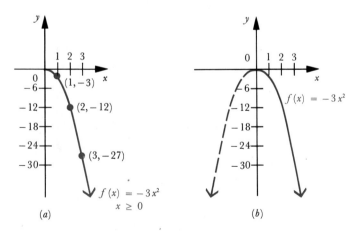

2 $f(x) = \sqrt{1 - x^2}$

SOLUTION. $f(-x) = \sqrt{1 - (-x)^2} = \sqrt{1 - x^2} = f(x)$, so that the graph of f is symmetric with respect to the y axis. Hence, we can get the graph of the function f by first locating points in the first quadrant (Figure 6a), and then reflecting this graph across the y axis (Figure 6b). f is an even function.

Figure 6

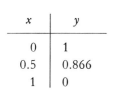

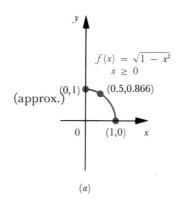

(a)

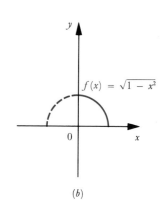

(b)

3   $f(x) = -5x^3$

SOLUTION. $f(-x) = -5(-x)^3 = -(-5x^3) = -f(x)$, so that the graph of f is symmetric with respect to the origin. Hence, we can graph f by first locating points in the second quadrant (Figure 7a), and then reflecting this graph across the origin (Figure 7b). f is an odd function.

Figure 7

x	y
0	0
-1	5
-2	40

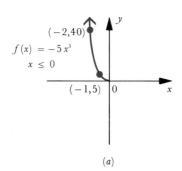

(a)

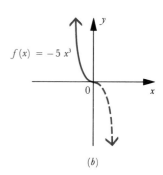

(b)

In summary, even and odd functions can be described geometrically by using symmetry. If the graph of the function f is symmetric with respect to the y axis, it is an even function, whereas if the graph of f is symmetric with respect to the origin, it is an odd function. Algebraically, a function f is said to be an *even function* if $f(x) = f(-x)$ for all x in the domain of f; a function f is said to be an *odd function* if $f(-x) = -f(x)$ for all x in the domain of f.

Some functions are neither even nor odd. For example, $f(x) = 2x + 1$ is neither even nor odd because

$$f(-x) = 2(-x) + 1$$
$$= -2x + 1$$

so that

$$f(x) \neq f(-x)$$
$$\| \qquad \|$$
$$2x + 1 \qquad -2x + 1$$

and

$$-f(x) \neq f(-x)$$
$$\| \qquad \|$$
$$-2x - 1 \qquad -2x + 1$$

This fact can also be seen from the graph of $f(x) = 2x + 1$ (Figure 8) in that there is no symmetry with respect to either the y axis or the origin.

Figure 8

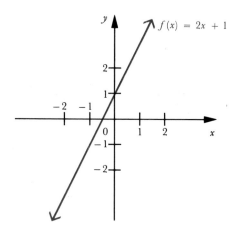

EXAMPLES

Determine whether the following functions are even, odd, or neither, and sketch the graph.

1 $f(x) = 4x^2$

SOLUTION. $f(-x) = 4(-x)^2 = 4x^2 = f(x)$, so that f is an even function. Note that the graph of f is symmetric with respect to the y axis (Figure 9).

Figure 9

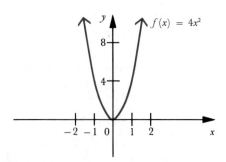

2 $f(x) = 5x^3$

SOLUTION. $f(-x) = 5(-x)^3 = -5x^3 = -f(x)$, so that f is an odd function and the graph of f is symmetric with respect to the origin (Figure 10).

Figure 10

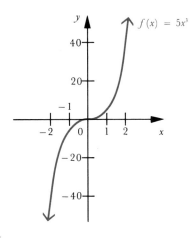

3 $f(x) = x^2 + 4x + 4$

SOLUTION. $f(-x) = (-x)^2 + 4(-x) + 4 = x^2 - 4x + 4 \neq f(x)$, so that f is not an even function. Also $f(-x) \neq -f(x)$ for all x; hence, f is not an odd function. Therefore, f is neither even nor odd (Figure 11).

Figure 11

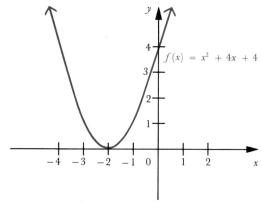

5 Increasing and Decreasing Functions

The concepts of increasing and decreasing functions can be introduced by considering the graphs of some functions. For example, consider the motion of a point traveling along the graph of f (Figure 1) from left to right. The point rises as its x coordinate increases from 0 to 3, and then declines as its x coordinate increases from 3 to 8. In this case we say that "*f is an increasing function* in the interval (0, 3)," and "*f is a decreasing function* in the interval (3, 8)."

Figure 1

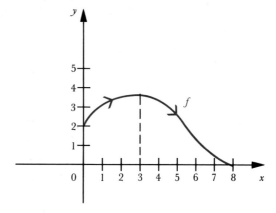

More formally, a function f is said to be a *(strictly) increasing function* in an interval I, if whenever a and b are two numbers in I such that $a < b$, we have $f(a) < f(b)$ (Figure 2).

Figure 2

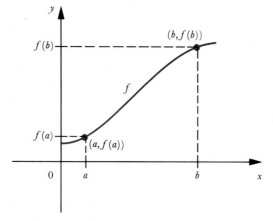

A function f is said to be a *(strictly) decreasing function* in an interval I, if, whenever a and b are two numbers in I such that $a < b$, we have $f(a) > f(b)$ (Figure 3).

Figure 3

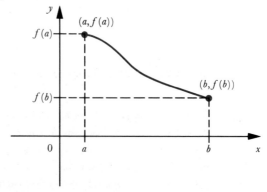

EXAMPLES

Determine whether each of the following functions are increasing or decreasing.

1   $f(x) = 3x + 2$

SOLUTION. The graph of f (Figure 4) indicates that f is an increasing function on R, for example, for -2 and 7 we have $-2 < 7$ and $f(-2) < f(7)$, since $3(-2) + 2 < 3(7) + 2$ or $-4 < 23$.

Figure 4

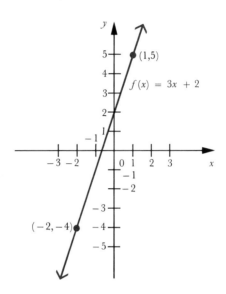

2   $f(x) = -2x^3$

SOLUTION. The graph of f (Figure 5) indicates that f is a decreasing function on R. For any two real numbers, say 1 and 2, $1 < 2$ implies $f(1) > f(2)$, since $-2 > -16$.

Figure 5

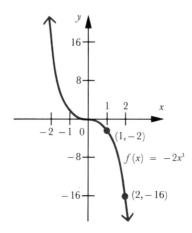

3 $f(x) = 3$

SOLUTION. The graph of f (Figure 6) indicates that f is neither an increasing nor a decreasing function on R. For example, for 1 and 5, $1 < 5$ but $f(1) = f(5)$. (Why?)

Figure 6

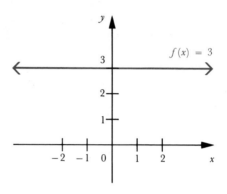

4 $f(x) = x^2$

SOLUTION. The graph of f (Figure 7) indicates that f is a decreasing function on the interval $(-\infty, 0)$ and f is increasing on the interval $(0, \infty)$.

Figure 7

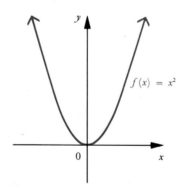

PROBLEM SET 3

1 Discuss the symmetry of each of the following functions to determine if f is symmetric with respect to the y axis or the origin. Use symmetry to sketch the graph.
   a) $f(x) = -2x$
   b) $f(x) = |x|$
   c) $f(x) = 5x^3$
   d) $f(x) = -\sqrt{4 - x^2}$
   e) $f(x) = -2x^4$
   f) $f(x) = 3x^2 + 1$

2 Give an example of two functions f and g that are symmetric with respect to the y axis. Indicate which of the following functions is symmetric with respect to the y axis.

a) $f + g$ b) $f - g$ c) $f \cdot g$ d) $\dfrac{f}{g}$

3 Indicate whether each of the given functions is even or odd or neither.
   a) $f(x) = x^3 + \dfrac{1}{x}$ b) $f(x) = x^4 + 1$
   c) $f(x) = x^3 + x^2$ d) $f(x) = \sqrt{x} + 3$
   e) $f(x) = (x^2 + 1)^3$ f) $f(x) = x^2 - 2x + 3$

4 Can a function be both even and odd? If so, give an example of one.

5 Let $f(x) = -x^3$. Sketch the graph of f for $x \in [1, 4]$. Is f increasing or decreasing in the interval $[1, 4]$?

6 Indicate the intervals where each of the given functions is increasing or decreasing. Sketch the graph of f.
   a) $f(x) = -5x + 7$ b) $f(x) = |x| - x$
   c) $f(x) = -5x^2$ d) $f(x) = |-3x|$
   e) $f(x) = x^3 + 1$ f) $f(x) = \sqrt[3]{x}$

7 Decide whether each of the given functions whose graph is given in Figure 8 is even, odd, or neither. Does the graph of the function have symmetry? Indicate those intervals where the given functions are either increasing or decreasing.

Figure 8

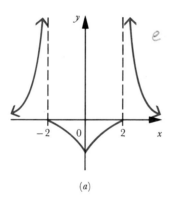

(a)

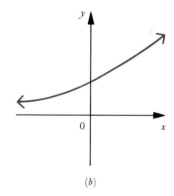

(b)

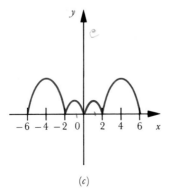

(c)

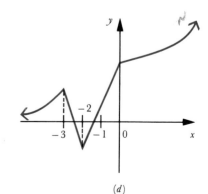

(d)

8 Given the two functions f and g defined by $f(x) = x^2 + 4x + 1$ and $g(x) = 2x + 4$.
  a) Is f even or odd? Is g even or odd?
  b) Is f increasing or decreasing? Is g increasing or decreasing?
  c) Show that $f(x) - 2g(x) = x^2 - 7$. Is $f(x) - 2g(x)$ even?

REVIEW PROBLEM SET

1 Which of the following relations are functions? Indicate the domain and range of each.
  a) $\{(4, 1), (3, -1), (5, 1), (-1, 1)\}$ b) $\{(1, 1), (2, 2), (3, 3), (4, 4)\}$
  c) $\{(0, 1), (0, -1), (1, 3), (2, 4)\}$ d) $\{(x, y) \mid y \leq 3x + 1\}$

2 Which of the graphs in Figure 9 represent a function?

Figure 9

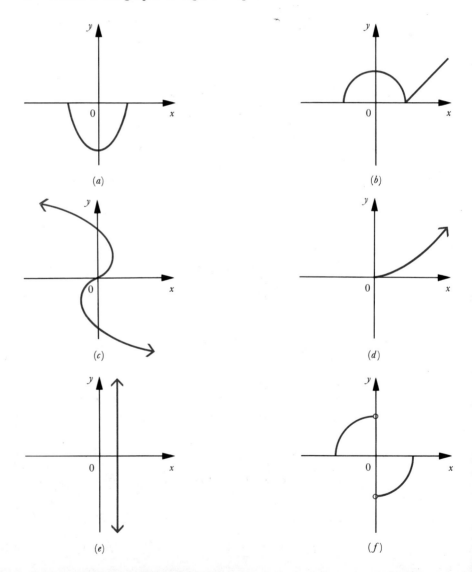

(a)

(b)

(c)

(d)

(e)

(f)

3 Let $f(x) = 2x^2 + 3x$. Find the domain of f; also find each of the following expressions.
a) $f(-2)$
b) $f(-1)$
c) $f(0)$
d) $f(\tfrac{1}{2})$
e) $f(a)$
f) $f(b)$
g) $f(a+b)$
h) $f(a+b) - f(a)$

4 Find the domain and range of each of the following functions; also find $f(1)$, $f(0)$, and $f(2)$.
a) $f(x) = \sqrt{4 - x^2}$
b) $f(x) = -2x$
c) $f(x) = -x - 1$
d) $f(x) = \dfrac{x+4}{x}$

5 Represent each of the following statements as a formula.
a) The volume of a sphere as a function of the radius.
b) The area of an equilateral triangle as a function of the length of the sides.
c) The radius of a sphere as a function of its surface area.

6 Let f be the function defined by $f(x) = 2x$ and g be the function defined by $g(x) = x^2 - 3$. Find
a) $f + g$
b) $f - g$
c) $f \cdot g$
d) $\dfrac{f}{g}$
e) $2f + g$
f) $2f - g$
g) $3f + f$
h) $2g \cdot g$
i) Graph $f + g$.

7 Discuss the symmetry for the following functions. Use your results to graph the functions.
a) $f(x) = -5x^2$
b) $f(x) = -x - 2$
c) $f(x) = \dfrac{-7}{x}$
d) $f(x) = |x - 2|$
e) $f(x) = -4x^3$
f) $f(x) = \sqrt{25 - x^2}$

8 Indicate whether each of the following functions is symmetric with respect to the y axis or the origin. Indicate the domain and range of f. Are the functions increasing or decreasing? Are the functions even or odd? Graph the functions.
a) $f = \{(x, y) \mid y = 2x - 1\}$
b) $f(x) = |x|^2$
c) $f(x) = |2x| - 2x$
d) $f = \{(x, y) \mid y = -\tfrac{1}{2}x + 1\}$
e) $f(x) = \dfrac{|x|}{x}$
f) $f(x) = \dfrac{3}{x}$
g) $f(x) = [\![3x]\!]$
h) $f(x) = 5 - x^2$
i) $f(x) = 3x^3 - 1$
j) $f(x) = \dfrac{3}{x^2}$

9 Find examples of functions for which the following statements are true.
   a) $f(x^2) = (f(x))^2$
   b) $f(|x|) = |f(x)|$
   c) $f(x + y) = f(x) + f(y)$
   d) $f([\![x]\!]) = [\![f(x)]\!]$

10 Let $f(x) = 3 - 2x$ and $g(x) = |x|$. Find
   a) $f(a - 1)$
   b) $f(b - 5)$
   c) $f[g(a)]$
   d) $g[f(b)]$

11 Express y as a function of $x [y = f(x)]$ in each of the following cases. Also graph the function.
   a) If y is directly proportional to x and if $y = 8$ when $x = 12$
   b) If y is directly proportional to x^2 and if $y = 18$ when $x = 3$
   c) If y is directly proportional to $\sqrt{x}$ and if $y = 16$ when $x = 16$
   d) If y is directly proportional to $\sqrt{x}$ and if $y = 9$ when $x = 16$
   e) If y is inversely proportional to x and if $y = 4$ when $x = 5$
   f) If y is inversely proportional to x and if $y = 12$ when $x = \frac{3}{4}$

12 Hook's law states that the extension of an elastic spring beyond its natural length is directly proportional to the force applied. If a weight of 8 pounds causes a spring to stretch from a length of 9 inches to a length of 9.5 inches, what weight will cause it to stretch to a length of 1 foot?

13 The power required to operate a fan is directly proportional to the speed of the fan. If 1 horsepower will drive a fan at a speed of 480 revolutions per minute,
   a) How fast will 8 horsepower drive it?
   b) What power will be required to give it a speed of 600 revolutions per minute?

CHAPTER 4

Linear and Quadratic Functions

CHAPTER 4

Linear and Quadratic Functions

1 Polynomial Functions

In Chapter 3 we considered the general properties of functions. This chapter deals with particular types of functions, polynomial functions. Any function with domain R whose values are found by means of a particular polynomial is called a *polynomial function*. The degree of the polynomial is also called the *degree of the polynomial function*. For example, $f(x) = 2x^3 + 5x^2 - 9x + 3$ is a polynomial function of degree 3. The numbers 2, 5, -9, and 3 are called the *coefficients* of the polynomial function. A function such as $f(x) = 4$ is called a *zero-degree polynomial* function, whereas $f(x) = 0$ is called the *zero polynomial* function and no degree is assigned to it.

EXAMPLES

Identify the degree and the coefficients of each of the polynomial functions.

1   $f(x) = -5x^3 + 7x^2 + 3x - 4$

SOLUTION. The degree of the polynomial function is 3. The coefficients of the polynomial function are -5, 7, 3, and -4.

2   $f(x) = x^4 + \sqrt{2}\, x^2 + 5$

SOLUTION. The degree of the polynomial function is 4 and the coefficients are 1, $\sqrt{2}$, and 5.

3   $f(x) = 2$

SOLUTION. The degree of the polynomial function is 0 and the coefficient is 2.

2 Linear Functions

A function defined by an equation of the form $f(x) = mx + b$, where m and b are constant numbers, is called a *linear function*. Such functions are called linear functions because their graphs are straight lines. For example, we can show that the graph of $f(x) = 2x + 1$ is a straight line by showing that any three points of its graph lie on the same straight line. In particular, consider the three points P_1: $(-1, -1)$, P_2: $(1, 3)$, and

Figure 1

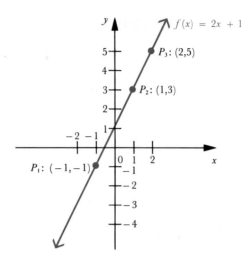

P_3: (2, 5) which belong to the graph of f (Figure 1). If the distance from P_1 to P_3 (denoted by $\overline{P_1P_3}$) equals the sum of the distances from P_1 to P_2 (denoted by $\overline{P_1P_2}$) and P_2 to P_3 (denoted by $\overline{P_2P_3}$), then the points P_1, P_2, and P_3 lie on a straight line, since the shortest distance between two points is a straight line. Now,

$$\overline{P_1P_3} = \sqrt{(2+1)^2 + (5+1)^2} = \sqrt{9+36} = \sqrt{45} = 3\sqrt{5}$$
$$\overline{P_1P_2} = \sqrt{(1+1)^2 + (3+1)^2} = \sqrt{4+16} = \sqrt{20} = 2\sqrt{5}$$

and

$$\overline{P_2P_3} = \sqrt{(2-1)^2 + (5-3)^2} = \sqrt{1+4} = \sqrt{5}$$

Since $3\sqrt{5} = 2\sqrt{5} + \sqrt{5}$, we have $\overline{P_1P_3} = \overline{P_1P_2} + \overline{P_2P_3}$, and the three points lie on the same straight line. Thus, we conclude that the graph of $f(x) = 2x + 1$ is a straight line. This argument can be applied to the general form of a linear function.

PROPERTY

The graph of a linear function $f(x) = mx + b$ is a straight line.

PROOF. Consider any three points on the graph of f and denote them by P_1: (x_1, y_1), P_2: (x_2, y_2), and P_3: (x_3, y_3) such that $x_1 < x_2 < x_3$ (Figure 2). These three points lie on the same straight line if the distance $\overline{P_1P_3}$ equals the sum of the two distances $\overline{P_1P_2}$ and $\overline{P_2P_3}$. Now

$$\begin{aligned}\overline{P_1P_3} &= \sqrt{(x_3 - x_1)^2 + (y_3 - y_1)^2} \\ &= \sqrt{(x_3 - x_1)^2 + [(mx_3 + b) - (mx_1 + b)]^2} \\ &= \sqrt{(x_3 - x_1)^2 + m^2(x_3 - x_1)^2} \\ &= \sqrt{(x_3 - x_1)^2(1 + m^2)}\end{aligned}$$

Figure 2

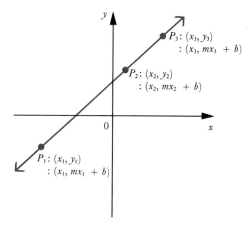

Since $x_3 > x_1$, $x_3 - x_1$ is positive so that

$$\overline{P_1P_3} = (x_3 - x_1)\sqrt{1 + m^2}$$

Similarly, we can show that

$$\overline{P_1P_2} = (x_2 - x_1)\sqrt{1 + m^2}$$

and

$$\overline{P_2P_3} = (x_3 - x_2)\sqrt{1 + m^2}$$

Therefore,

$$\begin{aligned}\overline{P_1P_2} + \overline{P_2P_3} &= (x_2 - x_1)\sqrt{1 + m^2} + (x_3 - x_2)\sqrt{1 + m^2} \\ &= [(x_2 - x_1) + (x_3 - x_2)]\sqrt{1 + m^2} \\ &= (x_3 - x_1)\sqrt{1 + m^2} \\ &= \overline{P_1P_3}\end{aligned}$$

Hence, any three points on the graph of f lie on the same straight line. That is, the graph of $f(x) = mx + b$ is a straight line.

Figure 3

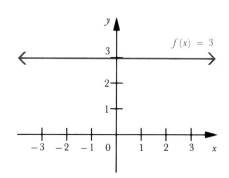

If $m = 0$, then $f(x) = b$ is called a *constant function*. Its graph is the set of all points with an ordinate of b; it is a line parallel or coincidental to the x axis through the point $(0, b)$. If, for example, $b = 3$, then $f(x) = 3$ has the set representation $f = \{(x, y) \mid y = 3\}$ (Figure 3).

EXAMPLES

1. Find the domain and the range of the following linear function and sketch the graph.

$$f(x) = 2x + 5$$

SOLUTION. Since for any real number x the calculation can be performed to determine a corresponding number $f(x)$, the domain of the function f is the set of real numbers R and the range of f is also the set of real numbers R. Since the graph of a linear function is a straight line, it is enough to determine the graph by locating two points (Figure 4).

Figure 4

x	$f(x)$
0	5
$-\frac{5}{2}$	0

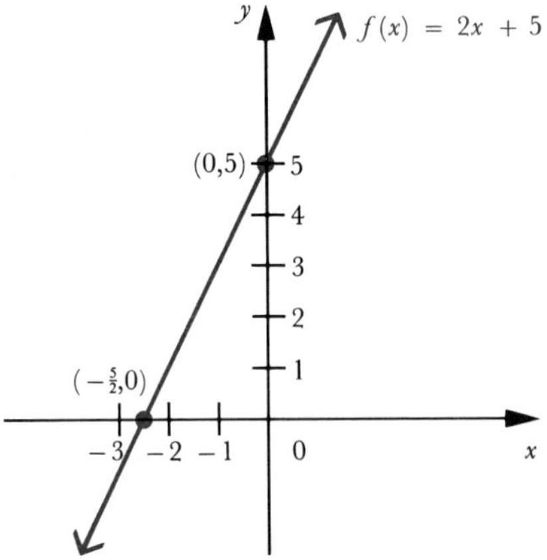

2. The temperature of a body is C degrees Celsius and the corresponding

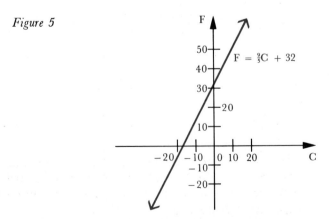

Figure 5

Fahrenheit temperature is denoted by F. Find the relation between these quantities if it is linear and if F = 32 when C = 0 and F = 212 when C = 100.

SOLUTION. Since the relationship is given to be linear, we have F = $mC + b$. Now F = 32 when C = 0, so that 32 = $m(0) + b$, or $b = 32$. Also, F = 212 when C = 100, so that 212 = $100m + 32$, or $m = \frac{9}{5}$. Therefore, the relationship between the two numbers can be expressed by the formula F = $\frac{9}{5}C + 32$ (Figure 5).

2.1 Slope of a Line

Figure 6 displays the graphs of the functions $y = x$, $y = 2x$, and $y = 5x$.

Figure 6

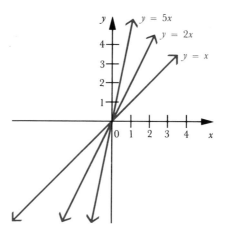

Note that each of these functions is increasing at a different "rate." For example, as

	x increases from 1 to 2	
$y = x$	increases from 1 to 2	(1-unit increase)
$y = 2x$	increases from 2 to 4	(2-unit increase)

and

$y = 5x$	increases from 5 to 10	(5-unit increase)

What we would like to do next is to derive some way of measuring the "inclination" or "rate of change" of lines and to relate such a measure to the equation of the line. The simplest way to do this is by what is called the *slope* of a line.

DEFINITION

Suppose that $P_1: (x_1, y_1)$ and $P_2: (x_2, y_2)$ are two points of a line (Figure 7). The number s that is defined by the equation

$$s = \frac{y_2 - y_1}{x_2 - x_1}$$

provided that $x_1 \neq x_2$ is called the *slope* of the line.

Note that the value of s will be the same regardless of the choice of the two points on the line.

Figure 7

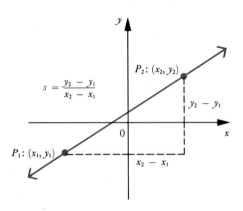

Since

$$\frac{y_1 - y_2}{x_1 - x_2} = \frac{-(y_1 - y_2)}{-(x_1 - x_2)} = \frac{y_2 - y_1}{x_2 - x_1}$$

the order in which we take the two points when subtracting the coordinates does not change the value.

For example, the slope of the line containing the points $(-2, 5)$ and $(3, -4)$ is

$$\frac{-4 - 5}{3 - (-2)} = -\frac{9}{5}$$

or, equivalently,

$$\frac{5 - (-4)}{-2 - 3} = -\frac{9}{5}$$

Accordingly, we can speak of $(y_2 - y_1)/(x_2 - x_1)$ as the slope of the line containing the two points (x_1, y_1) and (x_2, y_2), without specifying which comes first. If the line containing two points (x_1, y_1) and (x_2, y_2) is vertical, that is, if $x_1 = x_2$ for each distinct pair of points on the line, then $x_2 - x_1 = 0$, so that

$$s = \frac{y_2 - y_1}{x_2 - x_1} = \frac{y_2 - y_1}{0}$$

has no meaning in the algebra of real numbers, and we say that the slope s of the line is undefined. For example, the slope s of the line $x = 4$ is undefined, since $x_1 = x_2 = 4$ (Figure 8).

Figure 8

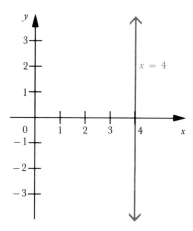

If $y_1 = y_2$ for each distinct pair of points on the line, then $y_2 - y_1 = 0$; therefore,

$$s = \frac{y_2 - y_1}{x_2 - x_1} = \frac{0}{x_2 - x_1} = 0$$

and the slope of the line is 0. For example, the slope of the line $y = 3$ is zero, since $y_1 = y_2 = 3$ (Figure 9).

Figure 9

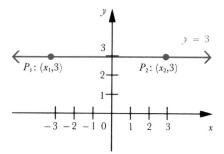

EXAMPLE

Determine the slope of the line containing the two points whose coordinates are given by

a) (6, 2) and (3, 7) b) (3, −2) and (5, −6)

SOLUTION

a) The slope of the line is given by $s = (y_2 - y_1)/(x_2 - x_1)$, so that for $P_1: (6, 2)$ and $P_2: (3, 7)$, we have

$$s = \frac{7-2}{3-6} = \frac{5}{-3} = -\frac{5}{3}$$

b) If $P_1: (3, -2)$ and $P_2: (5, -6)$, then

$$s = \frac{-6 - (-2)}{5 - 3} = \frac{-6 + 2}{5 - 3} = -\frac{4}{2} = -2$$

The graph of a linear function f is a straight line for which the slope is always defined. This can be shown by considering any two distinct points (x_1, y_1) and (x_2, y_2) on the graph of f. $x_1 \neq x_2$ (why?), so that $x_2 - x_1 \neq 0$, and thus $s = (y_2 - y_1)/(x_2 - x_1)$ is always defined.
The slope of $f(x) = mx + b$ can be computed as follows:

$$s = \frac{y_2 - y_1}{x_2 - x_1} = \frac{f(x_2) - f(x_1)}{x_2 - x_1} = \frac{(mx_2 + b) - (mx_1 + b)}{x_2 - x_1}$$

$$= \frac{mx_2 - mx_1}{x_2 - x_1} = \frac{m(x_2 - x_1)}{x_2 - x_1} = m$$

That is, *the slope of a linear function $f(x) = mx + b$ is the coefficient of the x term, namely m.* Furthermore, by the above computation, we see again that the value of m will be the same regardless of the choice of the two points on the graph of f. From now on, we will use m to represent the slope of a straight line; that is,

$$m = \frac{y_2 - y_1}{x_2 - x_1} \quad \text{where } x_1 \neq x_2$$

Thus, for example, the slope of the line $y = 2x + 1$ is 2; the slope of $y = -\frac{1}{2}x$ is $-\frac{1}{2}$.

The value of the slope of a line can be used to determine whether the linear function is increasing or decreasing. This can be derived as follows.

Let (x_1, y_1) and (x_2, y_2) be any two points on the graph of $y = mx + b$ such that $x_1 < x_2$ (Figure 10).
For $m > 0$, we have

$$m = \frac{y_2 - y_1}{x_2 - x_1} > 0$$

Since $x_1 < x_2$ implies that $x_2 - x_1 > 0$, we have $y_2 - y_1 > 0$ or $y_1 < y_2$. Thus, for $m > 0$, $y = mx + b$ is an increasing function and its graph rises from left to right (Figure 10).

Figure 10

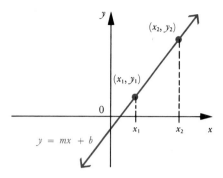

Similarly, for $m < 0$, we have

$$m = \frac{y_2 - y_1}{x_2 - x_1} < 0$$

Since $x_1 < x_2$ implies that $x_2 - x_1 > 0$, we have $y_2 - y_1 < 0$, or $y_1 > y_2$. Thus, for $m < 0$, $y = mx + b$ is a decreasing function and its graph falls from left to right (Figure 11).

Figure 11

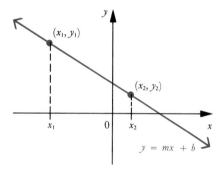

If $m = 0$, then $y = mx + b = 0x + b = b$, and $y = b$ is a constant function (Figure 12).

Figure 12

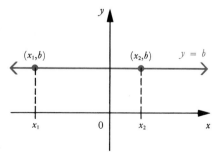

Thus, for $m \neq 0$, $y = f(x) = mx + b$ is either an increasing function if $m > 0$ or a decreasing function if $m < 0$. If $m = 0$, then f is neither increasing or decreasing.

EXAMPLES

1. Find the slope of each of the following lines, and determine if the function is increasing, decreasing, or constant. Assume that y represents the range variable.
   a) $2y + 3x = 4$
   b) $y - 2x - 6 = 0$
   c) $y = 4$

   SOLUTION

   a) $2y + 3x = 4$ is equivalent to $y = -\frac{3}{2}x + 2$. Hence, $m = -\frac{3}{2}$. The graph is decreasing (Figure 13).

   Figure 13

   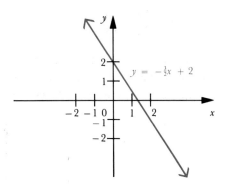

   b) $y - 2x - 6 = 0$ is equivalent to $y = 2x + 6$. Hence, $m = 2$ and the graph is increasing (Figure 14).

   Figure 14

   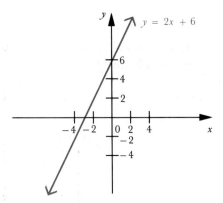

   c) $y = 4$ is a constant function with $m = 0$ and the graph neither increases nor decreases (Figure 15).

2. If a book salesman estimates the annual cost c of operating his car (in dollars) as $c = 0.08m + 1{,}500$ where m is his total mileage, should his

Figure 15

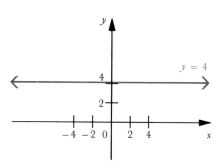

figures be questioned by the Internal Revenue Service, if the salesman reports that he drove 20,000 miles and claims a cost of $3,250?

SOLUTION

$$c = 0.08m + 1{,}500$$

If $m = 20{,}000$, then

$$\begin{aligned} c &= (0.08)(20{,}000) + 1{,}500 = 1{,}600 + 1{,}500 \\ &= 3{,}100 \end{aligned}$$

His cost is $3,100, and since he claims a cost of $3,250, his figures should be questioned.

2.2 Geometry of Lines

If two *distinct* lines are graphed on the same coordinate axis, they either intersect at one point or are parallel. We can determine whether two lines intersect or are parallel by examining the relationship between their slopes. Let $y = m_1 x + b_1$ and $y = m_2 x + b_2$ be the equations of two *distinct* lines. If their graphs intersect at one point (x, y), then we can find the x coordinate of this point by solving these two equations simultaneously. Thus,

$$\begin{aligned} y &= y \\ m_1 x + b_1 &= m_2 x + b_2 \\ m_1 x - m_2 x &= b_2 - b_1 \\ (m_1 - m_2)x &= b_2 - b_1 \\ x &= \frac{b_2 - b_1}{m_1 - m_2} \end{aligned}$$

From this solution we can see that the coordinate will exist, and thus a point of intersection will exist whenever $m_1 \neq m_2$. On the other hand, if

$m_1 = m_2$, there is no point of intersection (why?), and we have the following property.

PROPERTY 1

Two lines with slopes m_1 and m_2 are parallel if and only if $m_1 = m_2$.

For example, the line containing the points P_1: (3, 3) and P_2: (5, 6) is parallel to the line containing the points P_3: (−1, 1) and P_4: (1, 4), since their slopes are identical (Figure 16). That is,

$$m_1 = \frac{6-3}{5-3} = \frac{3}{2} \quad \text{and} \quad m_2 = \frac{4-1}{1+1} = \frac{3}{2}$$

Figure 16

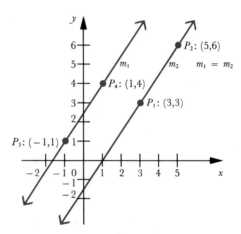

We can also determine when the graphs of two lines $y = m_1 x + b_1$ and $y = m_2 x + b_2$ are perpendicular from the relationship between their slopes. (See Problem 21, Problem Set 1.) Thus, we have the following result.

PROPERTY 2

Two lines with slopes m_1 and m_2 are perpendicular if and only if $m_1 m_2 = -1$.

For example, consider the line containing the points P_1: (3, 3), P_2: (5, 6), with slope

$$m_1 = \frac{6-3}{5-3} = \frac{3}{2}$$

and the line containing the points Q_1: (1, 4) and Q_2: (−2, 6), with slope

$$m_2 = \frac{6-4}{-2-1} = -\frac{2}{3}$$

Since $m_1 m_2 = (\frac{3}{2})(-\frac{2}{3}) = -1$, the two lines are perpendicular (Figure 17).

Figure 17

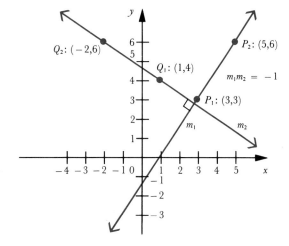

EXAMPLE

Given the points P_1: $(3, 2)$, P_2: $(4, 5)$, and P_3: $(0, 3)$, show that the line containing P_1 and P_2 is perpendicular to the line containing P_1 and P_3.

SOLUTION. The slope m_1 of the line containing P_1 and P_2 is

$$\frac{5-2}{4-3} = \frac{3}{1} = 3$$

The slope m_2 of the line containing P_1 and P_3 is

$$\frac{3-2}{0-3} = \frac{1}{-3} = -\frac{1}{3}$$

Thus, $m_1 m_2 = (3)(-\frac{1}{3}) = -1$, and the two lines are perpendicular.

2.3 Forms of Equations of Lines

A linear function of the form $y = mx + b$ will always contain the point $(0, b)$, for when $x = 0$, $y = m \cdot 0 + b = b$. Since the point $(0, b)$ is on the y axis, we call b the *y intercept* of the line. m is the slope. Hence, the form $y = mx + b$ is called the *slope-intercept form* of the line. Suppose that a line with slope m contains a point P_1: (x_1, y_1) (Figure 18). Then a point P: (x, y) different from P_1 is on the line if and only if the line containing P and P_1 also has slope m, that is, if and only if

$$m = \frac{y - y_1}{x - x_1} \qquad x \neq x_1$$

Figure 18

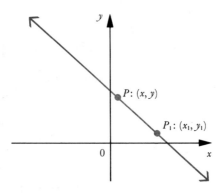

If this equation is written in the form $y - y_1 = m(x - x_1)$, it is called the *point-slope form* of the equation of the line containing the point (x_1, y_1), with slope m. Thus, the equation of the line containing the point $(-1, 2)$, with slope 3, is $y - 2 = 3(x + 1)$ (Figure 19).

Figure 19

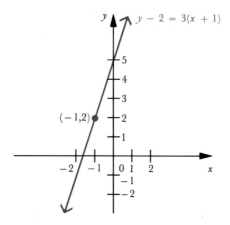

EXAMPLES

1 Find the equation of each of the following lines with the given slope m and the given point. Sketch the graph.
  a) $m = 5$, $P_1 = (1, 3)$
  b) $m = -1$, $P_1 = (-1, 2)$

SOLUTION

a) Using the point-slope form $y - y_1 = m(x - x_1)$, we get $y - 3 = 5(x - 1)$ (Figure 20).

Figure 20

x	y
1	3
2	8
0	−2

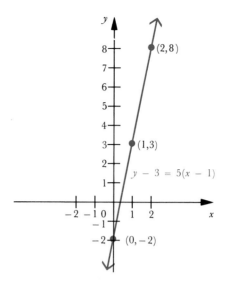

b) Using the point-slope form $y - y_1 = m(x - x_1)$, we have $y - 2 = -1(x + 1)$ (Figure 21).

Figure 21

x	y
−1	2
1	0
2	−1
3	−2

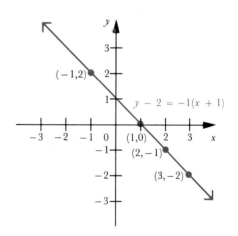

2   Find the equation of the line that contains the points $(-2, 5)$ and $(3, -4)$.

SOLUTION. The slope of the line containing the points $(-2, 5)$ and $(3, -4)$ is

$$\frac{5 - (-4)}{-2 - 3} = \frac{9}{-5}$$

so that the equation of the line is $y - 5 = -\frac{9}{5}(x + 2)$ or, equivalently,

$$y = -\frac{9}{5}x + \frac{7}{5} \quad \text{(Figure 22)}$$

Figure 22

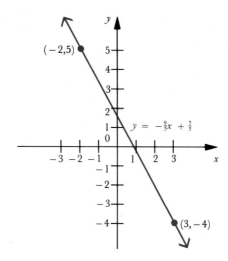

3   Determine the slope and the y intercept of each of the following lines. Also sketch the graph.
a) $y = -3x + 7$
b) $5x - y - 6 = 0$

SOLUTION

a) The slope of the line $y = -3x + 7$ is -3; and the y intercept is 7. (Why?) (Figure 23).

Figure 23

x	y
0	7
1	4
2	1
3	-2

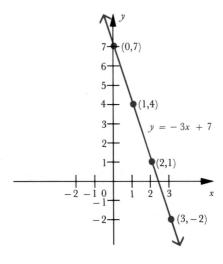

b) $5x - y - 6 = 0$ can be written as $y = 5x - 6$; therefore, the slope is 5 and the y intercept is -6 (Figure 24).

4   A builder can expect to sell 26,000 bricks if he charges 15 cents per brick, but only 8,000 bricks if he charges 19 cents per brick. Assuming

Figure 24

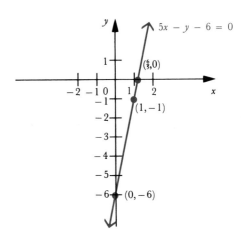

that the relationship is linear, find the equation of the line which relates the number of bricks the builder can expect to sell y to the price of the bricks x. How many of these bricks can he expect to sell if he charges 18 cents per brick?

SOLUTION. The number of bricks he can expect to sell is 26,000 if the price is 15 cents per brick, so that

$$x = 15 \quad \text{when } y = 26{,}000$$

and

$$x = 19 \quad \text{when } y = 8{,}000$$

The slope of the line between the points (15, 26,000) and (19, 8,000) is

$$\frac{8{,}000 - 26{,}000}{19 - 15} = -\frac{18{,}000}{4} = -4{,}500$$

Substituting into the equation $y - y_1 = m(x - x_1)$ for $(x_1, y_1) = (15, 26{,}000)$ and $m = -4{,}500$, we get

$$y - 26{,}000 = -4{,}500(x - 15)$$

If $x = 18$, then

$$\begin{aligned} y &= -4{,}500(18 - 15) + 26{,}000 \\ &= (-4{,}500)(3) + 26{,}000 \\ &= -13{,}500 + 26{,}000 \\ &= 12{,}500 \end{aligned}$$

He can expect to sell 12,500 bricks if he charges 18 cents per brick.

5   Find the equation of a line containing (3, −2) and parallel to
   a)   The x axis
   b)   The y axis

   SOLUTION

   a) The slope of the line containing the point (3, −2) and parallel to the x axis is 0. Hence, the equation of that line is $y = -2$ (Figure 25).
   b) The slope of the line containing the point (3, −2) and parallel to the y axis is undefined. Hence, the equation of that line is $x = 3$ (Figure 25).

Figure 25

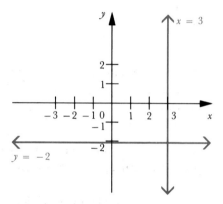

6   Find a value for k such that the line represented by $3x + ky = 5$ is parallel to the line $y = -2x + 1$.

   SOLUTION. $3x + ky = 5$ implies that

   $$y = -\frac{3}{k}x + \frac{5}{k}$$

   so that

   $$\frac{-3}{k} = -2$$

Figure 26

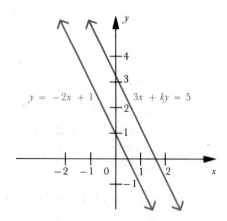

since the line $y = -(3/k)x + 5/k$ is parallel to the line $y = -2x + 1$ (Figure 26). Therefore,

$$k = \tfrac{3}{2}$$

7 a) Find the equation of the line containing the point $(3, 2)$ that is parallel to the line containing the points $(-4, -2)$ and $(-2, 2)$.
   b) Find the equation of the line containing $(3, 2)$ that is perpendicular to the line containing the points $(-4, -2)$ and $(-2, 2)$.

SOLUTION

a) Since the line we are seeking is parallel to the line containing the points $(-4, -2)$ and $(-2, 2)$, its slope is also

$$\frac{2 - (-2)}{-2 - (-4)} = 2$$

Hence, the equation of the line is $y - 2 = 2(x - 3)$ or, equivalently, $y = 2x - 4$ (Figure 27).

b) From part a, the slope of the line containing the points $(-4, -2)$ and $(-2, 2)$ is 2. Hence, a line with slope m will be perpendicular to this line if $2m = -1$, that is, if $m = -\tfrac{1}{2}$. Thus, the line we seek contains the point $(3, 2)$ and has a slope of $-\tfrac{1}{2}$. Therefore, the equation is $y - 2 = -\tfrac{1}{2}(x - 3)$ or, equivalently, $y = -\tfrac{1}{2}x + \tfrac{7}{2}$ (Figure 28).

Figure 27

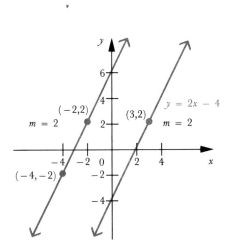

Figure 28

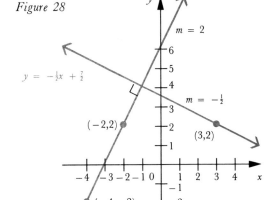

PROBLEM SET 1

1 Which of the following functions are polynomial functions? Indicate the degree of the polynomial functions.

a) $f(x) = 3x^2 + 5x - 13$ b) $f(x) = x^{-3} + x^2 + \dfrac{1}{x}$

c) $f(x) = 3x + \sqrt{5}x^5 + 3x^3 - 5$ d) $f(x) = \dfrac{1}{x^2} - 3x + 1$

e) $f(x) = 7^{-1}$ f) $f(x) = \tfrac{1}{3}x^4 - 2x^2 + 5$

2 Find a constant function whose graph contains the point (3, 5).

3 For what values of a, b, and c will $f(x) = ax^2 + bx + c$ be a constant function?

4 Indicate the x and y intercepts and sketch the graph of each of the following.
  a) $y = 2x$
  b) $y = 2x + 1$
  c) $y = -2x + 7$
  d) $y = 5x - 10$
  e) $y = 4x + 3$
  f) $y = 3x + 2$
  g) $2x + 3y + 6 = 0$
  h) $4x - 3y - 12 = 0$

5 Discuss the following linear functions, where $y = f(x)$. Indicate the domain and range. Is the function increasing or decreasing? Is the function even or odd? Find the slope and sketch the graph.
  a) $\{(x, y) \mid y = -3x + 5\}$ b) $3x - y + 1 = 0$
  c) $2x - 3y - 9 = 0$ d) $-x + 2y - 11 = 0$
  e) $f(x) = -7x + 2$ f) $f(x) = 3 + 5x$
  g) $-4x + 3y = 0$ h) $x = 4 - y$
  i) $\dfrac{x}{2} + \left(\dfrac{y}{-1}\right) = 1$ j) $y - 2 = -5(x - 1)$
  k) $f(x) = 3 - 2x$ l) $x + 4y - 3 = 0$
  m) $f(x) = -\tfrac{3}{4}x + \tfrac{1}{2}$

6 Find a linear function f whose slope is -3 and such that
  a) $f(1) = 2$ b) $f(-1) = 4$
  c) $f(0) = -3$ d) $f(7) = -1$

7 Find the slope of the linear function f if $f(1) = -2$ and
  a) $f(0) = 3$ b) $f(2) = 4$
  c) $f(3) = 3$ d) $f(-2) = 7$

8 Find the slope of the line containing each of the following pairs of points.
  a) $(2, -1)$ and $(-3, 4)$ b) $(1, -4)$ and $(2, 3)$
  c) $(1, 5)$ and $(-2, 3)$ d) $(4, 3)$ and $(-3, -4)$
  e) $(6, -1)$ and $(0, 2)$

9 Find the equations of the lines of Problem 8.

10 Determine a linear function f that satisfies the given conditions.

a) $f(1) = 2$ and $f(3) = 1$ b) $f(3x) = 3f(x)$
c) $f(3x + 5) = 3f(x) + 5$ d) $f(2) = -4$ and $f(-1) = -f(2)$

11 a) Show that if a, b, and c are constant real numbers such that a and b are not both zero, then the graph of the equation $ax + by + c = 0$ is a straight line. (This is called the *general form* of the equation of the line.)

b) Find the slope and the y intercept of the line whose equation is $3x - 2y + 5 = 0$.

12 a) Show that an equation of the line joining the two points (x_1, y_1) and (x_2, y_2), $x_1 \neq x_2$, is

$$y - y_1 = \frac{y_2 - y_1}{x_2 - x_1}(x - x_1)$$

(This is called the *two-point form* of the equation of the line.)

b) Find the equation of the line that contain the points $(-2, 4)$ and $(-1, -5)$.

13 a) Show that the equation of a line whose intercepts are $(a, 0)$ and $(0, b)$, with $a \neq 0$ and $b \neq 0$, is $x/a + y/b = 1$. (This is called the *intercept form* of the equation of the line.)

b) Find the equation of the line whose intercepts are $(5, 0)$ and $(0, -3)$.

14 Find the equation of the line with the given slope m and the given point (x_1, y_1). Sketch the graph.
a) $m = -3$, $(x_1, y_1) = (-1, 2)$
b) $m = 5$, $(x_1, y_1) = (3, 1)$
c) $m = \frac{22}{5}$, $(x_1, y_1) = (-1, 4)$
d) $m = 2$, $(x_1, y_1) = (1, 2)$
e) $m = 3$, $(x_1, y_1) = (3, 5)$
f) $m = 0$, $(x_1, y_1) = (-1, -5)$

15 The points $P_1: (x_1, y_1)$, $P_2: (x_1, y_2)$, and $P_3: (x_3, y_3)$ are collinear if they lie on the same straight line. This is the same as saying that the slope between P_1 and P_2 is the same as the slope between P_1 and P_3. Use this concept of slope to determine whether or not the following points are collinear.
a) $(1, 1)$, $(2, 4)$, and $(3, 2)$
b) $(0, 3)$, $(1, 1)$, and $(2, -1)$
c) $(1, -3)$, $(-1, -11)$, and $(-2, -15)$
d) $(1, 5)$, $(-2, -1)$, and $(-3, -3)$

16 Find the slope of each line that is (i) parallel to the given line; (ii) perpendicular to the given line.
a) $y = -2x + 3$ b) $y = 7x - 13$
c) $y = -5x + 10$ d) $y = 3$
e) $y = -x$

17 Show that
   a) The points $(-5, 4)$, $(7, -11)$, $(12, 25)$, and $(0, 40)$ are the vertices of a parallelogram (a four-sided figure with opposite sides parallel).
   b) The points $(2, 4)$, $(3, 8)$, $(5, 1)$, and $(4, -3)$ are the vertices of a parallelogram.

18 Find the equation of the line with $m = 2$ that contains the point $(2, 3)$. Then find the equation of the line that contains the point $(-2, 1)$, which is perpendicular to the line.

19 Show that the following points are vertices of a square: P_1: $(-2, 5)$, P_2: $(8, 2)$, P_3: $(-5, -5)$, P_4: $(5, -8)$.

20 A projectile fired straight up attains a velocity of v feet per second after t seconds of flight, and the relation between the numbers v and t is linear. If the projectile is fired at a velocity of 200 feet per second and reaches a velocity of 90 feet per second after 2 seconds of flight, express v as a function of t. How soon after it is fired does the projectile reach its highest point?

21 Show that two lines are perpendicular if and only if the product of their slopes is -1. [*Hint:* Let (x, y) be the point of intersection and consider points on each line whose abscissas are $x + 1$, and then apply the Pythagorean theorem to the resulting right triangle.]

22 Assume that the total amount of money spent annually on radio advertising grows at a constant rate. Given the information that the actual 1960 and 1970 expenditures for radio advertising where 545 and 889 million dollars, respectively, find the linear equation of the trend line that contains the two points, and use it to predict the amount that will be spent on radio advertising in the year 1974.

23 An apartment building was built in 1959 at a cost of \$350,000. What is its value (for tax purposes) in 1972, if it is being depreciated linearly over 40 years according to the formula $v = c - (c/N)n$, where c (in dollars) is the original cost of the property, N is the number of years (being linearly depreciated), and v is the value of the undepreciated balance at the end of n years?

3 Systems of Linear Equations

It was pointed out in Section 2 that if the slopes of two linear equations are the same, then either the lines are parallel or they coincide. For example, the lines $y = 3x + 2$ and $y = 3x - 1$ are parallel, since they both have the same slope 3 and have different y intercepts (Figure 1a).

The lines $y = 2x + 1$ and $2y - 4x = 2$ coincide, since they both have the same slope 2 and the same y intercept (Figure 1b).

Figure 1

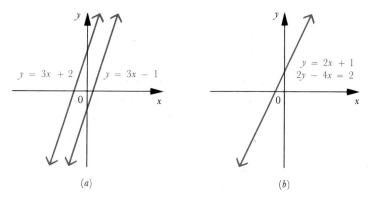

(a)   (b)

We also discovered in this section that if the slopes of two linear equations are different, then the two lines intersect. For example, the lines $y = 2x - 1$ and $y = -3x + 1$ intersect (Figure 2).

Figure 2

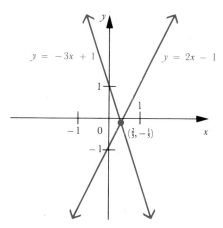

One purpose in this section is to develop algebraic techniques that determine the coordinates of the point of intersection of two lines if such a point exists. We will see how it is possible to generalize the techniques used to solve linear systems containing two unknowns to include linear systems containing more than two unknowns by using the *substitution* and *elimination* methods. In Chapter 8 we will investigate two different methods: the *row-reduction method* and *Cramer's rule*. We will see that row reduction is essentially the same as the elimination method, whereas Cramer's rule make use of determinants.

3.1 Substitution Method

One direct algebraic method for solving systems of linear equations is called the *method of substitution*. Let us see how this method works by returning to the above example:

$$\begin{cases} y = 2x - 1 \\ y = -3x + 1 \end{cases}$$

Geometrically, solving this linear system means to find the coordinates of the point of intersection of the two lines (Figure 2). In other words, we determine the algebraic solution by finding a pair of numbers (x, y) that satisfies both equations simultaneously.

Since we are trying to find one value for x that yields the same y value in both equations, we assume that y represents the same number in each equation. Hence,

$$y = y$$

or

$$2x - 1 = -3x + 1$$

from which we get

$$5x = 2$$

That is,

$$x = \tfrac{2}{5}$$

Hence,

$$y = 2x - 1 = 2(\tfrac{2}{5}) - 1 = -\tfrac{1}{5}$$

so that the point of intersection is $(\tfrac{2}{5}, -\tfrac{1}{5})$, and the solution set is $\{(\tfrac{2}{5}, -\tfrac{1}{5})\}$.

EXAMPLES

In each of the following examples, decide whether or not the linear system has a solution, graph the equations of each system on the same coordinate system, and then solve the system by substitution. Finally, check the solution by testing both equations.

1 $\begin{cases} y = 3x \\ y = 5x + 1 \end{cases}$

SOLUTION. Since the slopes are different, the lines intersect (Figure 3). Assuming the y values to be the same, we get

$$y = y$$

or

$$5x + 1 = 3x$$

Figure 3

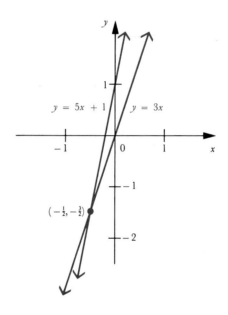

so that

$$2x = -1$$

or

$$x = -\tfrac{1}{2}$$

Hence, $y = 3(-\tfrac{1}{2}) = -\tfrac{3}{2}$, so that $\{(-\tfrac{1}{2}, -\tfrac{3}{2})\}$ is the solution set.

Check: $y = 3x$ and $y = 5x + 1$; hence

$$-\tfrac{3}{2} \stackrel{?}{=} 3(-\tfrac{1}{2}) \qquad -\tfrac{3}{2} \stackrel{?}{=} 5(-\tfrac{1}{2}) + 1$$
$$-\tfrac{3}{2} = -\tfrac{3}{2} \qquad -\tfrac{3}{2} = -\tfrac{3}{2}$$

2 $\begin{cases} x + y = 3 \\ 5y + 5x = 1 \end{cases}$

SOLUTION. First, we can rewrite the system by putting each of the two equations in slope-intercept form.

$$\begin{cases} y = -x + 3 \\ y = -x + \frac{1}{5} \end{cases}$$

Since the slopes are equal and the y intercepts are different, the lines are parallel. Consequently, the system does not have a solution (Figure 4). Hence, the solution set is empty.

Figure 4

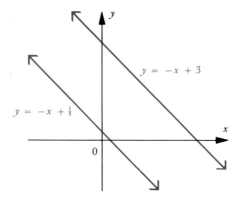

3 $\begin{cases} x - 6y = 1 \\ 3x - 3 = 18y \end{cases}$

SOLUTION. This system can be rewritten as

$$\begin{cases} y = \frac{1}{6}x - \frac{1}{6} \\ y = \frac{1}{6}x - \frac{1}{6} \end{cases}$$

Hence, the lines coincide (Figure 5). The solution of the system, the

Figure 5

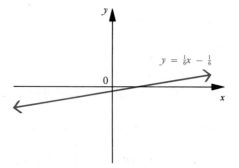

intersection of the lines, is the set $\{(x, y) \mid y = \frac{1}{6}x - \frac{1}{6}\}$, since all points on this line satisfy the equations of the system.

4. $\begin{cases} 5x + 2y = 1 \\ 3x - y = 2 \end{cases}$

SOLUTION. The graph of these equations shows that a solution exists (Figure 6). The second equation can be rewritten as $y = 3x - 2$, and

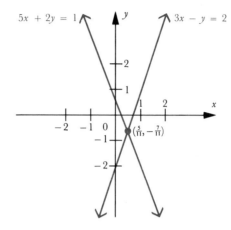

Figure 6

by substituting the latter equation into the first equation we have

$$5x + 2(3x - 2) = 1 \quad \text{or} \quad 5x + 6x - 4 = 1$$

or

$$11x = 5$$

so that

$$x = \tfrac{5}{11}$$

Hence,

$$y = 3(\tfrac{5}{11}) - 2 = \tfrac{15}{11} - \tfrac{22}{11} = -\tfrac{7}{11}$$

so that the solution set is $\{(\tfrac{5}{11}, -\tfrac{7}{11})\}$.

Check:

$$5\left(\frac{5}{11}\right) + 2\left(-\frac{7}{11}\right) = \frac{25 - 14}{11} = 1$$

and

$$3\left(\frac{5}{11}\right) - \left(-\frac{7}{11}\right) = \frac{15 + 7}{11} = 2$$

When systems of two linear equations in two variables are graphed on the same coordinate plane, three general relationships between the two lines can occur. Either they coincide (Figure 1b), they are parallel lines (Figure 1a), or they intersect at a point. (Figure 2). Hence, we have the following possibilities:

1. If the two lines coincide (Figure 1b), the coordinates of every point on the first line are solutions of the given system and all points on one line are points on the second line. In this case, we say that the system is *dependent*.

2. If the two lines are parallel (Figure 1a), they have no points in common. Hence, the solution is the empty set and the system is called *inconsistent*.

3. If the two lines intersect at one point (Figure 2), we call the system *independent*.

In summary, to solve an independent system of linear equations in two variables means to find the ordered pair that satisfies the equations simultaneously. We now turn our attention to linear systems with more than two variables.

EXAMPLES

Use the substitution method to solve each of the following systems.

1. $\begin{cases} x + y + z = 6 \\ 2x - y - z = 0 \\ x - y + 2z = 7 \end{cases}$

SOLUTION. We are seeking an ordered triple of numbers (x, y, z) that satisfies all three equations simultaneously. First, rewrite the equation $x + y + z = 6$ as $z = 6 - x - y$. Then, substituting in the remaining two equations, we get

$$\begin{cases} z = 6 - x - y \\ 2x - y - (6 - x - y) = 0 \\ x - y + 2(6 - x - y) = 7 \end{cases}$$

That is,

$$\begin{cases} z = 6 - x - y \\ 3x - 6 = 0 \\ -x - 3y + 12 = 7 \end{cases}$$

However, the latter two equations in the system can be solved as a linear system containing two unknowns. Here we get $x = 2$ and $y = 1$ (why?), so that $z = 6 - x - y = 6 - 2 - 1 = 3$. Hence, the solution set is $\{(2, 1, 3)\}$.

Check:

$$2 + 1 + 3 = 6$$
$$2(2) - 1 - 3 = 0$$
$$2 - 1 + 2(3) = 7$$

2. $\begin{cases} 2x - y - z = 13 \\ x + y + z = 2 \\ -x + y - 2z = -2 \end{cases}$

SOLUTION. From the first equation, we get $y = 2x - z - 13$, so that after substitution the system becomes

$$\begin{cases} y = 2x - z - 13 \\ x + (2x - z - 13) + z = 2 \\ -x + (2x - z - 13) - 2z = -2 \end{cases}$$

The latter two equations of the system yield the system

$$\begin{cases} 3x = 15 \\ x - 3z = 11 \end{cases}$$

so that

$$x = 5 \quad \text{and} \quad z = -2$$

Since $y = 2x - z - 13$, we have $y = 2(5) - (-2) - 13 = -1$. Hence, the solution set is $\{(5, -1, -2)\}$.

Check:

$$2(5) - (-1) - (-2) = 13$$
$$5 + (-1) + (-2) = 2$$
$$-(5) + (-1) - 2(-2) = -2$$

3.2 Elimination Method

Now we will investigate another algebraic procedure for solving a system of two linear equations. We will see that this method can also be extended to systems containing more than two linear equations and more than two variables.

Let us consider a specific example of the method by solving the following system:

$$\begin{cases} x - y = 1 \\ 3x + 2y = 8 \end{cases}$$

First, multiply the first equation by -3 and add it to the second equation, so that the system becomes

$$\begin{cases} x - y = 1 \\ 0 + 5y = 5 \end{cases}$$

Next, multiply the second equation by $\frac{1}{5}$ to get

$$\begin{cases} x - y = 1 \\ \phantom{x - {}} y = 1 \end{cases}$$

Substituting $y = 1$ into the first equation yields $x = 2$. Thus, the solution set is $\{(2, 1)\}$, since $x = 2$ and $y = 1$ satisfy each equation in the system simultaneously (Figure 7).

Figure 7

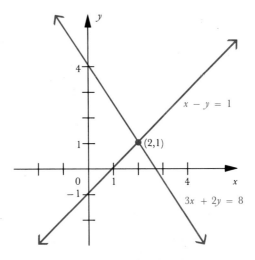

This procedure for solving a system of linear equations is called the *elimination method* or, equivalently, the *process of elimination*.

The process of elimination may also be applied to systems containing more than two unknowns. We illustrate this application in the following examples.

EXAMPLES

Use the elimination method to solve the following systems of linear equations:

1. $\begin{cases} 2x_1 + x_2 - 2x_3 = 10 \\ 3x_1 + 2x_2 + 2x_3 = 1 \\ 5x_1 + 4x_2 + 3x_3 = 4 \end{cases}$

SOLUTION

(A) $\begin{cases} 2x_1 + x_2 - 2x_3 = 10 \\ 3x_1 + 2x_2 + 2x_3 = 1 \\ 5x_1 + 4x_2 + 3x_3 = 4 \end{cases}$

First, replace the second equation by the sum of -3 times the first equation and 2 times the second equation to obtain the system (B):

(B) $\begin{cases} 2x_1 + x_2 - 2x_3 = 10 \\ x_2 + 10x_3 = -28 \\ 5x_1 + 4x_2 + 3x_3 = 4 \end{cases}$

The system (B) is *equivalent* to the system (A), since every solution of the system (B) is also a solution of the system (A). Next, replace the third equation of (B) by the sum of -5 times the first equation and 2 times the third to obtain the equivalent system (C):

(C) $\begin{cases} 2x_1 + x_2 - 2x_3 = 10 \\ x_2 + 10x_3 = -28 \\ 3x_2 + 16x_3 = -42 \end{cases}$

Now replace the third equation of (C) by the sum of -3 times the second equation and the third equation to get the equivalent system (D):

(D) $\begin{cases} 2x_1 + x_2 - 2x_3 = 10 \\ x_2 + 10x_3 = -28 \\ -14x_3 = 42 \end{cases}$

Next multiply the third equation by $-\frac{1}{14}$ and add -1 times the second equation to the first equation to obtain the equivalent system (E):

(E) $\begin{cases} 2x_1 - 12x_3 = 38 \\ x_2 + 10x_3 = -28 \\ x_3 = -3 \end{cases}$

Finally, multiply the first equation by $\frac{1}{2}$ to obtain the equivalent system (F):

(F) $\begin{cases} x_1 - 6x_3 = 19 \\ x_2 + 10x_3 = -28 \\ x_3 = -3 \end{cases}$

From the third equation, $x_3 = -3$, so that after substituting into the second equation we get $x_2 = 2$; then, after substituting into the first equation, we obtain $x_1 = 1$. Thus, the solution is $\{(1, 2, -3)\}$, since $x_1 = 1$, $x_2 = 2$, and $x_3 = -3$ satisfy each equation in the system simultaneously.

2. $\begin{cases} x_1 - 2x_2 + 3x_3 = -1 \\ 2x_1 - x_2 + 2x_3 = 2 \\ 3x_1 + x_2 + 2x_3 = 3 \end{cases}$

SOLUTION. ($\downarrow$ is used to indicate that the systems are equivalent.)

(A) $\begin{cases} x_1 - 2x_2 + 3x_3 = -1 \\ 2x_1 - x_2 + 2x_3 = 2 \\ 3x_1 + x_2 + 2x_3 = 3 \end{cases}$

(E) $\begin{cases} x_1 - 2x_2 + 3x_3 = -1 \\ x_2 - \frac{4}{3}x_3 = \frac{4}{3} \\ \frac{7}{3}x_3 = -\frac{10}{3} \end{cases}$

First, replace the second equation by the sum of -2 times the first equation and the second equation.

Next replace equation one with 2 times equation two plus equation one.
$\downarrow$

(B) $\begin{cases} x_1 - 2x_2 + 3x_3 = -1 \\ 3x_2 - 4x_3 = 4 \\ 3x_1 + x_2 + 2x_3 = 3 \end{cases}$

(F) $\begin{cases} x_1 + \frac{1}{3}x_3 = \frac{5}{3} \\ x_2 - \frac{4}{3}x_3 = \frac{4}{3} \\ \frac{7}{3}x_3 = -\frac{10}{3} \end{cases}$

Next, replace equation three with the sum of -3 times equation one and the third equation.

Multiply equation three by $\frac{3}{7}$.
$\downarrow$

(C) $\begin{cases} x_1 - 2x_2 + 3x_3 = -1 \\ 3x_2 - 4x_3 = 4 \\ 7x_2 - 7x_3 = 6 \end{cases}$

(G) $\begin{cases} x_1 + \frac{1}{3}x_3 = \frac{5}{3} \\ x_2 - \frac{4}{3}x_3 = \frac{4}{3} \\ x_3 = -\frac{10}{7} \end{cases}$

Multiply equation two by $\frac{1}{3}$.
$\downarrow$

Now replace equation one with the sum of $-\frac{1}{3}$ times equation three and equation one.
$\downarrow$

(D) $\begin{cases} x_1 - 2x_2 + 3x_3 = -1 \\ x_2 - \frac{4}{3}x_3 = \frac{4}{3} \\ 7x_2 - 7x_3 = 6 \end{cases}$

Now, replace the third equation with -7 times the second equation plus the third equation.

(H) $\begin{cases} x_1 & = \frac{15}{7} \\ x_2 - \frac{4}{3}x_3 & = \frac{4}{3} \\ x_3 & = -\frac{10}{7} \end{cases}$

| Finally, replace equation two with the sum of $\frac{4}{3}$ times equation three and equation two.

(I) $\begin{cases} x_1 & = \frac{15}{7} \\ x_2 & = -\frac{4}{7} \\ x_3 & = -\frac{10}{7} \end{cases}$

Hence, the solution is
$\{(\frac{15}{7}, -\frac{4}{7}, -\frac{10}{7})\}$.

PROBLEM SET 2

1 Decide whether or not each of the following linear systems has a solution. Graph the equations of each system on the same coordinate system and then solve the system by substitution. Finally, check the solution.

a) $2x - y = 0$
$x + y = 1$

b) $7x + y = 5$
$y - 2x = 3$

c) $-3x + y = 3$
$-2x - y = -5$

d) $x + y + 8 = 0$
$2x - 7y = 5$

e) $x + y = 0$
$x - y = 0$

f) $3y + 2z = 5$
$2y = -3z + 1$

g) $3x + y = 1$
$9x + 3y = -4$

h) $13r + 5s = 2$
$2 - 2s = 6r$

i) $x - y = 7$
$-5x + 5y = 35$

j) $x = 5 - y$
$y = 5 - x$

k) $2x - y = 9$
$5x - 3y = 14$

l) $5x - 2y = 35$
$x + 4y = 25$

m) $4x - 3y = 1$
$3x - 4y = 6$

n) $8 = x + 3y$
$x + 8y = 53$

o) $2x - y = 5$
$x + 2y = 25$

p) $\frac{1}{3}x - \frac{1}{4}y = 2$
$\frac{1}{4}x - \frac{1}{2}y = 7$

2 Solve each of the following systems using the elimination method.

a) $\frac{1}{2}x + \frac{1}{3}y = 13$
$\frac{1}{5}x + \frac{1}{8}y = 5$

b) $4x + 3y = 5$
$x + 4y = 6$

c) $3x - 2y = 6$
$6x - 3y = 1$

d) $3x + y = 3$
$x - 3y = 2$

e) $\begin{aligned} x + y &= 4 \\ 3x - y + 3z &= 7 \\ -5x + 7y - 2z &= 2 \end{aligned}$

f) $\begin{aligned} x + y + 2z &= 4 \\ -x + 5y - z &= -5 \\ 3x - 17 + 2z &= 0 \end{aligned}$

g) $\begin{aligned} x + y + z &= 80 \\ 1 - 2y &= 0 \\ x + y - 4z &= 0 \end{aligned}$

h) $\begin{aligned} x + y + z &= 2 \\ x + 2y - z &= 4 \\ 2x - y + z &= 0 \end{aligned}$

i) $\begin{aligned} 2x + y - 3z &= 9 \\ x - 2y + 4z &= 5 \\ 3x + y - 2z &= 15 \end{aligned}$

j) $\begin{aligned} 2x + 3y + z &= 6 \\ x - 2y + 3z &= 3 \\ 3x + y - z &= 8 \end{aligned}$

k) $\begin{aligned} x + y + z &= 6 \\ x - y + 2z &= 12 \\ 2x + y - z &= 1 \end{aligned}$

l) $\begin{aligned} x + y + 2z &= 4 \\ x + y - 2z &= 0 \\ x - y &= 0 \end{aligned}$

m) $\begin{aligned} 7x + y + 3z &= -6 \\ 4x - 5y + 6z &= -27 \\ x + 15y - 9z &= 64 \end{aligned}$

n) $\begin{aligned} x + 3y &= 4 \\ y + 3z &= 4 \\ z + 3x &= 4 \end{aligned}$

o) $\begin{aligned} x + 3y - 2z &= -21 \\ 7x - 5y + 4z &= 31 \\ 2x + y + 3z &= 17 \end{aligned}$

p) $\begin{aligned} 3x + 2y - z &= 4 \\ 2x + y - z &= 3 \\ x + 2y + 3z &= 6 \end{aligned}$

q) $\begin{aligned} 8x + 3y - 18z &= -76 \\ 10x + 6y - 6z &= -50 \\ 4x + 9y + 12z &= 10 \end{aligned}$

r) $\begin{aligned} 5x + 3y - 2z &= 1 \\ 2x + 4y + z &= 0 \\ 4x - 7y - 7z &= 6 \end{aligned}$

4 Quadratic Functions

Polynomial functions of degree 2 are called *quadratic functions*. Thus, $f(x) = ax^2 + bx + c$, where a, b, and c are real numbers, $a \neq 0$, is the general representation of a quadratic function. A quadratic function can also be written in set form as $\{(x, y) \mid y = ax^2 + bx + c, a \neq 0, a, b, c \in R\}$. In graphing any quadratic function we will locate the x intercepts of f; that is, we will identify those real number values of x, if any, for which $f(x) = 0$. Hence, it is necessary to solve equations of the form $ax^2 + bx + c = 0$. Such equations are called *quadratic equations*.

4.1 Quadratic Equations

An equation that is equivalent to an equation of the form $ax^2 + bx + c = 0$, where a, b, and c are real numbers with $a \neq 0$, is called a second-degree equation or *quadratic equation* in x. For example, $4x^2 - 9 = 5x$ and $-5x^2 = 13$ are quadratic equations because the first can be expressed as $4x^2 - 5x - 9 = 0$, and the second can be expressed as $-5x^2 + 0 \cdot x - 13 = 0$. The solution or solution set of a quadratic equation is the set of all possible values that satisfy the equation. For example, $\{-\frac{3}{2}, \frac{3}{2}\}$ is the solution set for the equation $4x^2 - 9 = 0$, so that the graph of $f(x) = 4x^2 - 9$ intersects the x axis at points $(-\frac{3}{2}, 0)$ and $(\frac{3}{2}, 0)$. Quadratic equations can be solved by using the following methods.

1 Factoring Method

We saw in Chapter 1 how to factor some second-degree polynomials. We shall use factoring as a tool for solving second-degree equations that are factorable. Thus, the equation $x^2 - 6x + 5 = 0$ can also be written as $(x - 1)(x - 5) = 0$. In Chapter 1, a property states that *the product of two real numbers is zero whenever one or both of the two numbers is zero.* That is, if a and b are real numbers, $ab = 0$ if and only if $a = 0$, or $b = 0$, or both.

This property of real numbers implies that the product

$$(x - 1)(x - 5) = 0$$

whenever $x - 1 = 0$, or $x - 5 = 0$, or both. Solving these two first-degree equations, we have $x = 1$ or $x = 5$, and both values of x satisfy the original equation. The factoring method can be illustrated as follows. Suppose we want to solve

$$x^2 + 4x - 5 = 0$$

By factoring, we have

$$(x - 1)(x + 5) = 0$$

Setting each factor equal to 0, we have

$$x - 1 = 0 \quad \text{or} \quad x + 5 = 0$$

so that

$$x = 1 \quad \text{or} \quad x = -5$$

Hence, for $x^2 + 4x - 5 = 0$, the solution set is $\{1, -5\}$.

2 Completing-the-Square Method

Suppose that we want to solve $x^2 - 4 = 0$. Although the equation $x^2 - 4 = 0$ can be solved by factoring, it could also be solved as follows:

$$x^2 - 4 = 0$$

is equivalent to

$$x^2 = 4$$

so that

$$x = 2 \quad \text{or} \quad x = -2$$

and the solution set is $\{2, -2\}$. The key step here is the *extraction of roots* after expressing the equation in the equivalent form

$$(\quad)^2 = \square$$

so that

$$(\quad) = \sqrt{\square} \quad \text{or} \quad (\quad) = -\sqrt{\square}$$

This idea works as follows. Suppose we want to solve

$$x^2 - 2x - 7 = 0$$

This equation can be rewritten as

$$x^2 - 2x = 7$$

Next we want to write the left-hand side as the square of a linear polynomial. This can be accomplished as follows.

We can add to each side the number 1 (which is $\frac{1}{2}$ the coefficient of x, -2, squared) to get

$$x^2 - 2x + 1 = 7 + 1$$

so that

$$(x - 1)^2 = 8$$

The extraction of roots yields

$$x - 1 = \sqrt{8} \quad \text{or} \quad x - 1 = -\sqrt{8}$$

Hence,

$$x = 1 + \sqrt{8} \quad \text{or} \quad x = 1 - \sqrt{8}$$

Since $\sqrt{8} = 2\sqrt{2}$, the solution set is $\{1 + 2\sqrt{2}, 1 - 2\sqrt{2}\}$.

This technique is called *completing the square*. Its use involves finding the constant necessary to form a perfect quadratic square. From the identity $(x \pm d)^2 = x^2 \pm 2dx + d^2$, it appears that in order to complete the square when the two terms $x^2 \pm 2dx$ are given, we must add the number

d^2. But $d^2 = [\frac{1}{2}(2d)]^2$; therefore, the term to be added is the square of one-half the coefficient of x. Hence, the quadratic expression $x^2 - 6x$ can be written as a complete quadratic square by adding $[\frac{1}{2}(6)]^2$ or $(3)^2 = 9$. Thus, $x^2 - 6x + 9 = (x - 3)^2$ is a perfect quadratic square.

In the case where the coefficient of the x^2 term is not 1, we first divide the equation by the coefficient of the x^2 term and then proceed as before. For example, to solve the equation $3x^2 - 2x - 2 = 0$, we proceed as follows.

Divide each side of $3x^2 - 2x - 2 = 0$ by 3 to get

$$x^2 - \tfrac{2}{3}x - \tfrac{2}{3} = 0$$

or

$$x^2 - \tfrac{2}{3}x = \tfrac{2}{3}$$

Completing the square, we have

$$x^2 - \tfrac{2}{3}x + [\tfrac{1}{2}(\tfrac{2}{3})]^2 = \tfrac{2}{3} + [\tfrac{1}{2}(\tfrac{2}{3})]^2$$

or

$$x^2 - \tfrac{2}{3}x + \tfrac{1}{9} = \tfrac{2}{3} + \tfrac{1}{9}$$

so that

$$(x - \tfrac{1}{3})^2 = \tfrac{7}{9}$$

Hence,

$$x - \tfrac{1}{3} = \pm\sqrt{\tfrac{7}{9}}$$

which implies that

$$x = \frac{1}{3} \pm \frac{\sqrt{7}}{3}$$

That is,

$$x = \frac{1}{3} + \frac{\sqrt{7}}{3} \quad \text{or} \quad x = \frac{1}{3} - \frac{\sqrt{7}}{3}$$

Hence, the solution set is

$$\left\{\frac{1 + \sqrt{7}}{3}, \frac{1 - \sqrt{7}}{3}\right\}$$

3 *The Quadratic Formula*

The method of completing the square can be generalized to develop a formula that enables us to solve any quadratic equation.

Consider the quadratic equation $ax^2 + bx + c = 0$, with a, b, and c real numbers and $a \neq 0$. By applying the method of completing the square, we can obtain its solution set in terms of a, b, and c as follows.

$$ax^2 + bx + c = 0$$

Dividing by a, we have

$$x^2 + \frac{b}{a}x + \frac{c}{a} = 0$$

or

$$x^2 + \frac{b}{a}x = -\frac{c}{a}$$

By adding $[\frac{1}{2}(b/a)]^2 = b^2/4a^2$ to both sides of the equation, we have

$$x^2 + \frac{b}{a}x + \frac{b^2}{4a^2} = \frac{b^2}{4a^2} - \frac{c}{a}$$

or

$$x^2 + \frac{b}{a}x + \frac{b^2}{4a^2} = \frac{b^2 - 4ac}{4a^2}$$

The left-hand side of the equation is the square of $x + b/2a$; therefore, the equation can be written as

$$\left(x + \frac{b}{2a}\right)^2 = \frac{b^2 - 4ac}{4a^2}$$

which implies that

$$x + \frac{b}{2a} = \pm\sqrt{\frac{b^2 - 4ac}{4a^2}}$$

or

$$x + \frac{b}{2a} = \frac{\pm\sqrt{b^2 - 4ac}}{2a}$$

so that

$$x = \frac{-b}{2a} \pm \frac{\sqrt{b^2 - 4ac}}{2a} = \frac{-b \pm \sqrt{b^2 - 4ac}}{2a}$$

Therefore,

$$x = \frac{-b + \sqrt{b^2 - 4ac}}{2a} \quad \text{or} \quad x = \frac{-b - \sqrt{b^2 - 4ac}}{2a}$$

That is, the solution set is

$$\left\{ \frac{-b + \sqrt{b^2 - 4ac}}{2a}, \frac{-b - \sqrt{b^2 - 4ac}}{2a} \right\}$$

The above result, which provides a formula for solving any quadratic equation, is called the *quadratic formula*.

In summary, if $ax^2 + bx + c = 0$, $a \neq 0$, then

$$x = \frac{-b \pm \sqrt{b^2 - 4ac}}{2a}$$

which is an abbreviated way of writing the two possible solutions. Note that if $b^2 - 4ac < 0$, then $\sqrt{b^2 - 4ac}$ is not defined in the real numbers, in which case the quadratic equation has no real solutions. Negative radicals will be handled in Chapter 5, when we cover complex numbers.

EXAMPLES

1 Solve $6x^2 + 5x - 4 = 0$ by the factoring method.

   SOLUTION

   $$6x^2 + 5x - 4 = 0$$

   Factoring the left side of the equation, we have

   $$(2x - 1)(3x + 4) = 0$$

   so that

   $$2x - 1 = 0 \quad \text{or} \quad 3x + 4 = 0$$

   Therefore,

   $$x = \tfrac{1}{2} \quad \text{or} \quad x = -\tfrac{4}{3}$$

   The solution set is $\{\tfrac{1}{2}, -\tfrac{4}{3}\}$.

2 Solve $x^2 + 4x + 2 = 0$ by using the completing-the-square method.

   SOLUTION

   $$x^2 + 4x + 2 = 0$$

so that

$$x^2 + 4x = -2$$

Now, if the square of one-half the coefficient of the first-degree term $[\frac{1}{2}(4)]^2$ is added to each side of the equation, we have

$$x^2 + 4x + [\tfrac{1}{2}(4)]^2 = -2 + [\tfrac{1}{2}(4)]^2$$

so that

$$x^2 + 4x + 4 = -2 + 4$$

or

$$x^2 + 4x + 4 = 2$$

The left-hand side of the equation is the square of $(x + 2)$; thus, the equation can be written as

$$(x + 2)^2 = 2$$

which implies that

$$x + 2 = \pm\sqrt{2}$$

so that

$$x = -2 + \sqrt{2} \quad \text{or} \quad x = -2 - \sqrt{2}$$

That is, the solution set is $\{-2 - \sqrt{2},\ -2 + \sqrt{2}\}$.

3 Solve $3x^2 - 4x + 1 = 0$ by using the quadratic formula.

SOLUTION

$$3x^2 - 4x + 1 = 0$$

We have $a = 3$, $b = -4$, and $c = 1$. Substituting into the quadratic formula

$$x = \frac{-b \pm \sqrt{b^2 - 4ac}}{2a}$$

we obtain

$$x = \frac{-(-4) \pm \sqrt{(-4)^2 - 4(3)(1)}}{2(3)}$$

or

$$x = \frac{4 \pm \sqrt{4}}{6}$$

That is,

$$x = \frac{4 \pm 2}{6}$$

Hence,

$$x = \frac{4 + 2}{6} = 1 \quad \text{or} \quad x = \frac{4 - 2}{6} = \frac{1}{3}$$

That is, the solution set is $\{\frac{1}{3}, 1\}$.

4.2 Properties of Quadratic Functions

The methods of solving quadratic equations can be used to help graph quadratic functions. First, we shall determine the graph of the quadratic function $f(x) = x^2$ by locating some of its points (Figure 1a), and then drawing the curve connecting these points (Figure 1b).

Figure 1

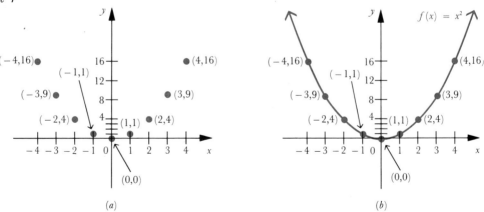

The graph of any quadratic function will have the same general shape as that illustrated in Figure 1, although the location of the graph will vary, depending upon the specific values of a, b, and c. Such graphs are called *parabolas*. Specific examples of quadratic functions are shown in Figure 2. Notice that the curve opens upward when the coefficient of the x^2 term is positive (Figures 2a and c), and downward when the coefficient of the x^2 term is negative (Figures 2b and d). In general, the graph of

Figure 2

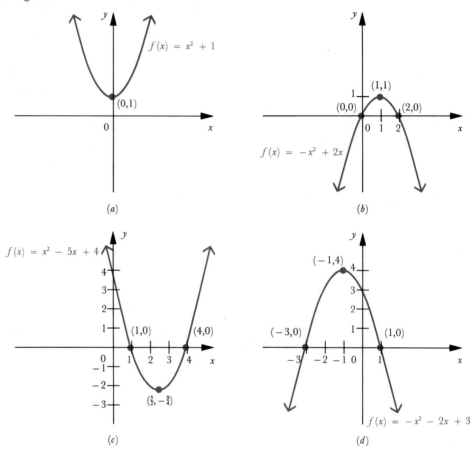

$f(x) = ax^2 + bx + c$ opens upward when $a > 0$ and downward when $a < 0$.

This property can be shown as follows. Factoring out a from the two terms involving x in $y = ax^2 + bx + c$, $a \neq 0$, we have

$$y = a\left(x^2 + \frac{b}{a}x\right) + c$$

Completing the square we obtain

$$y = a\left(x^2 + \frac{b}{a}x + \frac{b^2}{4a^2}\right) + c - \frac{b^2}{4a}$$

or

$$y = a\left(x + \frac{b}{2a}\right)^2 + c - \frac{b^2}{4a}$$

Since $(x + b/2a)^2 \geq 0$ for all values of x, $y \geq c - (b^2/4a)$ if $a > 0$ and $y \leq c - (b^2/4a)$ if $a < 0$. That is, the curve opens upward for $a > 0$ and opens downward for $a < 0$. Note that $(x + (b/2a))^2 = 0$ when $x = -b/2a$ so that the point $(-b/2a, c - (b^2/4a))$ is the *minimum point* when $a > 0$ and the *maximum point* when $a < 0$.

This point is also referred to as the *extreme point*, or the *vertex*, of the parabola. Locating this point will help in sketching the graphs of quadratic functions. For example, in order to graph $y = x^2 - 3x + 2$, notice that the solution of the quadratic equation $x^2 - 3x + 2 = 0$ is 1 or 2, so that the x intercepts are 1 and 2. By completing the square in the right-hand member of $y = x^2 - 3x + 2$, we obtain

$$y = (x^2 - 3x + \tfrac{9}{4}) - \tfrac{9}{4} + 2$$

or

$$y = (x - \tfrac{3}{2})^2 - \tfrac{1}{4}$$

From this equation we can see that the extreme point occurs when the term $(x - \tfrac{3}{2})^2$ is zero, that is, when $x - \tfrac{3}{2} = 0$ or $x = \tfrac{3}{2}$. Thus, the extreme point is $(\tfrac{3}{2}, -\tfrac{1}{4})$. Also, the y intercept is found by setting x equal to zero, so that $y = 2$. By connecting these points, we are able to sketch the parabola (Figure 3).

Figure 3

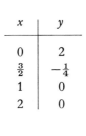

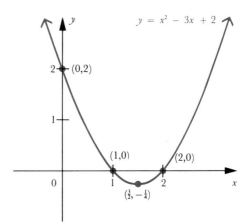

EXAMPLES

In Examples 1 to 3, find the domain, the x intercepts, the y intercepts, the extreme point, and the range of each of the following quadratic functions and sketch the graph.

1   $f(x) = -x^2 + 2x$

SOLUTION. The domain of f is the set of real numbers. In locating the x intercepts of the graph of the function f, let $f(x) = 0$, and then find values of x for which $-x^2 + 2x = 0$. This can be written as $-x(x - 2) = 0$, so that $x = 0$ or 2. The y intercept is found by assigning the value 0 to x, so that the y intercept is 0. We obtain the coordinates of the extreme point by completing the square. Thus, $y = -x^2 + 2x$ can be written as $y = -(x^2 - 2x + 1) + 1$ (why?), or $y = -(x - 1)^2 + 1$. For $x = 1$, $(x - 1)^2$ is zero, so that the extreme point is $(1, 1)$. The range of f is the set $\{y \mid y \leq 1\}$ (Figure 4).

Figure 4

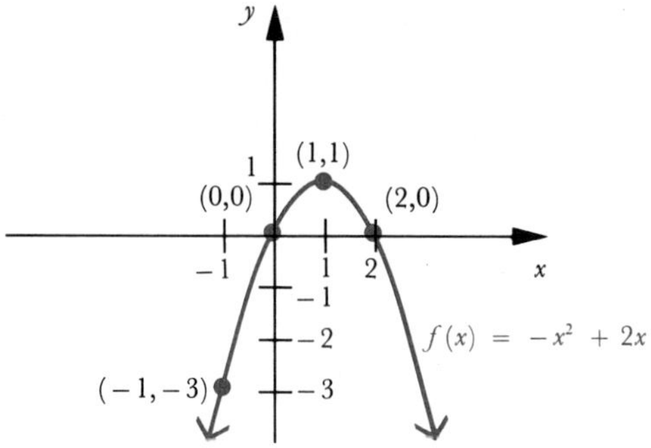

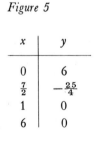

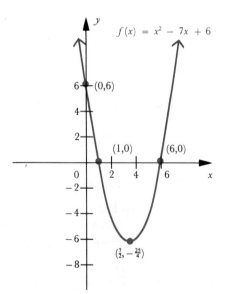

2 $f(x) = x^2 - 7x + 6$

SOLUTION. The domain of f is the set of real numbers. The x intercepts of f are the values of x such that $x^2 - 7x + 6 = 0$; that is,

$$(x - 6)(x - 1) = 0$$

where $x = 1$ or 6. The y intercept of f is $f(0) = 6$. To obtain the coordinates of the extreme point write

Figure 5

x	y
0	6
$\frac{7}{2}$	$-\frac{25}{4}$
1	0
6	0

$$y = f(x) = x^2 - 7x + 6$$

as

$$y = x^2 - 7x + (\tfrac{7}{2})^2 - (\tfrac{7}{2})^2 + 6$$

or

$$y = (x - \tfrac{7}{2})^2 - \tfrac{25}{4}$$

For $x = \tfrac{7}{2}$, y has an extreme value $-\tfrac{25}{4}$. (Why?) Thus, the extreme point is at $(\tfrac{7}{2}, -\tfrac{25}{4})$. The range of f is the set $\{y \mid y \geq -\tfrac{25}{4}\}$ (Figure 5).

3 $f(x) = 3x^2 - 2x - 5$

SOLUTION. The domain of f is the set of real numbers. The x intercepts of f are the values of x such that $3x^2 - 2x - 5 = 0$; that is, $(3x - 5) \times (x + 1) = 0$, where $x = -1$ or $\tfrac{5}{3}$. The y intercept of f is $f(0) = -5$. To obtain the coordinates of the extreme point, write

$$y = f(x) = 3x^2 - 2x - 5$$

as

$$y = 3(x^2 - \tfrac{2}{3}x) - 5$$

or

$$y = 3(x^2 - \tfrac{2}{3}x + \tfrac{1}{9}) - 5 - \tfrac{1}{3} = 3(x - \tfrac{1}{3})^2 - \tfrac{16}{3}$$

For $x = \tfrac{1}{3}$, y has an extreme value of $-\tfrac{16}{3}$. Thus, the extreme point is $(\tfrac{1}{3}, -\tfrac{16}{3})$. The range of f is the set $\{y \mid y \geq -\tfrac{16}{3}\}$ (Figure 6).

Figure 6

x	y
-1	0
0	-5
$\tfrac{1}{3}$	$-\tfrac{16}{3}$
$\tfrac{5}{3}$	0

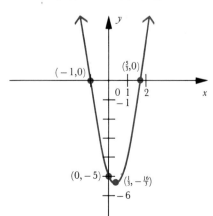

4 A projectile is fired from a balloon in such a way that it is h feet above the ground t seconds after the firing. If $h = 96t - 16t^2$, find
a) h when $t = 1$.
b) the maximum height reached by the projectile.
c) the graph of the function.

SOLUTION

a) At $t = 1$, $h = 96(1) - 16(1^2) = 80$ feet.
b) $h = 96t - 16t^2$
$h = -16(t^2 - 6t)$

Completing the square, we get

$$h = -16(t^2 - 6t + 9) + 144$$

or

$$h = -16(t - 3)^2 + 144$$

Hence, h has a maximum value of 144 feet when $t = 3$ seconds.

c) The t intercepts are found by setting $h = 0$. Thus,

$$96t - 16t^2 = 0$$
$$-16t(t - 6) = 0$$

or $t = 0$ and $t = 6$. Notice that the h intercept is also 0. The graph is determined by constructing the parabola that contains the points $(0, 0)$, $(6, 0)$, and $(3, 144)$ (Figure 7).

Figure 7

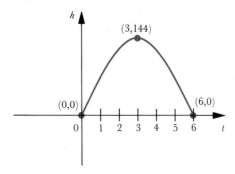

5 A survey showed that a manufacturer can expect to sell 26,000 boxes of detergent a week in a certain market area if he charges 19 cents per

box, 16,000 boxes if he charges 21 cents per box and 14,000 boxes if he charges 22 cents per box. The functional relationship between the number of boxes sold and the price per box is expressed by the equation $D = p^2 - 45p + 520$, where D is the weekly demand (in thousands of boxes) and p is the price per box (in cents). How many boxes of detergent might he sell if he charges 15 cents per box? Graph the function that relates the number of boxes sold to the price per box.

SOLUTION. Since $D = p^2 - 45p + 520$, substituting $p = 15$ in the equation, we find that $D = 15^2 - 45(15) + 520 = 70$, so that there would be a demand for 70,000 boxes. The graph is determined by constructing the parabola that contains the points (19, 26,000), (21, 16,000), and (22, 14,000) (Figure 8).

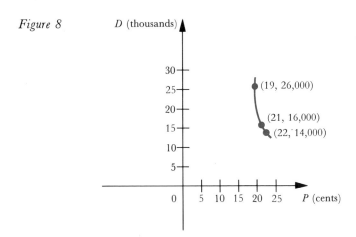

Figure 8

The graphs of quadratic functions can be used to illustrate the nature of the solutions of associated quadratic equations $ax^2 + bx + c = 0$, $a \neq 0$. The solutions of the quadratic equation $ax^2 + bx + c = 0$, $a \neq 0$, can be observed by inspecting the x intercepts of the associated function defined by $f(x) = ax^2 + bx + c$. Let us consider the graphs of $f(x) = x^2 - 4x + 4$, $f(x) = x^2 - 5x + 4$, and $f(x) = x^2 + 1$. The quadratic equation $x^2 - 4x + 4 = (x - 2)(x - 2) = 0$ has two equal solutions, $x = 2$ and $x = 2$, so that the graph of the associated function $f(x) = x^2 - 4x + 4$ intercepts the x axis at one point $x = 2$ (Figure 9a). The equation $x^2 - 5x + 4 = (x - 1)(x - 4) = 0$ has two different solutions, $x = 1$ and $x = 4$, so that the graph of the associated function $f(x) = x^2 - 5x + 4$ intercepts the x axis at two points, 1 and 4 (Figure 9b). Finally, the equation $x^2 + 1 = 0$ has no real solutions. Accordingly, the graph of the associated function $f(x) = x^2 + 1$ does not cross the x axis (Figure 9c).

Figure 9

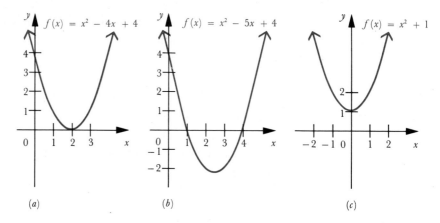

(a) (b) (c)

The solutions of any quadratic equation of the form $ax^2 + bx + c = 0$, where a, b, and c are real numbers, $a \neq 0$, are given by the quadratic formulas

$$x = \frac{-b + \sqrt{b^2 - 4ac}}{2a} \quad \text{or} \quad x = \frac{-b - \sqrt{b^2 - 4ac}}{2a}$$

We can see from these general forms that solutions that are not real numbers can only occur if the expression under the radical is negative, that is, if $b^2 - 4ac < 0$. For example, the solutions of the quadratic equation $2x^2 - x + 2 = 0$ are not real numbers, since $b^2 - 4ac = (-1)^2 - 4(2)(2) = -15$.

On the other hand, if the expression $b^2 - 4ac$ is positive or zero, then the solutions are real numbers. Thus, the solutions of the quadratic equation $x^2 + 6x + 5 = 0$ are real numbers because $b^2 - 4ac = 36 - 4(5) = 16$.

In the case where $b^2 - 4ac = 0$, the solutions are equal. For example, the solutions of the quadratic equation $x^2 - 6x + 9 = 0$ are equal, since $b^2 - 4ac = (-6)^2 - 4(9) = 0$.

Thus, we can determine whether or not the solutions of quadratic equations are real numbers simply by evaluating the expression $b^2 - 4ac$. This expression is called the *discriminant* of the quadratic equation.

We can restate the above results as follows:

1. If the discriminant $b^2 - 4ac$ of a quadratic equation $ax^2 + bx + c = 0$, $a \neq 0$, is zero, then the associated function $f(x) = ax^2 + bx + c$ intercepts the x axis at one point (Figure 9a) and there is one real root.

2. If the discriminant of a quadratic equation is positive, that is, if $b^2 - 4ac > 0$, then the associated function $f(x) = ax^2 + bx + c$, $a \neq 0$, intercepts the x axis at two different points (Figure 9b) and there are two real roots.

3 If the discriminant $b^2 - 4ac < 0$, then the associated function $f(x) = ax^2 + bx + c$, $a \neq 0$, does not intercept the x axis (Figure 9c) and there are no real roots.

PROBLEM SET 3

1 Solve each of the following equations by the factoring method.
   a) $3x^2 - 7x = 0$
   b) $2x^2 - 19x - 33 = 0$
   c) $5(x + 25) = 6x^2$
   d) $x(x - 2) = 9 - 2x$
   e) $x^2 - 6x = -8$
   f) $4x^2 + 11x + 6 = 0$
   g) $4x^2 - 16 = 0$
   h) $18x^2 + 61x - 7 = 0$
   i) $3x^2 - 2x = 5$
   j) $10x^2 - 31x - 14 = 0$

2 Solve each of the following equations using the method of completing the square.
   a) $3x^2 - 8x + 2 = 0$
   b) $x^2 - 12x + 35 = 0$
   c) $x^2 + 3x - 1 = 0$
   d) $x^2 - 3x + 2 = 0$
   e) $3x^2 - 7x - 3 = 0$
   f) $x^2 + 18x + 12 = 0$
   g) $x^2 - 13x + 3 = 0$
   h) $x^2 + 2x - 3 = 0$
   i) $x^2 - 8x + 7 = 0$
   j) $x^2 + 21x + 10 = 0$

3 Solve each of the following quadratic equations using the quadratic formula. Indicate the discriminant of each equation.
   a) $2x^2 - 5x + 1 = 0$
   b) $2x^2 + x - 1 = 0$
   c) $20 = 12x - x^2$
   d) $12x^2 + 29x - 11 = 0$
   e) $x^2 - x + 1 = 0$
   f) $5 + x = 6x^2$
   g) $x^2 - 18x + 56 = 0$
   h) $32x^2 - 4x + 21 = 0$
   i) $x^2 + 3x + 13 = 0$
   j) $4x^2 + 11x = 3$

4 For each of the following quadratic functions, determine the domain, the extreme point and the range, and the x and y intercepts. Sketch the graph.
   a) $f(x) = 2x^2 - 3$
   b) $f(x) = x^2 - 3$
   c) $f(x) = -x^2 - 2x - 1$
   d) $f(x) = (x - 5)^2$
   e) $f(x) = x^2 + 5x + 6$
   f) $f(x) = -x^2 - 1$
   g) $f(x) = 2x^2 - 3x$
   h) $f(x) = -(x + 1)^2$
   i) $f(x) = x^2 + 4x + 3$
   j) $f(x) = -x^2 + x - 5$

5 Graph on the same coordinate system $y = ax^2$, where $a \in \{-10, -5, -2, -1, 0, 1, 2, 5, 10\}$. Compare the graphs for different values of a. What do you notice?

6 Sketch the graph of each of the following functions on the same coordinate system.
   a) $f(x) = x^2 - 2$
   b) $f(x) = x^2 - 1$
   c) $f(x) = x^2 + 1$
   d) $f(x) = x^2 + 2$

7 A ball is thrown vertically upward from the edge of a roof in such a manner that it eventually falls to the street 112 feet below. If it moves so

that its distance h from the roof at time t seconds is given by $h = 96t - 16t^2$, find

a) h at $t = 2$.
b) the maximum height above the street reached by the ball.
c) the graph of the function.

8 A stone is projected vertically upward with initial velocity 112 feet per second and moves according to the law $h = 112t - 16t^2$, where h is the distance from the starting point. Find

a) h at $t = 2$.
b) The maximum height reached by the stone.
c) the graph of the function.

9 A travel agency advertises all-expenses-paid trips to the World Series for special groups. Transportation is by charter bus, which seats 48 passengers, and the charge per person is $80 plus an additional $2 for each empty seat. (Thus, if there are 4 empty seats each person has to pay $88; if there are 6 empty seats each person has to pay $92; and so on.) If there are x empty seats, how many passengers are there on the bus? How much does each passenger have to pay? What are the travel agency's total receipts? Graph the function that relates the travel agency's total receipts to the number of empty seats.

5 Quadratic Inequalities

Inequalities which are equivalent to those of the form $ax^2 + bx + c < 0$ or $ax^2 + bx + c > 0$, where $a \neq 0$, are called *quadratic inequalities*. Thus, $x^2 + 6 < 5x$ and $x^2 + x > 2$ are examples of quadratic inequalities, since the first can be expressed as $x^2 - 5x + 6 < 0$ and the second as $x^2 + x - 2 > 0$.

The graphs of quadratic functions can be used to solve quadratic inequalities. For example, consider the quadratic function $f(x) = x^2 - 3x + 2$. The graph of f intersects the x axis at the two points $(1, 0)$ and $(2, 0)$ (Figure 1). Notice, also, that if $x < 1$ or if $x > 2$, then $f(x) > 0$; if $1 < x < 2$, then $f(x) < 0$. Hence,

$$\{x \mid f(x) = x^2 - 3x + 2 > 0\} = \{x \mid x < 1\} \cup \{x \mid x > 2\}$$
$$= (-\infty, 1) \cup (2, \infty)$$

and

$$\{x \mid f(x) = x^2 - 3x + 2 < 0\} = \{x \mid 1 < x < 2\} = (1, 2)$$

Figure 1

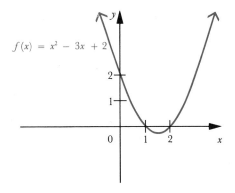

EXAMPLES

1 Use the graph of $f(x) = 3 + 2x - x^2$ to find the values of x for which $3 + 2x - x^2 < 0$.

SOLUTION. The inequality is satisfied by all those values of x on the graph of f that have ordinates below the x axis, that is, by all x such that $f(x) < 0$. The x intercepts of the graph of $f(x) = 3 + 2x - x^2$ are -1 and 3, so that

$$\{x \mid 3 + 2x - x^2 < 0\} = \{x \mid x < -1\} \cup \{x \mid x > 3\}$$
$$= (-\infty, -1) \cup (3, \infty) \qquad \text{(Figure 2)}$$

Figure 2

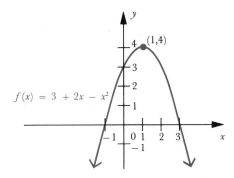

2 Use the graph of $f(x) = 2x^2 + 5x - 3$ to find the values of x for which $2x^2 + 5x - 3 > 0$.

SOLUTION. The inequality is satisfied by all values of x for which the graph of f lies above the x axis, that is, by all x such that $f(x) > 0$. The x intercepts of the graph of f are -3 and $\frac{1}{2}$. Hence,

$$\{x \mid 2x^2 + 5x - 3 > 0\} = \{x \mid x > \tfrac{1}{2} \text{ or } x < -3\}$$
$$= \{x \mid x < -3\} \cup \{x \mid x > \tfrac{1}{2}\}$$
$$= (-\infty, -3) \cup (\tfrac{1}{2}, \infty) \qquad \text{(Figure 3)}$$

Figure 3

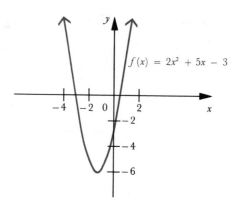

3 Use the graph of $f(x) = x^2 + 4$ to find the values of x for which $x^2 + 4 < 0$.

SOLUTION. Upon examining the graph of the associated function $f(x) = x^2 + 4$ (Figure 4), we notice that $x^2 + 4 > 0$ for all real numbers x, so that $\{x \mid x^2 + 4 < 0\} = \emptyset$.

Figure 4

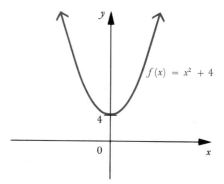

We have seen how to solve quadratic inequalities by using the graphs of the associated functions. Although this method is easy to follow, an alternate *algebraic method* based on the factors of the quadratic can also be used to solve these inequalities.

Consider the inequality $x^2 - 5x + 6 < 0$. By factoring the quadratic expression, we can rewrite the inequality as $(x - 2)(x - 3) < 0$. The solution set will be those values of x for which the product of $x - 2$ and $x - 3$ is a negative number. Recalling the rules for multiplying signed numbers, either $x - 2$ is a positive number and $x - 3$ is a negative number, or $x - 2$ is a negative number and $x - 3$ is a positive number. That is, either

i $x - 2 > 0$ and simultaneously $x - 3 < 0$

or

ii $x - 2 < 0$ and simultaneously $x - 3 > 0$

Let us consider these two cases:

i $x - 2 > 0$ and $x - 3 < 0$. Solving these first-degree inequalities, we have

$x > 2$ and $x < 3$

or

$\{x \mid x > 2\} \cap \{x \mid x < 3\}$

That is,

$2 < x < 3$ or $\{x \mid 2 < x < 3\}$

ii $x - 2 < 0$ and $x - 3 > 0$. Solving these first-degree inequalities, we have

$x < 2$ and $x > 3$

or

$\{x \mid x < 2\} \cap \{x \mid x > 3\}$

But x cannot be both less than 2 and at the same time greater than 3; that is,

$\{x \mid x < 2\} \cap \{x \mid x > 3\} = \emptyset$

Therefore, the solution set of the inequality $x^2 - 5x + 6 < 0$ is $\{x \mid 2 < x < 3\}$. These two cases are illustrated in Figure 5. In the

Figure 5

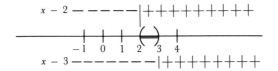

figure, the set of values of x for which the two expressions $x - 2$ and $x - 3$ are positive or negative are indicated, so that the values of x for which the product $(x - 2)(x - 3)$ is negative are obvious.

EXAMPLES

Solve each of the following inequalities and illustrate the solution graphically.

1   $x^2 + x > 2$

SOLUTION

$$x^2 + x > 2 \quad \text{or} \quad x^2 + x - 2 > 0$$

By factoring the left side, we get

$$(x + 2)(x - 1) > 0$$

Since the product of these two expressions is positive, we have either

i $x + 2 > 0$ and $x - 1 > 0$

or

ii $x + 2 < 0$ and $x - 1 < 0$

Solving these inequalities, we have

i $x + 2 > 0$ and $x - 1 > 0$

That is,

$x > -2$ and $x > 1$

or, equivalently,

$$\{x \mid x > -2\} \cap \{x \mid x > 1\} = \{x \mid x > 1\}$$

ii $x + 2 < 0$ and $x - 1 < 0$

$x < -2$ and $x < 1$

or, equivalently,

$$\{x \mid x < -2\} \cap \{x \mid x < 1\} = \{x \mid x < -2\}$$

Thus, the solution set for $x^2 + x > 2$ is $\{x \mid x > 1\} \cup \{x \mid x < -2\}$, as shown in Figure 6.

Figure 6

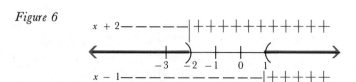

2 $x^2 - 6x < 7$

SOLUTION

$$x^2 - 6x < 7 \quad \text{or} \quad x^2 - 6x - 7 < 0$$

so that

$$(x - 7)(x + 1) < 0$$

Since the product is negative, we have either

i $x - 7 > 0$ and $x + 1 < 0$

so that

$x > 7$ and $x < -1$

That is,

$$\{x \mid x > 7\} \cap \{x \mid x < -1\} = \emptyset$$

or

ii $x - 7 < 0$ and $x + 1 > 0$

$x < 7$ and $x > -1$

That is,

$$\{x \mid x < 7\} \cap \{x \mid x > -1\} = \{x \mid -1 < x < 7\}$$

Thus, the values of x that satisfy $x^2 - 6x < 7$ are found in case 2; that is, the solution set is $\{x \mid -1 < x < 7\}$ (Figure 7).

Figure 7

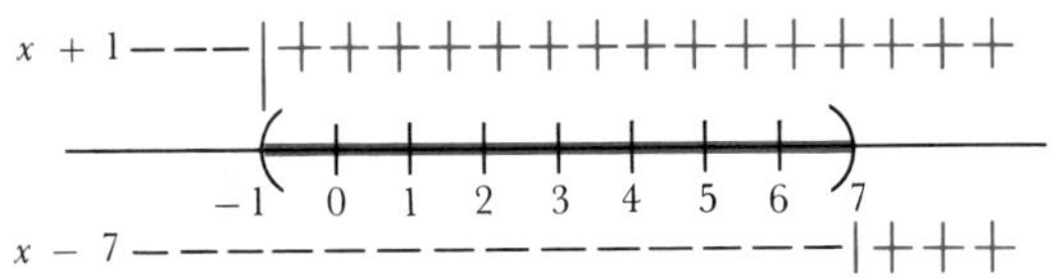

Inequalities involving fractions such as $(x - 1)/(x + 2) > 0$ can be solved in a similar fashion. Here we consider the possible ways for the expression to be positive. The quotient of two numbers is positive if both numbers are positive or if both numbers are negative. Hence, we can solve $(x - 1)/(x + 2) > 0$ by considering the two possibilities. Either

i $x - 1 > 0$ and $x + 2 > 0$

or

ii $x - 1 < 0$ and $x + 2 < 0$

Solving these two cases separately, we have

i $x - 1 > 0$ and $x + 2 > 0$

$x > 1$ and $x > -2$

or, equivalently,

$$\{x \mid x > 1\} \cap \{x \mid x > -2\} = \{x \mid x > 1\}$$

ii $x - 1 < 0$ and $x + 2 < 0$

$x < 1$ and $x < -2$

or, equivalently,

$$\{x \mid x < 1\} \cap \{x \mid x < -2\} = \{x \mid x < -2\}$$

Hence, the solution set is $\{x \mid x > 1\} \cup \{x \mid x < -2\}$.

PROBLEM SET 4

1. Sketch the graph of each of the following quadratic functions, and then use the graph to solve the inequality. Write the solution in interval form.
   a) $y = -2x^2 + 3x + 2$; $-2x^2 + 3x + 2 \geq 0$
   b) $y = 3x^2 - 5x + 2$; $3x^2 - 5x + 2 \leq 0$
   c) $y = (x - 2)^2$; $x^2 - 4x \leq -4$
   d) $y = 5x^2 - 9x + 4$; $5x^2 - 9x + 4 \geq 0$
   e) $y = x^2 - 5x + 4$; $x^2 - 5x + 4 \leq 0$
   f) $y = -2x^2 + 6x + 7$; $-2x^2 + 6x + 7 \geq 0$
   g) $y = -4x^2 + 9x - 3$; $-4x^2 + 9x - 3 < 0$
   h) $y = -3x^2 + 2x - 9$; $-3x^2 + 2x - 9 > 0$

2. Solve each of the following inequalities using factoring and illustrate the solutions on a number line.
   a) $2x^2 + x < 1$
   b) $6x^2 - x - 7 < 0$
   c) $2x^2 + 9x - 5 < 0$
   d) $x^2 - 4 > 0$
   e) $x^2 > x + 2$
   f) $x^2 + 2x < 3$

g) $x^2 + 5x + 6 > 0$ h) $x^2 + x - 6 \leq 0$
i) $x^2 < -4x - 4$ j) $40 - 3x - x^2 < 0$

6 Quadratic Forms and Radical Equations

Some equations that are not quadratic equations will be solved in this section by converting the equation to the quadratic form $au^2 + bu + c = 0$, where u is some expression involving another variable. For example, $3x^4 + 2x^2 - 1 = 0$, $x - \sqrt{x} + 2 = 0$, and $(x^2 - 1)^2 + 2(x^2 - 1) - 3 = 0$ are equations that are quadratic in form. We can solve equations of this type by first making a substitution to put them in the form $au^2 + bu + c = 0$ and then solving for u. This technique is illustrated in the following examples.

EXAMPLES

Solve each of the following equations.

1 $x^4 - 5x^2 + 6 = 0$

SOLUTION. We first write $x^4 - 5x^2 + 6 = 0$ as

$$(x^2)^2 - 5(x^2) + 6 = 0$$

Letting $u = x^2$, we have the quadratic in the form

$$u^2 - 5u + 6 = 0$$

Solving for u, we get

$$u^2 - 5u + 6 = 0$$
$$(u - 2)(u - 3) = 0$$
$$u - 2 = 0 \quad \text{or} \quad u - 3 = 0$$

That is,

$$u = 2 \quad \text{or} \quad u = 3$$

Now, since $u = x^2$ and $u = 2$ or $u = 3$, we have

$$x^2 = 2 \quad \text{or} \quad x^2 = 3$$

Thus,

$$x = \pm\sqrt{2} \quad \text{or} \quad x = \pm\sqrt{3}$$

By substituting these values in the original equation, we can verify that the solution set is $\{-\sqrt{2}, \sqrt{2}, -\sqrt{3}, \sqrt{3}\}$.

2. $x^{-2} + x^{-1} - 6 = 0$

   SOLUTION. Since $x^{-2} + x^{-1} - 6 = (x^{-1})^2 + (x^{-1}) - 6 = 0$, the equation is quadratic in form. Letting $u = x^{-1}$, we have

   $$u^2 + u - 6 = 0$$
   $$(u + 3)(u - 2) = 0$$
   $$u + 3 = 0 \quad \text{or} \quad u - 2 = 0$$
   $$u = -3 \quad \text{or} \quad u = 2$$

   Replacing u by x^{-1}, we have

   $$x^{-1} = -3 \quad \text{or} \quad x^{-1} = 2$$
   $$\frac{1}{x} = -3 \quad \text{or} \quad \frac{1}{x} = 2$$

   so that

   $$-3x = 1 \quad \text{or} \quad 2x = 1$$
   $$x = -\tfrac{1}{3} \quad \text{or} \quad x = \tfrac{1}{2}$$

   The solution set is $\{-\tfrac{1}{3}, \tfrac{1}{2}\}$.

3. $\left(x - \dfrac{8}{x}\right)^2 + \left(x - \dfrac{8}{x}\right) = 42$

   SOLUTION. Let $u = [x - (8/x)]$ in the equation $[x - (8/x)]^2 + [x - (8/x)] = 42$. We have

   $$u^2 + u = 42$$

   or

   $$u^2 + u - 42 = 0$$

   so that

   $$(u + 7)(u - 6) = 0$$

   which implies that

   $$u = -7 \quad \text{or} \quad u = 6$$

   so that

   $$x - \frac{8}{x} = -7 \quad \text{or} \quad x - \frac{8}{x} = 6$$

Therefore,

$$x^2 + 7x - 8 = 0 \quad \text{or} \quad x^2 - 6x - 8 = 0$$

so that

$$x = -8 \quad \text{or} \quad x = 1 \quad \text{or} \quad x = 3 - \sqrt{17} \quad \text{or} \quad x = 3 + \sqrt{17}$$

The solution set is $\{-8, 1, 3 - \sqrt{17}, 3 + \sqrt{17}\}$.

4. $x + 2 + \sqrt{x + 2} - 2 = 0$

SOLUTION

$$x + 2 + \sqrt{x + 2} - 2 = 0 \quad \text{or} \quad (\sqrt{x + 2})^2 + \sqrt{x + 2} - 2 = 0$$

Letting $u = \sqrt{x + 2}$ in the equation, we have

$$u^2 + u - 2 = 0$$

so that

$$(u - 1)(u + 2) = 0$$

That is,

$$u - 1 = 0 \quad \text{or} \quad u + 2 = 0$$

Therefore,

$$u = 1 \quad \text{or} \quad u = -2$$

Replacing u by $\sqrt{x + 2}$, we get

$$\sqrt{x + 2} = 1 \quad \text{or} \quad \sqrt{x + 2} = -2$$
$$(x + 2) = 1 \quad \text{or} \quad (x + 2) = 4 \quad \text{(squaring both sides)}$$
$$x = -1 \quad \text{or} \quad x = 2$$

Check: For $x = -1$,

$$(x + 2) + \sqrt{x + 2} - 2 = 0$$
$$[(-1) + 2] + \sqrt{(-1) + 2} - 2 = 1 + \sqrt{1} - 2 = 2 - 2 = 0$$

Therefore, $x = -1$ is a solution.
For $x = 2$,

$$[(2) + 2] + \sqrt{2 + 2} - 2 = 4 + 2 - 2 \neq 0$$

Therefore, $x = 2$ is not a solution. Hence, the solution set is $\{-1\}$.

It is important to notice that in the above example the process of squaring both sides of an equation in order to remove the radicals introduced an "apparent solution," which later proved to be invalid. If we had not checked our solutions, this erroneous solution would not have been detected. Whenever an equation is squared on both sides, the resulting equation is not necessarily equivalent to the equation being squared. In the example above, this occurred when $\sqrt{x+2} = -2$ was squared to get $x + 2 = 4$. Here an equation with no real number solution was converted to one that was solvable. Therefore, it is necessary to check all solutions that result from this process.

Some equations involve radicals that are not quadratic in form. These equations are solved by first raising both sides to some power in order to remove the radicals. The resulting equation is then solved by the appropriate method. Care must be taken, however, to check the validity of the solutions obtained. The invalid solutions, if any, are called *extraneous solutions*.

EXAMPLES

Solve and check the following equations.

1   $\sqrt{2x + 5} = 3$

SOLUTION. We first eliminate the radical by squaring both sides to get

$$(\sqrt{2x+5})^2 = 3^2$$
$$2x + 5 = 9$$

Solving this latter equation, we have

$$2x = 4 \quad \text{or} \quad x = 2$$

Check: We are to determine if $x = 2$ is a solution of the given equation $\sqrt{2x+5} = 3$. Substituting 2 for x, we have

$$\sqrt{2(2)+5} = \sqrt{4+5} = \sqrt{9} = 3$$

Hence, $\{3\}$ is the solution set. (*Note:* $\sqrt{9} = 3$, not ± 3, because of the definition of principal square roots.)

2   $\sqrt{1 - 5x} + \sqrt{1 - x} = 2$

SOLUTION. Before squaring, we add $-\sqrt{1-x}$ to both sides of the equation in order to simplify the subsequent steps.

$$\sqrt{1 - 5x} + \sqrt{1 - x} = 2$$
$$\sqrt{1 - 5x} = 2 - \sqrt{1 - x}$$

Squaring both sides, that is,

$$(\sqrt{1 - 5x})^2 = (2 - \sqrt{1 - x})^2$$

we have

$$1 - 5x = 4 - 4\sqrt{1 - x} + 1 - x$$

or

$$-4 - 4x = -4\sqrt{1 - x}$$

or

$$1 + x = \sqrt{1 - x}$$

Again, squaring both sides, we have

$$(1 + x)^2 = (\sqrt{1 - x})^2 \quad \text{or} \quad 1 + 2x + x^2 = 1 - x$$

That is,

$$x^2 + 3x = 0$$

Solving for x, we have

$$x(x + 3) = 0$$
$$x = 0 \quad \text{or} \quad x + 3 = 0 \quad \text{or} \quad x = -3$$

Check: For $x = 0$, we have

$$\sqrt{1 - 5(0)} + \sqrt{1 - 0} \stackrel{?}{=} 2$$
$$\sqrt{1} + \sqrt{1} \stackrel{?}{=} 2$$
$$1 + 1 = 2$$

Therefore, $x = 0$ is a solution. For $x = -3$, we have

$$\sqrt{1 - 5(-3)} + \sqrt{1 - (-3)} = \sqrt{16} + \sqrt{4} = 4 + 2 = 6 \neq 2$$

Therefore, $x = -3$ is an extraneous root, so that the solution set is $\{0\}$.

3 $\sqrt[4]{x^2 - 5x + 6} = \sqrt{x + 4}$

SOLUTION. Raising both sides of the equation to the fourth power, we get

$$x^2 - 5x + 6 = (x + 4)^2$$

or

$$x^2 - 5x + 6 = x^2 + 8x + 16$$

so that

$$-5x - 8x = 16 - 6 \quad \text{or} \quad -13x = 10$$

Therefore,

$$x = -\tfrac{10}{13}$$

Check: For $x = -\tfrac{10}{13}$, we have

$$\sqrt[4]{(-\tfrac{10}{13})^2 - 5(-\tfrac{10}{13}) + 6} \stackrel{?}{=} \sqrt{(-\tfrac{10}{13}) + 4}$$

Therefore,

$$\sqrt[4]{\tfrac{100}{169} + \tfrac{50}{13} + 6} \stackrel{?}{=} \sqrt{-\tfrac{10}{13} + \tfrac{52}{13}}$$

or

$$\sqrt[4]{\frac{100 + 650 + 1{,}014}{169}} \stackrel{?}{=} \sqrt{\frac{42}{13}}$$

or

$$\sqrt[4]{\frac{1{,}764}{169}} = \sqrt{\frac{42}{13}}$$

Therefore, the solution set is $\{-\tfrac{10}{13}\}$.

PROBLEM SET 5

Solve each of the following equations by reducing them to quadratic form, and check the solutions.

1. $x^4 - 13x^2 + 36 = 0$
2. $x^{-4} - 9x^{-2} + 20 = 0$
3. $x^{-8} - 17x^{-4} + 16 = 0$
4. $x^{1/3} - 1 - 12x^{-1/3} = 0$
5. $x^{-6} + 63x^{-3} - 64 = 0$
6. $x^{1/4} + 2 - 8x^{-1/4} = 0$
7. $x + 7 - \sqrt{x+7} - 2 = 0$
8. $\sqrt{x+20} - 4\sqrt[4]{x+20} + 3 = 0$
9. $x^2 + 3x + \sqrt{x^2 + 3x - 2} = 22$

10    $(x^2 + 1)^2 - 3(x^2 + 1) + 2 = 0$

11    $3(x + 3) + \sqrt{x + 3} = 2$ 12 $(x^2 + x)^2 - 18(x^2 + x) + 72 = 0$

13    $2x - 9\sqrt{x + 2} + 14 = 0$ 14 $(2x^2 - x)^2 - 3(2x^2 - x) + 2 = 0$

15    $x^2 + x + \dfrac{56}{x^2 + x} = 15$ 16 $3x - 1 - 2\sqrt{3x - 1} - 3 = 0$

17    $\left(3x - \dfrac{2}{x}\right)^2 + 6\left(3x - \dfrac{2}{x}\right) + 5 = 0$

18    $\dfrac{x^2}{x + 1} + \dfrac{2(x + 1)}{x^2} = 3$

19    $\dfrac{x + 1}{x} + 2 = 3\left(\dfrac{x}{x + 1}\right)$ 20 $\dfrac{x^2 + 1}{x} + \dfrac{4x}{x^2 + 1} - 4 = 0$

Solve each of the following equations and check for extraneous roots.

21    $\sqrt{2x + 5} = 4$ 22 $\sqrt{8x - 7} - x = 0$

23    $\sqrt{6x - 3} = 27$ 24 $\sqrt{3x + 1} = x - 1$

25    $\sqrt{3x + 1} = \sqrt{x + 3}$ 26 $2\sqrt{4x + 5} = \sqrt{8 - x} - 1$

27    $\sqrt{11 - x} - \sqrt{x + 6} = 3$ 28 $\sqrt{3 - x} - \sqrt{2 + x} = 3$

29    $\sqrt{x^2 + 6x} = x + \sqrt{2x}$ 30 $\sqrt{2x + \sqrt{7 + x}} = 3$

31    $\sqrt{x} = \sqrt{x + 16} - 2$ 32 $\sqrt{3x + 1} - 1 = \sqrt{3x - 8}$

33    $\sqrt{x + 4} + 1 = \sqrt{x + 11}$ 34 $\sqrt{x + 7} = 5 + \sqrt{x - 2}$

35    $\sqrt{7x - 6} = \sqrt{7x + 22} - 2$ 36 $\sqrt{2x^2 + 4} + 2 = 2x$

37    $\sqrt{39 - x} + \sqrt{11 - x} = 6$ 38 $\sqrt{x} + \sqrt{x - 6} = \dfrac{3}{\sqrt{x - 6}}$

39    $\sqrt{5 - 2x} + \sqrt{7 - 2x} = 4$ 40 $\sqrt{\sqrt{x + 16} - \sqrt{x}} = 2$

REVIEW PROBLEM SET

1   Find a linear function f such that $f(1) = 3$ and $f(2) = 5$.

2   Sketch the graph of each of the linear functions that are defined by the following equations. Find the domain and the range, and the slope of the graph and its x and y intercepts.
     a) $f(x) = -3x + 5$ b) $f(x) = 5x + 1$
     c) $f(x) = 2(x - 2) + 1$ d) $f(x) = \tfrac{1}{4}x + 1$
     e) $f(x) = -\tfrac{3}{4}x + 1$ f) $f(x) = -3x$
     g) $f(x) = -1$

3   Let f be a linear function. Give conditions so that each of the following equations are true for all real numbers.

a) $5f(x) = f(5x)$
b) $f(x + 7) = f(x) + f(7)$
c) $f(3x + 4) = 3f(x) + f(4)$
d) $f(3) = 4$ and $f(5) = 6$
e) $7f(x) = f(7x + 1)$
f) $f(3x + 2) = 3f(x) + 2$

4 Find a linear function f such that $f(1) = 3$ and such that the graph of f is parallel to the graph of the line whose slope is determined by the points $(-2, 1)$ and $(3, 2)$.

5 Let f be a linear function. Is $f(3t + 2)$ linear? Prove your answer.

6 If f is a constant function, find $f(2)$ if
a) $f(1) = -2$
b) $f(3) = 5$
c) $f(8) = -3$
d) $f(5) = 4$

7 Find the slope of the linear function f if $f(2) = -3$ and
a) $f(0) = 3$
b) $f(4) = 4$
c) $f(-2) = 1$
d) $f(5) = -10$

8 Find a function whose graph is the line joining the given points; also sketch the graph.
a) $(1, 2)$ and $(2, 4)$
b) $(1, 3)$ and $(4, 5)$
c) $(4, 1)$ and $(-2, 4)$
d) $(4, -7)$ and $(0, -5)$

9 Solve each of the following quadratic equations by the factoring method (a and b are constants).
a) $6x^2 - x - 1 = 0$
b) $6x^2 + bx = 2b^2$
c) $x^2 + 2ax = b^2 - a^2$
d) $(ax - bx)^2 = x(b - a)$
e) $4x^2 - 8x + 3 = 0$
f) $10x^2 + 11x - 6 = 0$
g) $ax^2 + 2ax + a = 2x + 2$
h) $25x^2 - 10x + 1 = 0$

10 Solve each of the following quadratic equations by completing the square, and check your solution by using the quadratic formula (a and b are constants).
a) $x^2 - 2ax + b^2 = 0$
b) $x^2 + x + k = kx$
c) $x^2 - x + 2 = 0$
d) $6x^2 - 31x + 21 = 0$
e) $x^2 + 10x + 13 = 0$
f) $3x^2 - 8x + 1 = 0$
g) $4x^2 - x - 1 = 0$
h) $16 - 6x - 3x^2 = 0$
i) $2x^2 + 5x - 17 = 0$
j) $49x^2 - 98x + 3 = 0$

11 Graph each of the following quadratic functions; determine the domain, the range, the extreme point, and the x and y intercepts.
a) $f(x) = 6x^2 - 5x - 4$
b) $f(x) = 2x^2 - x - 6$
c) $f(x) = x^2 + 6x + 9$
d) $f(x) = x^2 - 8x + 16$
e) $f(x) = -3 - 10x - 8x^2$
f) $f(x) = 10 + 3x - x^2$

12 Sketch the graph of each of the following quadratic functions, and then use the graph to solve the inequality. Write the solution in interval form.
a) $f(x) = 2x^2 + 5x - 12$; $2x^2 + 5x - 12 > 0$
b) $f(x) = 3x^2 + 5x + 2$; $3x^2 + 5x + 2 \leq 0$

c) $f(x) = -5x^2 + 7x - 6$; $-5x^2 + 7x - 6 \geq 0$
d) $f(x) = 4x^2 + 11x - 3$; $4x^2 + 11x - 3 < 0$

13 Solve each of the following equations.
a) $x^{-3/2} - 26x^{-3/4} - 27 = 0$
b) $\sqrt{x} - \sqrt[4]{x} - 2 = 0$
c) $x^2 - 6x - \sqrt{x^2 - 6x - 3} = 5$
d) $\dfrac{2}{\sqrt{x}} = \sqrt{x} + \sqrt{x-1}$
e) $3x^2 - 4x + \sqrt{3x^2 - 4x - 6} = 18$

14 Graph the equations of each system on the same coordinate system; then solve the system by substitution if possible.
a) $x + y = 5$
$x - y = 3$
b) $3x + y = 4$
$3x - y = -10$
c) $2x + y = 13$
$3x + y = 17$
d) $6x - y = 22$
$2x + y = 2$
e) $2x + 3y = -3$
$3x - 4y = 38$
f) $4x - 7y = 7$
$6x - 5y = 7$
g) $15x + 17y = 32$
$8x - 19y = -11$
h) $6x - 9y = 11$
$8x - 11y = 3$

15 Solve each of the following systems by the elimination method.
a) $x + y + z = 16$
$x - 4y + 3z = 42$
$x + 6y + 2z = 14$
b) $x + 2y + 3z = 4$
$x - 5y - 2z = -30$
$x + 4y - 3z = -33$
c) $9x + 12y - 4z = 16$
$10x - 4y + 7z = -11$
$12x + 13y - 11z = 24$
d) $x - 2y + 4z = -3$
$3x + y - 2z = 12$
$2x + y - 3z = 11$

CHAPTER 5

Roots of Polynomials and Complex Numbers

CHAPTER 5

Roots of Polynomials and Complex Numbers

1 Introduction

In this chapter we will consider functions similar to these examples: $f(x) = x^3 - 4x$, $f(x) = x^3 + x^2 - 10x + 8$, and $f(x) = x^4 - 2x^3 - 5x^2 + 6x + 3$. Such functions are called polynomial functions of degree higher than 2. In graphing such functions, it is helpful at times to determine the x intercepts, that is, the points where the graph crosses the x axis. These points can be found by solving for the "roots" of the equation $f(x) = 0$. The roots are also called the *zeros* of the function f. For example, the zeros of the polynomial function $f(x) = x^3 - 4x$ can be determined by solving the equation $x^3 - 4x = 0$ to get $x = 0$, $x = -2$, or $x = 2$. Other topics covered in this chapter include synthetic division, rational functions, complex numbers, and complex zeros of polynomial functions.

2 Division of Polynomials

Given the polynomial expressions $3x^2 + 5x + 1$ and $x^2 - 3x + 5$; the sum $(3x^2 + 5x + 1) + (x^2 - 3x + 5)$, the difference $(3x^2 + 5x + 1) - (x^2 - 3x + 5)$, and the product $(3x^2 + 5x + 1)(x^2 - 3x + 5)$ are polynomials. In general, the sum, difference, and product of any two polynomials always result in another polynomial. This is not always the case with the division of polynomials.

For example, the product

$$(x^2 + 3x + 9)(x - 3) = x^3 - 27$$

suggests that

$$\frac{x^3 - 27}{x - 3} = x^2 + 3x + 9$$

and we say that $x^3 - 27$ is "divisible by $x - 3$" with a quotient $x^2 +$

$3x + 9$. On the other hand, the equation

$$(x^2 + 3x + 9)(x - 3) + 1 = x^3 - 26$$

suggests that $x^3 - 26$ is *not* divisible by $x - 3$ in the sense described above, however, we can use the latter equation to write

$$\frac{x^3 - 26}{x - 3} = x^2 + 3x + 9 + \frac{1}{x - 3}$$

Here $x^2 + 3x + 9$ is the quotient and 1 is the remainder. In the first example the division resulted in a polynomial, in the second it did not. What we need in general in order to handle the problem of dividing polynomials is a theorem of algebra called the *division algorithm*. This theorem is stated as follows without proof.

If $f(x)$ and $D(x)$ are polynomials such that the degree of $D(x)$ is less than or equal to the degree of $f(x)$, with $D(x) \neq 0$, then there are unique polynomials $Q(x)$ and $R(x)$, where $R(x)$ is of degree less than or equal to $D(x)$ or $R(x) = 0$, such that $f(x) = Q(x)D(x) + R(x)$. $D(x)$ is called the *divisor*, $Q(x)$ is called the *quotient*, $f(x)$ is called the *dividend*, and $R(x)$ is called the *remainder*.

If $f(x) = 3x^3 - 2x^2 + x - 5$ and $D(x) = x - 2$, then the division algorithm guarantees the existence and uniqueness of polynomials $Q(x)$ and $R(x)$ which satisfy $3x^3 - 2x^2 + x - 5 = Q(x)(x - 2) + R(x)$. $Q(x)$ and $R(x)$ can be found as follows:

```
                 3x² + 4x  + 9
         ─────────────────────────
x - 2 )3x³ - 2x² + x  - 5
         3x³ - 6x²
         ─────────
               4x² +  x
               4x² - 8x
               ────────
                    9x -  5
                    9x - 18
                    ───────
                         13
```

Hence, $Q(x) = 3x^2 + 4x + 9$ and $R(x) = 13$.

EXAMPLE

Suppose that $f(x) = x^3 + 5x^2 + 6x + 3$ and $D(x) = x + 4$. Find $Q(x)$ and $R(x)$ such that $f(x) = (x + 4)Q(x) + R(x)$, where $Q(x)$ and $R(x)$ satisfy the division algorithm.

SOLUTION. The division can be arranged as follows:

$$\begin{array}{r}
x^2 + x + 2 \\
x + 4 \overline{\smash{\big)}x^3 + 5x^2 + 6x + 3} \\
\underline{x^3 + 4x^2} \\
x^2 + 6x \\
\underline{x^2 + 4x} \\
2x + 3 \\
\underline{2x + 8} \\
-5
\end{array}$$

Hence, $Q(x) = x^2 + x + 2$ and $R(x) = -5$.

2.1 Synthetic Division

The process of long division can be shortened considerably in cases where the divisor is of the form $x - r$. Consider the example of dividing $3x^3 - 2x^2 + x - 5$ by $x - 2$, as shown previously.

By inspecting the coefficients of the quotient we observe that they were obtained, with the exception of the first term, by multiplying the preceding coefficient of the quotient by -2 and subtracting this product from the corresponding coefficient in the dividend. The coefficients of the quotient were obtained as follows: 3 is the coefficient of the highest power term of the dividend; 4 is obtained by subtracting the product $(-2)(3)$ from -2, the coefficient of the second term of the dividend; 9 is obtained by subtracting the product $(-2)(4)$ from 1, the coefficient of the third term of the dividend; the remainder 13 is obtained by subtracting the product $(-2)(9)$ from -5, the last term of the dividend.

Since a pattern for determining the coefficients and remainder of the quotient exists, we can apply it to find the quotient when dividing by $x - r$ without showing all the steps that were previously used in long division. We shall show the application of this method to the same problem in the following way:

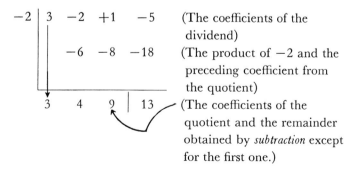

In this latter form we have shown only those numbers that are actually used to determine the quotient and remainder. We have only to interpret the results to complete the process. Since we are dividing a third-degree polynomial by a first-degree polynomial, the quotient will be a second-degree polynomial. Therefore, the numbers in the quotient row are the coefficients and remainder of a second-degree polynomial. That is, the quotient is $3x^2 + 4x + 9$ and the remainder is 13.

We shall make one further modification in this process. Instead of subtracting to get the last row, we can add the numbers in the columns above if we first change the sign of each number in the second row. We do this by changing the sign of the constant term of the divisor; then we can add in this process instead of subtracting. With these changes, the above problem can be shown as follows:

(Divisor with sign changed) $\boxed{2}$ $\boxed{3 \quad -2 \quad 1 \quad -5}$ (Coefficients of the dividend)

$ 6 \quad 8 \quad 18$

$ \boxed{3 \quad 4 \quad 9} \quad \boxed{13}$ (Remainder)

(Coefficients of the quotient)

The latter shortened form of division is known as *synthetic division*.

EXAMPLES

1. Use synthetic division to divide $2x^5 - 6x^3 + 3x - 5$ by $x + 3$.

    SOLUTION. We begin by representing the dividend in descending powers of the variable as follows:

    $$2x^5 - 6x^3 + 3x - 5 = 2x^5 + 0x^4 - 6x^3 + 0x^2 + 3x - 5$$

    Next arrange the coefficients in the form discussed. (Remember to change the sign of the constant of the divisor.)

    $-3 \ \big|\ 2 \quad 0 \quad -6 \quad 0 \quad 3 \quad -5$

    The first coefficient of the quotient is the same as the first number inside the division bracket:

    $-3 \ \big|\ 2 \quad 0 \quad -6 \quad 0 \quad 3 \quad -5$

    $ 2$

    The second number in the quotient is found by multiplying this number 2 by -3 (the number outside the bracket) and adding it to the second number inside the bracket:

$$\begin{array}{r|rrrrrr} -3 & 2 & 0 & -6 & 0 & 3 & -5 \\ & & -6 & & & & \quad [= 2(-3)] \\ \hline & 2 & -6 & & & & \quad [= 0 + (-6)] \end{array}$$

We continue this process until we have exhausted all the numbers of the dividend:

$$\begin{array}{r|rrrrrr} -3 & 2 & 0 & -6 & 0 & 3 & -5 \\ & & -6 & 18 & -36 & 108 & -333 \\ \hline & 2 & -6 & 12 & -36 & 111 & \big| \;-338 \end{array}$$

Hence,

$$2x^5 - 6x^3 + 3x - 5$$
$$= (2x^4 - 6x^3 + 12x^2 - 36x + 111)(x + 3) + (-338)$$

2 Use synthetic division to divide $3x^3 - 2x^2 + 1$ by $x - 2$.

SOLUTION

$$\begin{array}{r|rrrr} 2 & 3 & -2 & 0 & 1 \\ & & 6 & 8 & 16 \\ \hline & 3 & 4 & 8 & \big| \; 17 \end{array}$$

Hence, $Q(x) = 3x^2 + 4x + 8$ and the remainder is 17, so that $3x^3 - 2x^2 + 1 = (3x^2 + 4x + 8)(x - 2) + 17$.

Now consider $f(x) = 2x^3 - 5x^2 - x + 10$. We use synthetic division to divide $f(x)$ by $x - r$, where $r = 2$ as follows:

$$\begin{array}{r|rrrr} 2 & 2 & -5 & -1 & 10 \\ & & 4 & -2 & -6 \\ \hline & 2 & -1 & -3 & \big| \; 4 \end{array}$$

Hence,

$$f(x) = 2x^3 - 5x^2 - x + 10$$
$$= (2x^2 - x - 3)(x - 2) + 4$$

Using this result to find $f(2)$, we have $f(2) = 0 + 4 = 4$. Notice that the function value $f(2)$ is the same as the remainder we get when dividing $f(x)$ by $x - 2$.

This example can be generalized as follows:

THEOREM 1 (THE REMAINDER THEOREM)

If a polynomial $f(x)$ of degree $n > 0$ is divided by $x - r$, the remainder R is a constant and is equal to the value of the polynomial when r is substituted for x; that is, $f(r) = R$.

PROOF. Let $Q(x)$ be the quotient, so that, by the division algorithm, $f(x) = (x - r)Q(x) + R(x)$. Since the remainder $R(x)$ is of degree less than the divisor $x - r$, it must be constant and we will denote it as R. The equation $f(x) = (x - r)Q(x) + R$ holds for all x, and if we set $x = r$, we find that

$$f(r) = (r - r)Q(r) + R = 0 \cdot Q(r) + R = R$$

so that $f(r) = R$.

If the remainder $R = f(r)$ is zero, then the divisor $x - r$ and the quotient $Q(x)$ are factors of $f(x)$. Hence, we have a second theorem.

THEOREM 2 COROLLARY (THE FACTOR THEOREM)

If the value of $f(x)$ at the number r is zero, then $x - r$ is a factor of the polynomial $f(x)$ of degree $n > 0$; and, conversely, if $x - r$ is a factor, r is a zero; that is, $f(r) = 0$.

PROOF. Let $f(x) = Q(x)(x - r) + R$. If $f(r) = 0$, then $R = 0$ and $f(x) = Q(x)(x - r)$.

Conversely, if $x - r$ is a factor of $f(x)$, then $f(x) = Q(x)(x - r)$ and $f(r) = Q(r) \cdot 0 = 0$.

Thus, in order to find the value of the polynomial $f(x)$ at $x = r$ we can apply the remainder theorem together with synthetic division. The value of $f(r)$ is the same as the remainder we get when dividing $f(x)$ by $x - r$.

For example, if $f(x) = x^3 - 7x^2 + 3x - 2$, then $f(2)$ can be determined as follows:

$$\begin{array}{r|rrrr} 2 & 1 & -7 & 3 & -2 \\ & & 2 & -10 & -14 \\ \hline & 1 & -5 & -7 & -16 \end{array}$$

so that $f(2) = -16$.

EXAMPLES

1. Use synthetic division to find the quotient $Q(x)$ and the remainder R if $f(x) = 3x^3 - 6x^2 + x - 8$ is divided by $x - 3$. Also, find $f(3)$.

SOLUTION

$$
\begin{array}{r|rrrr}
3 & 3 & -6 & 1 & -8 \\
  &   & 9 & 9 & 30 \\
\hline
  & 3 & 3 & 10 & 22
\end{array}
$$

so that $Q(x) = 3x^2 + 3x + 10$, $R = 22$, and by the remainder theorem, $f(3) = 22$.

2 Find the remainder if $f(x) = 8x^4 - 28x^3 - 62x^2 + 7x + 15$ is divided by $x + 1$ and find $f(-1)$.

SOLUTION

$$
\begin{array}{r|rrrrr}
-1 & 8 & -28 & -62 & 7 & 15 \\
   &   & -8 & 36 & 26 & -33 \\
\hline
   & 8 & -36 & -26 & 33 & -18
\end{array}
$$

so that the remainder is -18 and, by the remainder theorem, $f(-1) = -18$. Note that in some cases it may be easier to compute $f(r)$ by direct substitution rather than by synthetic division, as shown in the following example.

3 Find the remainder if $f(x) = 3x^{73} - x^{37} - 1$ is divided by $x - 1$.

SOLUTION. By the remainder theorem, we know that the remainder $R = f(r) = f(1)$. Also, by substitution we have

$$f(1) = 3(1)^{73} - (1)^{37} - 1 = 3 - 1 - 1 = 1$$

That is, $f(1) = 1$ and the remainder R is 1.

4 Use synthetic division to determine the value of each of the following functions at the indicated value of x.
  a) $f(3)$ if $f(x) = 2x^3 - 6x^2 + x - 5$
  b) $f(2)$ if $f(x) = 3x^3 + 4x^2 - 10x - 15$

SOLUTION
a) Using synthetic division, we have

$$
\begin{array}{r|rrrr}
3 & 2 & -6 & 1 & -5 \\
  &   & 6 & 0 & 3 \\
\hline
  & 2 & 0 & 1 & -2
\end{array}
$$

Hence, $f(3) = -2$.

b)

2	3	4	−10	−15
		6	20	20
	3	10	10	5

Hence, $f(2) = 5$.

5 Show that $x - 6$ is a factor of $f(x) = x^3 - 6x^2 + x - 6$ and find $Q(x)$.

SOLUTION

6	1	−6	1	−6
		6	0	6
	1	0	1	0

Since $f(6) = 0$, $x - 6$ is a factor of $x^3 - 6x^2 + x - 6$, where $Q(x) = x^2 + 1$. That is, $x^3 - 6x^2 + x - 6 = (x - 6)(x^2 + 1)$.

PROBLEM SET 1

1 Find all the zeros of each of the following polynomial functions.
   a) $f(x) = -3x + 2$
   b) $f(x) = 3x^2 + x - 2$
   c) $f(x) = (x - 1)(x^2 - 3x + 2)$
   d) $f(x) = (x - 1)(x - 2)(x - 5)$
   e) $f(x) = x^4 - 16$
   f) $f(x) = (x - 1)^2 - 9$
   g) $f(x) = x^3 - 9x$
   h) $f(x) = x^4 - x^3$

2 Use long division to perform each of the following divisions.
   a) $5x^3 - 2x^2 + 3x - 4$ by $x - 3$
   b) $2x^4 + 3x^3 - 5x^2 + 2x - 1$ by $x + 1$
   c) $5x^5 - 3x^4 + 2x^3 + x^2 - 7x + 3$ by $x - 2$
   d) $2x^4 - 3x^3 + 5x^2 + 6x - 3$ by $x + 2$
   e) $-4x^6 - 5x^3 + 3x^2 + x + 7$ by $x - 1$
   f) $2x^4 + 3x^3 - 3x^2 + x - 1$ by $x + 4$

3 Use synthetic division to perform the divisions in Problem 2.

4 Use synthetic division to find $Q(x)$ and $f(r)$ so that $f(x) = (x - r)Q(x) + R$.
   a) $f(x) = 3x^3 + 6x^2 - 10x + 7$ and $r = 2$
   b) $f(x) = 3x^3 + 4x^2 - 7x + 16$ and $r = -1$
   c) $f(x) = 2x^3 - 5x^2 + 5x + 11$ and $r = \frac{1}{2}$
   d) $f(x) = -2x^4 + 3x^3 + 5x - 13$ and $r = 3$
   e) $f(x) = -3x^4 - 3x^3 + 3x^2 + 2x - 4$ and $r = -2$

5 If $f(x) = x^3 + 2x^2 - 13x + 10$, use synthetic division to determine $f(-5), f(-4), f(-3), f(-1), f(0), f(1), f(2), f(3), f(4)$, and $f(5)$. What are the factors of $f(x)$?

6 If $f(x) = 2x^3 + x^2 - 5x + 2$, use synthetic division to find $f(-2), f(-1), f(0), f(\frac{1}{2}), f(1)$, and $f(2)$. What are the factors of $f(x)$?

7 If $f(x) = 2x^3 - 6x^2 + x + k$, find k so that $f(3) = -2$.

8 Find k so that $x - 2$ is a factor of $f(x) = 3x^3 + 4x^2 + kx - 20$.

9 a) Show that $x - 1$ is a factor of $f(x) = 14x^{99} - 65x^{56} + 51$.
b) Show that $x + 4$ is a factor of $f(x) = 2x^2 + 13x + 20$.

10 For what values of n, where n is a positive integer, is each of the following true?
a) $x^n + a^n$ is divisible by $x + a$.
b) $x^n + a^n$ is divisible by $x - a$.
c) $x^n - a^n$ is divisible by $x + a$.
d) $x^n - a^n$ is divisible by $x - a$.

3 Graphs of Polynomial Functions of Degree Greater Than 2

Synthetic division often simplifies the process of graphing polynomial functions of degree greater than 2. For example, to graph the polynomial function $f(x) = 2x^3 - 7x^2 - 10x + 20$, we prepare a table of values of x and $f(x)$ by using synthetic division and then plot the points whose coordinates are $(x, f(x))$. From the table, we observe that the points $(x, f(x))$ to be plotted are $(-3, -67)$, $(-2, -4)$, $(-1, 21)$, $(0, 20)$, $(1, 5)$, $(2, -12)$, $(3, -19)$, $(4, -4)$, and $(5, 45)$ (Figure 1). Now the

Figure 1

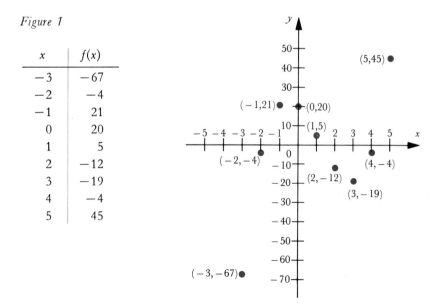

problem is how to draw the graph. Here we shall assume that the polynomial functions are *continuous* functions. Hence, if $a < b$ and $f(a) <$

$0 < f(b)$, then there is a zero, say c, such that $a < c < b$ and $f(c) = 0$ (Figure 2).

Figure 2

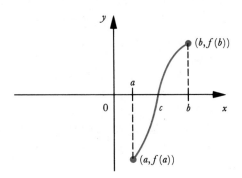

The question still remains whether the points we have already plotted are sufficient to give us a fairly accurate sketch of the graph, or whether there may be hidden "peaks" not shown thus far. We are *not* in a position to answer this question at present, but we can plot more points between those already located to get a rough sketch of the graph (Figure 3).

Figure 3

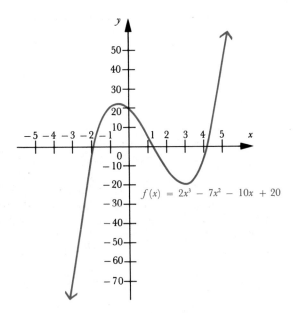

$f(x) = 2x^3 - 7x^2 - 10x + 20$

EXAMPLES

1 Sketch the graph of $f(x) = 2x^3 - 3x^2 - 12x + 13$.

SOLUTION. First we prepare a table of values of x and $f(x)$ by using synthetic division.

SECTION 3 GRAPHS OF POLYNOMIAL FUNCTIONS 253

x	$f(x)$
-3	-32
-2	9
-1	20
0	13
1	0
2	-7
3	4
4	45

These points whose coordinates $(x, f(x))$ appear in the table are plotted and the curve suggested by these points is drawn (Figure 4).

Figure 4

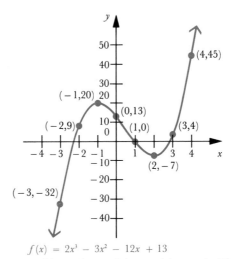

$f(x) = 2x^3 - 3x^2 - 12x + 13$

2 Sketch the graph of $f(x) = (x - 3)(x - 1)(x + 2)$. Use the graph to solve the inequality $(x - 3)(x - 1)(x + 2) < 0$.

SOLUTION. The x intercepts of the graph of f are 3, 1, and -2. Some additional points on the graph of f are given in the table. Notice that the portion of the graph below the x axis exhibits the values of x which

Figure 5

x	$f(x)$
-3	-24
-1	8
2	-4
4	18

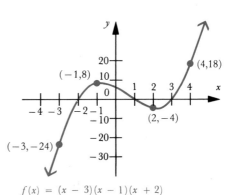

$f(x) = (x - 3)(x - 1)(x + 2)$

satisfy the inequality (Figure 5). Thus, the solution of the inequality is given by $\{x \mid (x - 3)(x - 1)(x + 2) < 0\} = (-\infty, -2) \cup (1, 3)$.

3 Sketch the graph of $f(x) = (x - 4)(x - 2)^2(x + 1)$. Use the graph to solve the inequality $(x - 4)(x - 2)^2(x + 1) > 0$.

SOLUTION. The x intercepts of the graph of f are 4, 2, and -1. Additional points on the graph of f are given on the table. Notice that the portion of the graph (Figure 6) above the x axis suggests the values of x which satisfy the inequality, so that $\{x \mid (x - 4)(x - 2)^2(x + 1) > 0\} = (-\infty, -1) \cup (4, \infty)$.

Figure 6

x	$f(x)$
-2	96
1	-6
3	-4
5	54

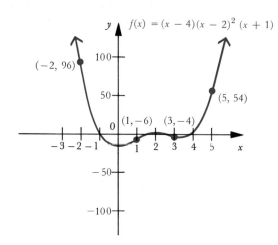

3.1 Rational Zeros

So far we have seen how to determine the zeros of first- and second-degree polynomial functions. This is a relatively easy process. However, it is more difficult to determine the zeros of polynomial functions of degree greater than 2. We shall now develop a method for determining zeros of polynomial functions with integral coefficients which are rational numbers. Such zeros are called *rational zeros*. To do so, we need the following property of integers: *Each composite integer has a unique set of prime factors.* For example, $24 = 2^3 \cdot 3$, and there is no other set of prime factors of 24.

THEOREM 1 (THE RATIONAL-ROOT THEOREM)

If $f(x) = a_n x^n + a_{n-1} x^{n-1} + \cdots + a_1 x + a_0$, and the coefficients are *integers* and p/q is a root, with p and q being integers and p/q reduced to lowest terms, then p is a divisor of a_0 and q is a divisor of a_n.

PROOF. (This proof is given for completeness sake, not because the student is expected to know it.) Since

$$a_n\left(\frac{p}{q}\right)^n + a_{n-1}\left(\frac{p}{q}\right)^{n-1} + \cdots + a_1\left(\frac{p}{q}\right) + a_0 = 0$$

it follows that

$$a_n p^n + a_{n-1} p^{n-1} q + \cdots + a_1 p q^{n-1} + a_0 q^n = 0$$

so that

(1) $\quad a_n p^n + a_{n-1} p^{n-1} q + \cdots + a_1 p q^{n-1} = -a_0 q^n$

or

(2) $\quad a_{n-1} p^{n-1} q + \cdots + a_1 p q^{n-1} + a_0 q^n = -a_n p^n$

Now, since both sides of Equation 1 are integers, p is a divisor of the left side and therefore, also of the right side; but p and q have no common factors since p/q is in lowest terms. Hence, every prime factor of p must be a factor of a_0 and the first part of the proof is finished.

Similarly, in Equation (2), q is a factor of the left side, hence of the right side. As before, q has no factors in common with p so q must be a divisor of a_n.

EXAMPLES

1 Find the rational roots of $f(x) = x^3 - 2x^2 - x + 2$.

SOLUTION. Assume that p/q represents the rational roots of $f(x) = 0$. By the rational-root theorem, p is a divisor of 2 and q is a divisor of 1. Hence, p can be any of the integers -1, 1, -2, or 2 and q can be either -1 or 1. Therefore, the possible values of p/q are -1, 1, -2, or 2. Using synthetic division to find $f(2)$ we have

$$\begin{array}{r|rrrr} 2 & 1 & -2 & -1 & 2 \\ & & 2 & 0 & -2 \\ \hline & 1 & 0 & -1 & 0 \end{array}$$

so that $f(2) = 0$ by the remainder theorem. Hence, 2 is a root and $x^3 - 2x^2 - x + 2 = (x - 2)(x^2 - 1)$. Since $Q(x) = x^2 - 1 = (x - 1)(x + 1)$, we see that the remaining roots are 1 and -1.

2 Find the rational roots of $f(x) = x^4 + 2x^3 - 7x^2 - 8x + 12$.

SOLUTION. By the rational-root theorem, if p/q is a rational root of $f(x) = 0$, then p is a divisor of 12 and q is a divisor of 1. Hence, p is any of the integers ±1, ±2, ±3, ±4, ±6, or ±12 and q is ±1, so that p/q is ±1, ±2, ±3, ±4, ±6, or ±12. Using synthetic division and the remainder theorem, we find that $f(1) = 0$, $f(-2) = 0$, $f(2) = 0$, and $f(-3) = 0$. Thus, the rational roots of $f(x) = x^4 + 2x^3 - 7x^2 - 8x + 12$ are 1, −2, 2, and −3.

3 Find the rational roots of $f(x) = 6x^3 - 11x^2 - 10x + 7$. Also sketch the graph of f.

SOLUTION.   Assume that p/q is a rational root of $f(x)$. Then p is a divisor of 7 and q is a divisor of 6. Hence, p can be any of the integers ±1 or ±7 and q can be ±1, ±2, ±3, or ±6. Thus, p/q can be ±1, ±$\frac{1}{2}$, ±$\frac{1}{3}$, ±$\frac{1}{6}$, ±7, ±$\frac{7}{2}$, ±$\frac{7}{3}$, and ±$\frac{7}{6}$. Using synthetic division and the remainder theorem, we find that $f(-1) = 0$, $f(\frac{1}{2}) = 0$, and $f(\frac{7}{3}) = 0$. Hence the rational roots of $f(x) = 6x^3 - 11x^2 - 10x + 7$ are −1, $\frac{1}{2}$, and $\frac{7}{3}$. The rational roots −1, $\frac{1}{2}$, and $\frac{7}{3}$ of f are the x intercepts of the graph of the function f (Figure 7).

Figure 7

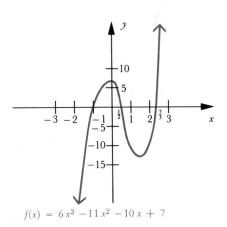

$f(x) = 6x^3 - 11x^2 - 10x + 7$

It is important to realize that the rational-root theorem has only very limited usefulness in determining roots. For example, the roots of the polynomial equation $x^2 - 2 = 0$ cannot be determined by the method used in the above examples since the roots of the equation are not rational numbers.

PROBLEM SET 2

1 Sketch the graph of each of the following polynomial functions. Also determine the x and y intercepts.
a) $f(x) = x^3 - 3x^2 + 4$
b) $f(x) = x^3 - 2x^2 - 5x + 6$
c) $f(x) = x(x - 1)(x + 2)$
d) $f(x) = 2x^3 - 3x^2 - 3x + 2$

e) $f(x) = (2x + 1)^4$
f) $f(x) = (x + 1)^3(x - 2)^2$

2 Sketch the graph of the following polynomial functions and then use the graph to solve the inequality. Write the solution in interval form.
  a) $f(x) = x(x - 1)(x + 2);\ x(x - 1)(x + 2) > 0$
  b) $f(x) = (x - 2)^2(x + 1);\ (x - 2)^2(x + 1) < 0$
  c) $f(x) = (x - 1)(x + 1)(x + 2);\ (x - 1)(x + 1)(x + 2) < 0$
  d) $f(x) = (x - 1)^3(x + 1)^2;\ (x - 1)^3(x + 1)^2 > 0$
  e) $f(x) = (x - 1)^2 x^3(x + 1);\ (x - 1)^2 x^3(x + 1) < 0$
  f) $f(x) = (x + 1)^3 x^2(x - 1);\ (x + 1)^3 x^2(x - 1) > 0$

3 Write down all rational numbers that might be roots of the following polynomial functions. Use synthetic division and the remainder theorem to test the possibilities to determine which of them are roots.
  a) $f(x) = 3x^3 - 7x^2 + 8x - 2$
  b) $f(x) = x^3 + 2x - 12$
  c) $f(x) = 5x^3 - 12x^2 + 17x - 10$
  d) $f(x) = x^3 - x^2 - 14x + 24$
  e) $f(x) = 2x^4 + 5x^3 + 2x^2 - 7x - 30$

4 Rational Functions

We have indicated in Section 2 that if $f(x)$ and $g(x)$ are polynomials, then $f(x) + g(x)$, $f(x) - g(x)$, and $f(x) \cdot g(x)$ are also polynomials. For the quotient $f(x)/g(x)$, the situation is different, since the quotient of two polynomials is not a polynomial nor can it necessarily be reduced to an equivalent expression which is a polynomial. The quotients of polynomials are called *rational expressions*. The functions such as $h(x) = x/(x + 1)$, $k(x) = (x + 2)/(x - 3)$, and $u(x) = (x^2 + 1)/(x^2 - 3)$ which are formed from rational expressions are called *rational functions*.

It should be noted here that the domain of a rational function $R(x) = f(x)/g(x)$ does not contain the zeros of the polynomial function g, since we cannot divide by zero. If a is a real number such that $f(a)$ and $g(a)$ are both zero, then the polynomials $f(x)$ and $g(x)$ have a common factor, $x - a$, because of the factor theorem. It follows that

$$R(x) = \frac{f(x)}{g(x)} = \frac{f_1(x)(x - a)}{g_1(x)(x - a)} = \frac{f_1(x)}{g_1(x)} \qquad x \neq a$$

For example,

$$R(x) = \frac{x^2 - 9}{x - 3} = \frac{(x - 3)(x + 3)}{x - 3} = x + 3 \qquad \text{for } x \neq 3$$

Observe that $R(3)$ is not defined, since 3 is not in the domain of R (Figure 1).

Figure 1

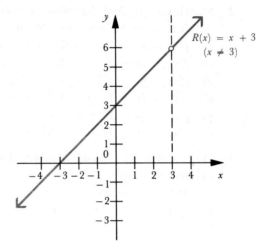

Next, consider a rational function $R(x) = f(x)/g(x)$, where $g(a) = 0$ and $f(a) \neq 0$. For example, for $R(x) = (x - 1)/(x - 2)$ the domain of R is $\{x \mid x \neq 2\}$ and the x intercept of R is 1, whereas the y intercept is $\frac{1}{2}$. Examining the behavior of the graph as x gets closer to 2 "from the right," we see that the corresponding value of $R(x)$ (see Table 1) becomes very large.

Table 1

x	5	4	3	$2\frac{1}{2}$	$2\frac{1}{8}$	$2\frac{1}{100}$	$2\frac{1}{1,000}$
$R(x)$	$\frac{4}{3}$	$\frac{3}{2}$	$\frac{2}{1}$	3	9	101	1,001

This situation results from the fact the denominator of the rational function R, $x - 2$, is becoming very close to zero, while the numerator of R, $x - 1$, is getting closer to 1 in value. Upon examining the behavior of the graph of R as x gets closer to 2 "from the left," the corresponding values of $R(x)$ (Table 2), which are negative numbers, become very large in absolute value.

Table 2

	5	-4	-3	-2	-1	0	1	$1\frac{1}{2}$	$1\frac{99}{100}$
$R(x)$	$\frac{6}{7}$	$\frac{5}{6}$	$\frac{4}{5}$	$\frac{3}{4}$	$\frac{2}{3}$	$\frac{1}{2}$	0	-1	-99

Both situations discussed in Tables 1 and 2 are described by saying that the graph of $R(x) = (x - 1)/(x - 2)$ is getting closer to the line $x = 2$

"asymptotically." The line $x = 2$ is called an *asymptote*; in this case, a *vertical asymptote* (Figure 2).

Figure 2

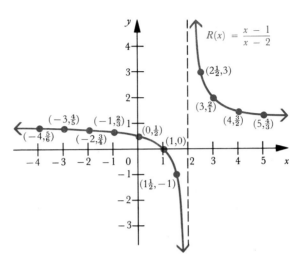

In general, if $g(a) = 0$ and $f(a) \neq 0$, then the graph of $h(x) = f(x)/g(x)$ has the line $x = a$ for a vertical asymptote. In addition, if $h(x) = f(x)/g(x)$ has as a vertical asymptote the line $x = a$, then the value of g at a is zero.

EXAMPLES

Find the domain, the x intercept, the y intercept, and the vertical and horizontal asymptotes, and sketch the graph of each of the following rational functions.

1 $f(x) = \dfrac{2}{x-3}$

Figure 3

x	$f(x)$
-2	$-\frac{2}{5}$
0	$-\frac{2}{3}$
2	-2
$2\frac{1}{2}$	-4
$2\frac{5}{6}$	-12
$2\frac{99}{100}$	-200
$3\frac{1}{100}$	200
$3\frac{1}{5}$	10
$3\frac{1}{2}$	4
4	2
5	1

CHAPTER 5 ROOTS OF POLYNOMIALS AND COMPLEX NUMBERS

$f(x) = \dfrac{2}{x-3}$

SOLUTION. The domain of f is $\{x \mid x \neq 3\}$. There is no x intercept and the y intercept is $-\tfrac{2}{3}$. A vertical asymptote occurs when $x - 3 = 0$, that is, when $x = 3$. Observe that as x gets closer to 3, $|f(x)|$ becomes very large. That is, as x gets closer to 3 "from the left," the values of $f(x)$ become "very large" negative numbers, and as x gets closer to 3 "from the right," the values of $f(x)$ become very large positive numbers (Figure 3).

2 $f(x) = \dfrac{x}{x+1}$

SOLUTION. The domain of f is $\{x \mid x \neq -1\}$. The x intercept is found by letting $x/(x+1) = 0$, so that $x = 0$ is the x intercept. The y intercept is $f(0) = 0/(0+1) = 0$. The vertical asymptote is $x + 1 = 0$ or, equivalently, $x = -1$. Note that as x approaches -1 from the left, the corresponding values of $f(x)$ are becoming positively larger numbers, whereas as x approaches -1 from the right, the corresponding values of $f(x)$ are becoming negatively "large" (Figure 4).

Dividing x by $x + 1$, $f(x) = x/(x+1)$ can also be expressed as $f(x) = 1 - 1/(x+1)$. Now, as x increases in the positive direction, the rational expression $1/(x+1)$ approaches 0. Also, as x decreases in the negative

Figure 4

x	$f(x)$
-3	$\tfrac{3}{2}$
-2	2
$-1\tfrac{1}{2}$	3
$-1\tfrac{1}{100}$	101
$-\tfrac{99}{100}$	-99
$-\tfrac{1}{2}$	-1
0	0
1	$\tfrac{1}{2}$
2	$\tfrac{2}{3}$
3	$\tfrac{3}{4}$

direction, $1/(x+1)$ again approaches 0. Thus, in both cases, $f(x) = 1 - 1/(x+1)$ approaches 1 and we say that $f(x)$ approaches 1 "asymptotically." This behavior can also be observed from the table associated with Figure 4. The line $y = 1$ is an asymptote, in this case, a *horizontal asymptote*.

3 $f(x) = \dfrac{x-1}{x^2-4}$

SECTION 4 RATIONAL FUNCTIONS 261

SOLUTION. The domain of f is $\{x \mid x \neq \pm 2\}$. The x intercept is determined by $(x - 1)/(x^2 - 4) = 0$, or $x = 1$ since it is true in general that if $a/b = 0$, then $a = 0$ and $b \neq 0$. The y intercept is $f(0) = (0 - 1)/(0^2 - 4) = \frac{1}{4}$. The vertical asymptotes are those values of x such that $x^2 - 4 = 0$, or $x = \pm 2$. Rewriting $f(x) = (x - 1)/(x^2 - 4)$ as

$$f(x) = \frac{\dfrac{1}{x} - \dfrac{1}{x^2}}{1 - \dfrac{4}{x^2}}$$

we can observe that as $|x|$ becomes infinitely large, $f(x)$ approaches 0 asymptotically so that $y = 0$ is a horizontal asymptote. When $x < -2$, $x - 1 < 0$ and $x^2 - 4 > 0$, so that $f(x) < 0$. When $-2 < x < 1$, $x - 1 < 0$ and $x^2 - 4 < 0$, so that $f(x) > 0$. When $1 < x < 2$, $x - 1 > 0$ and $x^2 - 4 < 0$, so that $f(x) < 0$. When $x > 2$, $x - 1 > 0$ and $x^2 - 4 > 0$, so that $f(x) > 0$ (Figure 5).

Figure 5

x	$f(x)$
-3	$-\frac{4}{5}$
$-2\frac{1}{2}$	$-\frac{14}{9}$
$-\frac{1}{2}$	$\frac{2}{5}$
0	$\frac{1}{4}$
1	0
$1\frac{1}{2}$	$-\frac{2}{7}$
$2\frac{1}{4}$	$\frac{20}{17}$
3	$\frac{2}{5}$

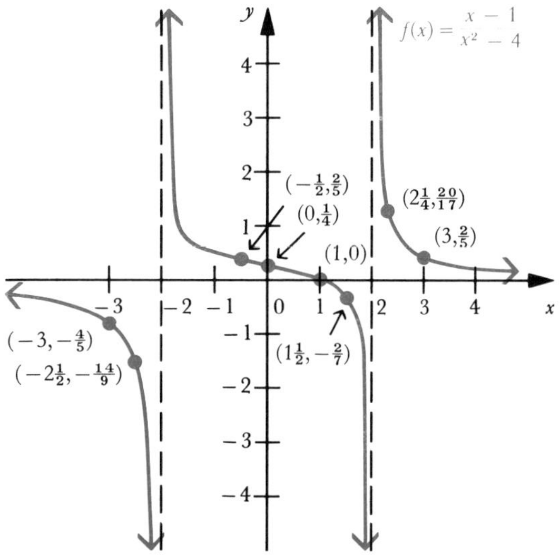

4 $f(x) = \dfrac{x^2}{x + 3}$

SOLUTION. The domain of f is $\{x \mid x \neq -3\}$. The x intercept is 0, the y intercept is 0, and the vertical asymptote is $x = -3$. $f(x) = x^2/(x + 3)$ can be expressed equivalently as $f(x) = x - 3 + 9/(x + 3)$, so that as x becomes infinitely larger, $f(x)$ approaches asymptotically the line $y = x - 3$. (Why?) Since $x^2 \geq 0$, $f(x)$ will be positive whenever $x + 3 > 0$, or $x > -3$, and $f(x)$ will be negative whenever $x + 3 < 0$, or $x < -3$ (Figure 6).

Figure 6

x	f(x)
−7	−12¼
−6	−12
−5	−12½
−2	4
−1	½
0	0
1	¼
3	1½
5	3⅜

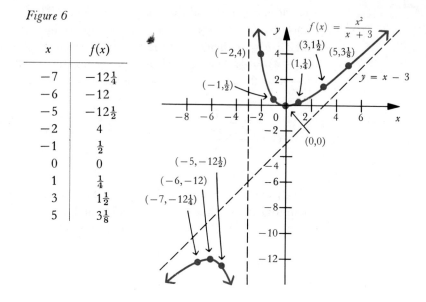

5 $f(x) = \dfrac{x^2 + 1}{x^2 + 2}$

SOLUTION. The domain of f is $\{x \mid x \in R\}$. The y intercept is

$$f(0) = \frac{0^2 + 1}{0^2 + 2} = \frac{1}{2}$$

There are no x intercepts, since $x^2 + 1 = 0$ has no real solutions. Also, there are no vertical asymptotes, since $x^2 + 2 = 0$ has no real solutions. Rewriting

$$f(x) = \frac{x^2 + 1}{x^2 + 2} \quad \text{as} \quad f(x) = 1 - \frac{1}{x^2 + 2}$$

Figure 7

x	f(x)
−4	17/18
−3	10/11
−2	5/6
−1	2/3
0	1/2
1	2/3
2	5/6
3	10/11
4	17/18

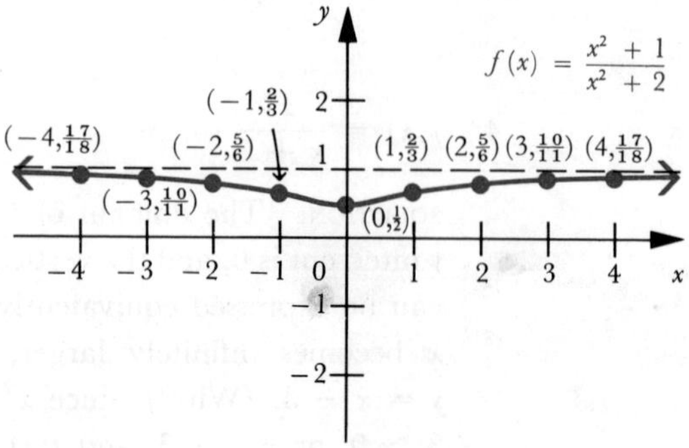

we see that $f(x)$ approaches 1 asymptotically as $|x|$ becomes infinitely large. (Figure 7). Thus $y = 1$ is a horizontal asymptote.

6 $f(x) = \dfrac{x^2 - 3x - 4}{x^2 + x - 6}$

SOLUTION. The domain of f is $\{x \mid x^2 + x - 6 \neq 0\} = \{x \mid x \neq 2 \text{ or } x \neq -3\}$. The x intercepts are -1 and 4, the roots of $x^2 - 3x - 4 = 0$, and the y intercept is $f(0) = -4/-6 = \frac{2}{3}$. The vertical asymptotes are $x = 2$ and $x = -3$, the roots of $x^2 + x - 6 = 0$. Expressing

$$f(x) = \dfrac{x^2 - 3x - 4}{x^2 + x - 6} \quad \text{as} \quad f(x) = \dfrac{1 - \dfrac{3}{x} - \dfrac{4}{x^2}}{1 + \dfrac{1}{x} - \dfrac{6}{x^2}}$$

we see that $f(x)$ approaches 1 asymptotically as $|x|$ becomes infinitely large (Figure 8).

Figure 8

x	$f(x)$
-5	$2\frac{4}{7}$
-2	$-1\frac{1}{2}$
-1	0
0	$\frac{2}{3}$
1	$\frac{3}{2}$
3	$-\frac{2}{3}$
4	0
5	$\frac{1}{4}$

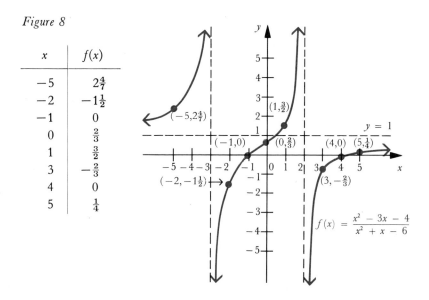

PROBLEM SET 3

Find all the horizontal and vertical asymptotes and sketch the graph of each of the following rational functions.

1 $f(x) = \dfrac{-2}{x - 3}$

2 $f(x) = \dfrac{x^2}{x^2 - 4}$

3 $f(x) = \dfrac{3x - 4}{2x^2 - 5x}$

4 $f(x) = \dfrac{x^2 - 9}{x^2 - x}$

5 $f(x) = \dfrac{x^2 + 1}{2x - 4}$

6 $f(x) = \dfrac{2x^2 - 3x}{x^2 - 2x - 3}$

7 $f(x) = \dfrac{x - \frac{1}{2}}{x^2 - 1}$

8 $f(x) = \dfrac{4}{x^2 + 3}$

9 $f(x) = \dfrac{5x}{x-2}$ **10** $f(x) = \dfrac{5x+2}{3x-9}$

11 Find all the horizontal and vertical asymptotes and sketch the graph of the function $f(x) = (x^2 + 1)/(x^2 - x)$. Does the graph of f intersect the horizontal asymptotes?

5 Complex Numbers

In Chapter 4, we discussed the solution of a quadratic equation, an equation that has an equivalent form $ax^2 + bx + c = 0$, where $a \neq 0$ and a, b, and c are real numbers. The solutions of such equations are real if the discriminant, $b^2 - 4ac$, is nonnegative. If the discriminant is negative, it is impossible to solve the quadratic equation *in the real number system*. For example, $\{x \mid x^2 + 1 = 0\} = \emptyset$ if we restrict ourselves to R. Our purpose here is to extend the real number system to a "new" system—the *complex number system*—which contains numbers that satisfy the equations such as $x^2 + 1 = 0$. An ordered pair of real numbers (a, b), which shall be denoted as $a + bi$, is called a *complex number*. The algebra of complex numbers is defined in terms of the algebra of real numbers as follows.

Assume that $z_1 = a_1 + b_1 i$ and $z_2 = a_2 + b_2 i$ are complex numbers. Then

1 *Equality.* $z_1 = z_2$ if and only if $a_1 = a_2$ and $b_1 = b_2$.

2 *Addition.* $z_1 + z_2 = (a_1 + a_2) + (b_1 + b_2)i$.

3 *Multiplication.* $z_1 \cdot z_2 = (a_1 a_2 - b_1 b_2) + (a_1 b_2 + a_2 b_1)i$.

For $z = a + bi$, a is called the *real part* of z and b is called the *imaginary part* of z. It would perhaps be better to identify a as the "non-i part" and b as the "i part" of the complex number, but the choice of words "real part" and "imaginary part" is accepted today for historical reasons. Hence, two complex numbers are equal if and only if the real and imaginary parts are equal; the real part of the sum of two complex numbers is the sum of the real parts and the imaginary part of the sum of two complex numbers is the sum of the imaginary parts.

The set of complex numbers will be denoted by C, so that in set notation, we have $C = \{a + bi \mid a, b \in R\}$.

If the imaginary part of a complex number is 0, we will consider the number to be real; hence, $a + 0i$ will be considered to be the real number a; and, in this sense, R can be considered to be a proper subset of C.

EXAMPLES

Find the sum and product of each of the following pairs of complex numbers and identify the real and imaginary parts of each result.

1 i, i

2 $5 + 6i, 9 + 3i$

3 $4 - 2i, -3 + i$

SOLUTION

1 $i + i = 2i$, so that the real part of $2i$ is 0 and the imaginary part is 2.

$$\begin{aligned} i \cdot i = i^2 &= (0 + 1i)(0 + 1i) \\ &= (0 \cdot 0 - 1 \cdot 1) + (0 \cdot 1 + 1 \cdot 0)i \quad \text{(why?)} \\ &= -1 \end{aligned}$$

Hence, the symbol i in $z = a + bi$ is sometimes denoted as $\sqrt{-1}$. The real part of -1 is -1 and the imaginary part is 0.

2 $(5 + 6i) + (9 + 3i) = 14 + 9i$; 14 is the real part and 9 is the imaginary part of $14 + 9i$.

$$\begin{aligned} (5 + 6i)(9 + 3i) &= (45 - 18) + (54 + 15)i \\ &= 27 + 69i \end{aligned}$$

27 is the real part and 69 is the imaginary part of $27 + 69i$.

3 $(4 - 2i) + (-3 + i) = 1 - i$; here, 1 is the real part and -1 is the imaginary part of $1 - i$.

$$\begin{aligned} (4 - 2i)(-3 + i) &= (-12 + 2) + (6 + 4)i \\ &= -10 + 10i \end{aligned}$$

The real part is -10 and the imaginary part is 10.

We have in Example 1 one of the properties which does not hold in R but holds in C: If $x \in R$, $x^2 \geq 0$, whereas it is possible to have $z \in C$ such that $z^2 < 0$ ($i^2 = -1$).

If the order properties of R did hold in C, then, by trichotomy,

$$i = 0 \quad \text{or} \quad i < 0 \quad \text{or} \quad i > 0$$

But, if $i = 0$, then $i \cdot i = 0$ implies that $-1 = 0$; hence, $i \neq 0$. On the other hand, if $i > 0$, then $i^2 > 0$ implies that $-1 > 0$ since $i^2 = -1$; hence, $i \not> 0$. Finally, if $i < 0$, then $i^2 > 0$, so that $-1 > 0$ since $i^2 = -1$; hence, $i \not< 0$. Consequently, trichotomy does *not* hold in C; that is, the order relation which exists in R does not exist in C.

The domain of functions can be extended to the set of complex numbers. For example, consider the function f whose domain is the set of complex numbers and whose rule of correspondence is $f(z) = 3z + 1$; then $f(i) = 3i + 1$ and $f(1 - i) = 3(1 - i) + 1 = 4 - 3i$.

5.1 Properties of Addition and Multiplication

Since we defined the operations of addition and multiplication on C in terms of the corresponding operations on R, it is not surprising that the properties of addition and multiplication on C are the same as the properties of addition and multiplication on R. (These properties are listed in Chapter 1.)

Assume that $z_1, z_2, z_3 \in C$; then the following properties hold.

1 CLOSURE OF ADDITION AND MULTIPLICATION:

i $z_1 + z_2 \in C$
ii $z_1 \cdot z_2 \in C$

2 COMMUTATIVITY OF ADDITION AND MULTIPLICATION:

i $z_1 + z_2 = z_2 + z_1$
ii $z_1 z_2 = z_2 z_1$

3 ASSOCIATIVITY OF ADDITION AND MULTIPLICATION:

i $z_1 + (z_2 + z_3) = (z_1 + z_2) + z_3$
ii $z_1 \cdot (z_2 \cdot z_3) = (z_1 \cdot z_2) \cdot z_3$

4 DISTRIBUTIVE PROPERTIES:

i $z_1 \cdot (z_2 + z_3) = (z_1 \cdot z_2) + (z_1 \cdot z_3)$
ii $(z_1 + z_2) \cdot z_3 = (z_1 \cdot z_3) + (z_2 \cdot z_3)$

5 IDENTITY:

i There exists $0 \in C$ such that $z + 0 = 0 + z = z$ for every $z \in C$.
ii There exists $1 \in C$ such that $z \cdot 1 = 1 \cdot z = z$ for every $z \in C$.

6 INVERSE:

i If $z \in C$, then there exists $-z \in C$ such that $z + (-z) = (-z) + z = 0$.
ii If $z \in C$, $z \neq 0$, then there exists $z^{-1} \in C$ such that $z \cdot z^{-1} = z^{-1} \cdot z = 1$.

PROOF OF PROPERTY 6i: Suppose that $z = a + bi$. Then, by letting $-z = -a - bi$, we have $(a + bi) + (-a - bi) = 0 + 0i = 0$. The proof of 6ii is given in Example 6.

For the proofs of the other properties, see Problem 6 on page 271.

5.2 Subtraction of Complex Numbers

If $z_1, z_2 \in C$, then $z_1 - z_2$, that is, the *difference* of z_1 and z_2, is defined as $z_1 - z_2 = z_1 + (-z_2)$.

EXAMPLES

1 If $z_1 = 7 + 4i$ and $z_2 = 3 + 5i$, then

$$z_1 - z_2 = (7 + 4i) - (3 + 5i) = (7 + 4i) + (-3 - 5i)$$
$$= 4 - i$$

2 If $z_1 = 4 - 5i$ and $z_2 = -5 + 7i$, then

$$z_1 - z_2 = (4 - 5i) - (-5 + 7i) = (4 - 5i) + (5 - 7i)$$
$$= 9 - 12i$$

5.3 Division of Complex Numbers

The *conjugate* of a complex number $z = a + bi$, written $\bar{z}$ (read z conjugate, or z bar for short), is defined as $\bar{z} = a - bi$. If $z_1 = a_1 + b_1 i$, $z_1 \neq 0$, and $z_2 = a_2 + b_2 i$, then the *quotient* $z_2 \div z_1$ is given by

$$\frac{z_2}{z_1} = \frac{z_2 \cdot \bar{z}_1}{z_1 \cdot \bar{z}_1}$$

or, equivalently,

$$\frac{a_2 + b_2 i}{a_1 + b_1 i} = \frac{(a_2 + b_2 i)(a_1 - b_1 i)}{(a_1 + b_1 i)(a_1 - b_1 i)}$$
$$= \frac{(a_2 a_1 + b_2 b_1) + (a_1 b_2 - b_1 a_2)i}{a_1^2 + b_1^2}$$

Notice that the result is a complex number of the form $x + yi$, with

$$x = \frac{a_2 a_1 + b_2 b_1}{a_1^2 + b_1^2} \quad \text{and} \quad y = \frac{a_1 b_2 - b_1 a_2}{a_1^2 + b_1^2}$$

and the denominator, $a_1^2 + b_1^2$, is a real number.

EXAMPLES

1. Find the conjugate of each of the following complex numbers.
   a) $3 + 3i$
   b) -4
   c) $5i$
   d) $-1 - i$

   SOLUTION
   a) $3 - 3i$
   b) -4
   c) $-5i$
   d) $-1 + i$

2. Show that $z\bar{z} \in R$ and $z\bar{z} \geq 0$.

   SOLUTION. Let $z = a + bi$; then $\bar{z} = a - bi$, so that

   $$\begin{aligned} z\bar{z} &= (a + bi)(a - bi) \\ &= (a^2 + b^2) + (ab - ab)i \\ &= (a^2 + b^2) + (0 \cdot i) \\ &= a^2 + b^2 \in R \end{aligned}$$

   Since $a^2 \geq 0$ and $b^2 \geq 0$ for any $a, b \in R$, we have $a^2 + b^2 \geq 0$.

3. Write each of the following quotients in the form $a + bi$.
   a) $\dfrac{1}{3 - 2i}$
   b) $\dfrac{1 + i}{1 - i}$
   c) $\dfrac{6 + 9i}{1 - 2i}$

   SOLUTION
   a) $\dfrac{1}{3 - 2i} = \dfrac{1}{3 - 2i} \cdot \dfrac{3 + 2i}{3 + 2i} = \dfrac{3 + 2i}{9 + 4} = \dfrac{3}{13} + \dfrac{2i}{13}$
   b) $\dfrac{1 + i}{1 - i} = \dfrac{1 + i}{1 - i} \cdot \dfrac{1 + i}{1 + i} = \dfrac{(1 + i)^2}{2} = \dfrac{2i}{2} = i$
   c) $\dfrac{6 + 9i}{1 - 2i} = \dfrac{6 + 9i}{1 - 2i} \cdot \dfrac{1 + 2i}{1 + 2i} = \dfrac{-12 + 21i}{5} = \dfrac{-21}{5} + \dfrac{21}{5}i$

5.4 Geometric Representation of Complex Numbers

Each ordered pair of real numbers (a, b) can be associated with the complex number $z = a + bi$, and each complex number $z = a + bi$ can be associated with the ordered pair of real numbers (a, b). Because of this one-to-one correspondence between the set of complex numbers and the set of ordered pairs of real numbers, we use the points in the plane associated with the ordered pairs of real numbers to represent the complex numbers. For example, the ordered pairs $(2, -3)$, $(5, 2)$, and (e, π) are used to represent complex numbers $z_1 = 2 - 3i$, $z_2 = 5 + 2i$, and $z_3 = e + \pi i$, respectively, as points in the plane (Figure 1). The plane on which the complex numbers are represented is called the *complex plane;* the horizontal axis (x axis) is called the *real axis,* and the vertical axis (y axis) is called the *imaginary axis.* Thus, complex numbers of the form $z = bi$ are represented by points of the form $(0, b)$, that is, by points on the imaginary axis, whereas complex numbers of the form $z = a$ are represented by points of the form $(a, 0)$, that is, by points on the real axis.

Figure 1

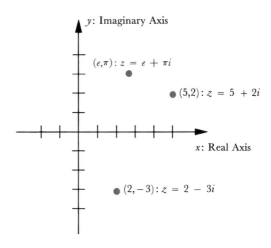

5.5 The Modulus of a Complex Number

If $z = a + bi$, then the *absolute value* or *length* or *modulus* of z, written $|z|$, is defined by

$$|z| = \sqrt{a^2 + b^2}$$

The modulus of $z = a + bi$ is the distance between the origin and the point (a, b) (Figure 2). Notice that $z \cdot \bar{z}$ is a positive real number and that $|z| = \sqrt{z \cdot \bar{z}}$ (see Problem 11).

Figure 2

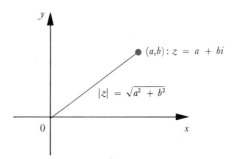

EXAMPLES

1 Let $z = 1 + \sqrt{3}\, i$. Find $|z|$ and show that $z\bar{z} = |z|^2$.

SOLUTION

$$|z| = \sqrt{1^2 + (\sqrt{3})^2} = \sqrt{1 + 3} = \sqrt{4} = 2$$

also

$$z\bar{z} = (1 + \sqrt{3}\,i)(1 - \sqrt{3}\,i) = 1^2 + (\sqrt{3})^2$$
$$= 4 = |z|^2 \quad \text{(Figure 3)}$$

Figure 3

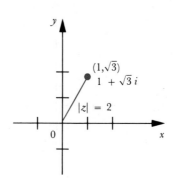

2. Let $z_1 = 4 + 3i$ and $z_2 = \sqrt{3} - i$. Find
   a) $|z_1|$ b) $|z_2|$ c) $|z_1 z_2|$

SOLUTION

a) $|z_1| = \sqrt{16 + 9} = \sqrt{25} = 5$
b) $|z_2| = \sqrt{(\sqrt{3})^2 + (-1)^2} = \sqrt{4} = 2$
c) $|z_1 z_2| = |(4 + 3i)(\sqrt{3} - i)|$
$= |(4\sqrt{3} + 3) + (3\sqrt{3} - 4)i|$
$= \sqrt{(4\sqrt{3} + 3)^2 + (3\sqrt{3} - 4)^2} = 10$

3. Let z_1 and z_2 be complex numbers; then $|z_1 z_2| = |z_1|\,|z_2|$.

PROOF. (This is a generalization of our result in Example 2.)

$$|z_1 z_2| = \sqrt{(z_1 z_2)(\overline{z_1 z_2})} = \sqrt{(z_1 \bar{z_1})(z_2 \bar{z_2})} \quad \text{(Problem 4)}$$
$$= \sqrt{|z_1|^2 |z_2|^2} = |z_1|\,|z_2|$$

4. Show that if z is a complex number, $z + \bar{z}$ is a real number.

SOLUTION. Let $z = a + bi$. Then $\bar{z} = a - bi$, so that

$$z + \bar{z} = (a + a) + (b - b)i$$
$$= 2a \in R$$

5. Let $z = 3 + 5i$. Find $\bar{z}$, $\bar{z} + z$, and $(z - \bar{z})/i$.

SOLUTION. Since $z = 3 + 5i$, $\bar{z} = 3 - 5i$, so that

$$\bar{z} + z = (3 - 5i) + (3 + 5i) = 6$$

and

$$\frac{z - \bar{z}}{i} = \frac{(3 + 5i) - (3 - 5i)}{i} = \frac{10i}{i} = 10$$

6 If $z \neq 0$, show that $1/z = z^{-1}$, that is, show that $1/z$ is the multiplicative inverse of z using the definition of division.

SOLUTION. If $z = a + bi \neq 0$, then

$$\frac{1}{z} = \frac{1}{a + bi} \cdot \frac{a - bi}{a - bi} = \frac{a - bi}{a^2 + b^2}$$

so that

$$z \cdot \frac{1}{z} = (a + bi) \cdot \frac{a - bi}{a^2 + b^2} = \frac{a^2 + b^2}{a^2 + b^2} = 1$$

PROBLEM SET 4

1 Perform the indicated operations, writing the answer in the form $a + bi$.
 a) $(5 + 6i) + (3 - 4i) - (2 + 7i)$
 b) $(6 + i)(5 - 3i)$
 c) $(3 + 2i)(2 - 5i)(1 - i)$
 d) $\dfrac{3 - 2i}{2 - i}$
 e) $\dfrac{(5 - i)^2}{1 + i}$
 f) $i^{27} + i^5 - i^9$
 g) $\dfrac{(-1 - \sqrt{2}\, i)^3}{5 - 3i}$
 h) $(3 + \sqrt{2}\, i)(3 - \sqrt{2}\, i)$
 i) $\dfrac{6}{7i}$
 j) $(3 + i)^3$
 k) $(4 + 3i)^{-1}$
 l) $(3 + 2i) - (7 - 3i)$
 m) $\dfrac{7 - 3i}{5i}$
 n) $\dfrac{4}{3 + 2i}$
 o) $\dfrac{35 + 8i}{4i}$
 p) $\dfrac{3 - i}{2 + 3i}$
 q) $\dfrac{-3}{5i^3}$
 r) $(2 - 3i)^4$
 s) $(2 - 7i)(2 + 3i)$
 t) $4i^{-13}$
 u) $\dfrac{2i}{(1 + i)^4}$
 v) $i^{18} - 3i^7$
 w) $\dfrac{3 + 5i}{4 - 3i}$

2 Find the real numbers x and y to satisfy each of the following equations.
 a) $x - 3 + 2iy = 8i$
 b) $3x - y + ix - 2iy = 6 - 3i$
 c) $3x + 2yi = 6 + 11i$

3. Find $\bar{z}$, the real part of z, the imaginary part of z, and $1/z$ for each of the following numbers.
   a) $z = 2 + \sqrt{3}\, i$
   b) $z = 1 - \tfrac{1}{2}i$
   c) $z = (2 + \sqrt{3}\, i)(1 - \tfrac{1}{2}i)^2$
   d) $z = 2i$

4. Prove each of the following statements.
   a) If $\bar{z} = z$, then z is real.
   b) $z + \bar{z} = 0$ if and only if the real part of z is 0. both ways
   c) $\overline{z_1 + z_2} = \bar{z}_1 + \bar{z}_2$
   d) $\overline{z_1 z_2} = \bar{z}_1 \bar{z}_2$
   e) $\overline{z_1/z_2} = \bar{z}_1/\bar{z}_2,\ z_2 \neq 0$
   f) $\bar{\bar{z}} = z$

5. Let z_1 and z_2 be complex numbers. Show that

$$\operatorname{Re}\left(\frac{z_1}{z_1 + z_2}\right) + \operatorname{Re}\left(\frac{z_2}{z_1 + z_2}\right) = 1$$

where $\operatorname{Re}(z)$ indicates the real part of z.

6. Assume the properties of addition and multiplication on R that are listed in Chapter 1.
   a) Prove that C is closed under addition and multiplication.
   b) Prove that commutativity of addition and multiplication on C.
   c) Prove the associativity of addition and multiplication on C.
   d) Prove the distributive properties on C.
   e) Prove the identity properties of addition and multiplication on C.

7. Let f be a function whose domain is the set of complex numbers and whose rule of correspondence is given by $f(z) = z^2 + 5z + i$. Find
   a) $f(2 - i)$
   b) $f(2 + i)$
   c) $f(1 - i)$
   d) $f(1 + i)$

8. We are given the set $T = \{1, -1, i, -i\}$.
   a) Construct a *multiplication* table as shown:

$\bullet$	1	-1	i	$-i$
1				
-1				
i				
$-i$				

   b) Is the set T closed under multiplication?
   c) Is multiplication on T commutative?

d) Is multiplication on T associative?
e) What is the identity element?
f) What are the reciprocals of each of the elements of T? (Two elements are reciprocals of each other if their product equals 1.)

9 If $z_1 = 3 + 4i$ and $z_2 = -3 + 2i$, find
   a) $|z_1|$ b) $|z_2|$ c) $|z_1 z_2|$ d) $|z_1| |z_2|$
   e) $\left|\dfrac{z_1}{z_2}\right|$ f) $\dfrac{|z_1|}{|z_2|}$

10 Let $z_1 = 1 + 2i$ and $z_2 = -3 + 5i$; find and simplify the following.
   a) $|z_1 + z_2|^2$ b) $|z_1 - z_2|^2$
   c) $|z_1 + z_2|^2 + |z_1 - z_2|^2$

11 Prove that $|z| = \sqrt{z \cdot \bar{z}}$ for any complex number z.

6 Complex Zeros of Polynomial Functions

We have seen (Chapter 4) that a polynomial function with real coefficients does not always have real number zeros. In particular, a quadratic polynomial function $f(x) = ax^2 + bx + c$, $a, b, c \in R$, $a \neq 0$, has real zeros if and only if the discriminant, $b^2 - 4ac$, is nonnegative. These real roots can be found by the quadratic formula

$$x = \frac{-b \pm \sqrt{b^2 - 4ac}}{2a}$$

If the zeros of the quadratic polynomial function are not real numbers, that is, if $b^2 - 4ac < 0$, the zeros are complex numbers and the quadratic formula can still be used. For example, the zeros of $f(x) = 2x^2 + x + 1$ can be found by using the quadratic formula; hence,

$$x = \frac{-1 \pm \sqrt{1 - 8}}{4} = \frac{-1 \pm \sqrt{-7}}{4} = \frac{-1 \pm \sqrt{7}i}{4}$$

The polynomial function $f(x) = x^3 - 6x^2 + 13x - 10$ can be factored as $f(x) = (x - 2)(x^2 - 4x + 5)$, so that after using the quadratic formula, we find that the zeros of f are 2, $2 - i$, and $2 + i$, and

$$f(x) = (x - 2)[x - (2 - i)][x - (2 + i)]$$

In fact, exactly the same formula holds if a, b, and c are allowed to be any complex numbers, instead of real numbers. In general, all polynomial functions are factorable as the product of linear factors in the complex domain, even those of degree greater than 2.

Notice that

$$2x^2 + x + 1 = 2\left(x - \frac{-1+\sqrt{7}i}{4}\right)\left(x - \frac{-1-\sqrt{7}i}{4}\right)$$

This result follows as a corollary of the *fundamental theorem of algebra*, whose proof depends upon methods generally considered beyond the scope of this text. We shall state the theorem without proof.

6.1 Fundamental Theorem of Algebra

If $f(x)$ is a polynomial of degree $n \geq 1$ with complex coefficients, then there is a complex number r such that $f(r) = 0$.

Assuming this fundamental theorem, we can prove the following theorems on factoring polynomials.

THEOREM 1 (THE FACTORIZATION THEOREM)

If $f(x) = a_n x^n + a_{n-1} x^{n-1} + \cdots + a_1 x + a_0$ and $a_n \neq 0$, n a positive integer, then

$$f(x) = a_n(x - r_1)(x - r_2) \cdots (x - r_n)$$

where the numbers r_j are complex numbers.

PROOF. By the fundamental theorem of algebra, $f(x) = 0$ has a root r_1, so that by the factor theorem (see Theorem 2, page 248)

$$f(x) = (x - r_1)Q_1(x)$$

$Q_1(x)$ is a polynomial of degree $n - 1$, so that it has a zero r_2 if $n - 1 \geq 1$, and, as above,

$$Q_1(x) = (x - r_2)Q_2(x)$$

so that

$$f(x) = (x - r_1)(x - r_2)Q_2(x)$$

where $Q_2(x)$ has degree $n - 2$. Continuing the process, we get

$$f(x) = (x - r_1)(x - r_2) \cdots (x - r_n)Q_n(x)$$

where $Q_n(x)$ has degree 0, that is, $Q_n(x)$ is a constant. Multiplying out this expression for $f(x)$, it is seen that the coefficient of x^n is Q_n; hence $Q_n = a_n$ and the theorem is proved.

THEOREM 2 (COROLLARY)

If $f(x)$ is a polynomial function of degree n, $n \neq 0$, then $f(x) = 0$ has exactly n roots, one for each of the n linear factors. (Note that some of the n roots may be the same.)

PROOF. By the factorization theorem,

$$f(x) = a_n(x - r_1)(x - r_2) \cdots (x - r_n)$$

Clearly, the numbers $r_1, r_2, \ldots, r_n$ are roots of $f(x) = 0$. Moreover, if $f(r) = 0$ for $r \neq r_i$, $i = 1, \ldots, n$, then

$$f(x) = a_n(x - r_1)(x - r_2) \cdots (x - r_n)(x - r)$$

so that the degree of $f(x)$ is $n + 1$ (why?), which contradicts our assumption that $f(x)$ is a polynomial of degree n.

Notice that the roots need not be distinct. For example, $x^2 - 4x + 4 = 0$ has two roots, both of which are equal to 2 and we say that $f(x) = x^2 - 4x + 4$ has $x = 2$ as a double root. In general, if

$$f(x) = (x - r)^s Q(x) \quad \text{and} \quad Q(r) \neq 0$$

we say that r is a zero of *multiplicity* s. For example,

$$f(x) = (x - 1)^2(x - 2)$$

has $x = 1$ as a root of multiplicity 2 and $x = 2$ as a root of multiplicity 1.

THEOREM 3 (CONJUGATE ROOT THEOREM)

If a polynomial of degree n, $n \neq 0$, has *real* coefficients, and $f(z_0) = 0$, where $z_0 = a + bi$, then $f(\overline{z_0}) = 0$.

PROOF. Let $f(z) = a_n z^n + a_{n-1} z^{n-1} + \cdots + a_1 z + a_0$ be a polynomial with real coefficients. Since $z_0 = a + bi$ is a zero of $f(z)$, then

$$f(z_0) = a_n z_0^n + a_{n-1} z_0^{n-1} + \cdots + a_1 z_0 + a_0 = 0$$

so that

$$\overline{f(z_0)} = \overline{a_n z_0^n + a_{n-1} z_0^{n-1} + \cdots + a_1 z_0 + a_0} = \overline{0} = 0$$

However,

$$\overline{a_n z_0^n + a_{n-1} z_0^{n-1} + \cdots + a_1 z_0 + a_0}$$
$$= \overline{a_n z_0^n} + \overline{a_{n-1} z_0^{n-1}} + \cdots + \overline{a_1 z_0} + \overline{a_0}$$

because the conjugate of the sum of two complex numbers is the same as the sum of the conjugate of two complex numbers (see Problem Set 4,

Problem 4c). Also,

$$\overline{a_n z_0^n} = \overline{a_n} \overline{z_0}^n,$$
$$\overline{a_{n-1} z_0^{n-1}} = \overline{a_{n-1}} \overline{z_0}^{n-1}, \ldots, \overline{a_1 z_0} = \overline{a_1}\, \overline{z_0}$$

and

$$\overline{a_n} \overline{z_0}^n = \overline{a_n} \overline{z_0}^n,$$
$$\overline{a_{n-1}}\, \overline{z_0}^{n-1} = \overline{a_{n-1} z_0^{n-1}}, \ldots, \overline{a_1} \overline{z_0} = \overline{a_1 z_0}$$

Since the conjugate of the real number is the real number itself, we have $\overline{a_0} = a_0, \overline{a_1} = a_1, \ldots, \overline{a_n} = a_n$. Hence,

$$f(\overline{z_0}) = a_n \overline{z_0}^n + a_{n-1} \overline{z_0}^{n-1} + \cdots + a_1 \overline{z_0} + a_0 = \overline{f(z_0)} = 0$$

That is, $\overline{z_0}$ is also a zero of $f(z)$.

For example, the polynomial function $f(x) = x^2 - 4x + 5$ has two zeros, one of which is the complex number $2 + i$. By the conjugate-root theorem, $2 - i$ is also a zero of $f(x) = x^2 - 4x + 5$, as shown by the following multiplication.

$$[x - (2 + i)][x - (2 - i)]$$
$$= x^2 - (2 + i)x - (2 - i)x + (2 + i)(2 - i)$$
$$= x^2 - (2 + i + 2 - i)x + 2^2 + 1^2$$
$$= x^2 - 4x + 5$$

EXAMPLES

1 Find the third-degree polynomial function f that has 1 and $1 - 2i$ as zeros.

SOLUTION. By the conjugate-root theorem, $1 + 2i$ is the third zero; so that

$$f(x) = (x - 1)[x - (1 - 2i)][x - (1 + 2i)]$$
$$= (x - 1)(x^2 - 2x + 5)$$
$$= x^3 - 3x^2 + 7x - 5$$

2 Form a polynomial $f(x)$ which has the following numbers as zeros: $-\frac{1}{2}, 1 + i$, and 1 as a double zero.

SOLUTION. Since $1 + i$ is a root of $f(x) = 0$, it follows from the conjugate-root theorem that $1 - i$ is also a root; therefore,

$$f(x) = (x + \tfrac{1}{2})(x - 1)^2[x - (1 + i)][x - (1 - i)]$$

has the given roots. Simplifying the equation, we get

$$f(x) = x^5 - \tfrac{7}{2}x^4 + 5x^3 - \tfrac{5}{2}x^2 - x + 1$$

3   Determine the multiplicity of the zeros of the polynomial

$$f(x) = x^4 - 4x^3 + 5x^2 - 4x + 4$$

SOLUTION.   Using synthetic division, the polynomial can be factored as $x^4 - 4x^3 + 5x^2 - 4x + 4 = (x - 2)^2(x + i)(x - i)$, so that 2 is a double zero and i and $-i$ are each zeros of multiplicity 1.

PROBLEM SET 5

1   Show that the equation

$$\frac{1}{x-3} + \frac{1}{x-2} - \frac{x-2}{x-3} = 0$$

has no solutions. Why doesn't this contradict the fundamental theorem of algebra?

2   Determine whether the given numbers are zeros of the given polynomial functions. If they are, find their multiplicities.
a) $f(x) = x^4 - x^3 - 18x^2 + 52x - 40$, $x = 2$
b) $f(x) = 4x^6 + 4x^5 + 9x^4 + 8x^3 + 6x^2 + 4x + 1$, $x = i$
c) $f(x) = 9x^4 - 12x^3 + 13x^2 - 12x + 4$, $x = \tfrac{2}{3}$

3   Use the roots to write each of the following polynomial functions in factored form.
a) $\{(x, f(x)) \mid f(x) = 2x^2 - x - 2\}$
b) $\{(x, f(x)) \mid f(x) = x^3 - 7x + 6\}$

4   a)   Given that $-1 + i$ is a zero of $f(x) = x^4 + 2x^3 - 4x - 4$, find all other zeros of $f(x)$.
b)   Given that i is a double zero of $f(x) = 2x^6 + x^5 + 2x^3 - 6x^2 + x - 4$, find all other zeros of $f(x)$.

5   Find polynomials having the following numbers as their zeros.
a) $2, 3 - i$ b) $2, 2, 1 + i, 1 - i$
c) $1 - 3i, 1 - 3i, 1 + 3i, 1 + 3i$ d) $i, i, 0, 1, 2i$
e) $2, 2, \tfrac{1}{2}(-1 + i\sqrt{3})$

REVIEW PROBLEM SET

1   Find the zeros of each of the following functions.
a) $f(x) = 3x^2 - 7x + 2$

b) $f(x) = 4x^2 - 16$
c) $f(x) = (x-1)(x-2)(x-3)$
d) $f(x) = (2x-1)(x-3)(2x+5)$
e) $f(x) = x^4 - 2x^3 + 8x - 16$
f) $f(x) = (3x-1)(x-2)(x^2 - 4x + 3)$

2 Use synthetic division to find the quotient $Q(x)$ and the remainder R when $D(x)$ is the divisor.
a) $f(x) = 3x^4 - 5x^3 - 4x^2 + 3x - 2$ and $D(x) = x - 2$
b) $f(x) = x^4 - 16$ and $D(x) = x + 2$
c) $f(x) = x^4 + 8x^3 - 8x^2 + 24x - 8$ and $D(x) = x + 1$
d) $f(x) = 3x^5 - 3x^2 + 6x + 7$ and $D(x) = x - 1$
e) $f(x) = 2x^4 + 5x^3 + 3x^2 + 10x - 7$ and $D(x) = x + 2$

3 Find k in each of the following cases.
a) If $x + \frac{1}{3}$ is a factor of $f(x) = 3x^4 - 2x^3 - 7x^2 + kx + 3$
b) If $x - 2$ is a factor of $f(x) = kx^3 + 3x^2 - 5x + 18$
c) If $x - 3$ is a factor of $f(x) = x^3 + 3x^2 - 12x - k$
d) If $x - k$ is a factor of $f(x) = x^2 - 3x + 2$

4 Use the remainder theorem to find each of the following values.
a) $f(-3)$ and $f(2)$ if $f(x) = 3x^3 - 5x^2 + 5x + 7$
b) $f(-2)$ and $f(1)$ if $f(x) = 81x^4 + 3x^3 - 2x^2 + 5x + 7$
c) $f(2)$ and $f(3)$ if $f(x) = 5x^3 - 11x^2 - 14x + 5$

5 Express each of the following numbers in the form of $a + bi$.
a) $-4(2 - \sqrt{-5})$
b) $\sqrt{-3}(3 + \sqrt{-6})$
c) $(3 + \sqrt{-5})^2$
d) $(6 + \sqrt{5}\,i)^2$
e) $(12 + \sqrt{-36}) + (8 - \sqrt{-16})$
f) $(5 + \sqrt{-16}) + (6 + \sqrt{-9})$
g) $(6 + 3i) + (8 + 5i)$
h) $(-7 + 6i) + (3 + 2i)$
i) $(3 + \sqrt{7}\,i)(5 + \sqrt{7}\,i)$
j) $(6 + 9i)(2 + 8i)$
k) $(-3 - 8i)(2 - 11i)$
l) $(5 - 2i)(5 + 2i)$
m) $(3 - 2\sqrt{7}\,i)(3 + 2\sqrt{7}\,i)$
n) $(5\sqrt{6} + 6\sqrt{5}\,i)(5\sqrt{6} - 6\sqrt{5}\,i)$
o) $(1 + i)(2 + 3i)(4 + 5i)$
p) $\dfrac{5 + i}{2 + 3i}$
q) $\dfrac{9 - 7i}{8 + 5i}$
r) $\dfrac{6 - 7i}{8 - 9i}$
s) $\dfrac{-3 - 8i}{2 - 11i}$
t) $\dfrac{\sqrt{3} + \sqrt{2}\,i}{2\sqrt{3} + 3\sqrt{2}\,i}$

6 Form the polynomial functions of the lowest possible degree with integral coefficients having the given numbers as their zeros.
a) $-1, 2,$ and -2
b) $-2, 3,$ and -4

c) 1, a double zero, and 3 $(x-1)^2$
d) 2, a double zero, and -2, a double zero $(x+2)^2(x-2)^2$
e) -1 and $1+i$
f) $-1+i$ and 2 and -3
    $-1-i$

7   Determine whether the given numbers are zeros of the given polynomial functions. If they are, find their multiplicities.
a) $f(x) = x^3 - 5x^2 + 8x - 4$; $x = 1$
b) $f(x) = 3(x-4)^4$; $x = 4$
c) $f(x) = 6x^4 + 25x^3 + 38x^2 + 25x + 6$; $x = -1$
d) $f(x) = ix^3 - 3x^2 - 4$; $x = i$
e) $f(x) = x^3 - 2x^2 + 4x - 8$; $x = -2i$

8   Sketch the graph of each of the following functions. Also determine the x and y intercepts.
a) $f(x) = (x-1)(x-2)(2x+1)$
b) $f(x) = x^3 - 2x^2 - 4x + 8$
c) $f(x) = x^3 - 8$
d) $f(x) = x^4 + 3x^3 - 8x^2 - 24x$
e) $f(x) = (3x+1)(2x-3)(x+5)$

9   Sketch the graph of the following polynomial functions and then use the graph to solve the corresponding inequalities. Write the solution in interval form.
a) $f(x) = (x-1)^2(x+1)$; $(x-1)^2(x+1) < 0$
b) $f(x) = (2-x)^2(1+x)(2+x)$; $(2-x)^2(1+x)(2+x) > 0$
c) $f(x) = (x-1)(x+1)(x+3)$; $(x-1)(x+1)(x+3) > 0$
d) $f(x) = -2(x-1)^2(x+2)$; $-2(x-1)^2(x+2) \le 0$
e) $f(x) = 3(x+1)^2(x-2)(x-3)$; $3(x+1)^2(x-2)(x-3) < 0$

10   Find all the asymptotes and sketch the graph of each of the following rational functions.   p. 264 #7
a) $f(x) = \dfrac{3}{x+1}$
b) $f(x) = \dfrac{x^2-4}{x^2+x}$
c) $f(x) = \dfrac{-4}{x-1}$
d) $f(x) = \dfrac{2x-3}{3x^2-4}$
e) $f(x) = \dfrac{x^2+3}{3x-9}$
f) $f(x) = \dfrac{x-\frac{1}{3}}{x^2-4}$

11   Let $z = a + bi$. Show that each of the following equations are true.
a) $\left|\dfrac{a+bi}{a-bi}\right|^2 = 1$
b) $\left|\dfrac{(a+bi)^3}{(a-bi)^2}\right| = |z|$

12   Which of the following statements is true? If the statement is false, give a counterexample.
a) $z + \bar{z} = 0$ if and only if Re $(z) = 0$, where Re (z) denotes the real part of z.

b) $z + 1/z$ is real if and only if Im $z = 0$ or $|z| = 1$, where Im z denotes the imaginary part of z.

c) If Im $z \neq 0$, then $z/(1 + z^2)$ is real if and only if $|z| = 1$.

13   Find all the rational zeros of each of the following polynomial functions.

a) $f(x) = x^3 - 3x^2 - 6x - 2$
b) $f(x) = x^3 + 5x^2 + 8x + 6$
c) $f(x) = x^3 - 8x^2 + 5x + 14$
d) $f(x) = x^3 - 2x^2 - 7x - 4$
e) $f(x) = x^4 + 4x^3 + 8x^2 + 8x + 3$
f) $f(x) = 8x^5 - 27x^2$

CHAPTER 6

Exponential and Logarithmic Functions

CHAPTER 6

Exponential and Logarithmic Functions

1 Introduction

Our objective in this chapter is to investigate two other types of functions, exponential functions and logarithmic functions. Logarithmic functions will evolve as "inverses" of exponential functions. Consequently, we will begin with a study of the concept of *inverse functions*.

2 Inverse Functions

Suppose that f is the function defined by $f(x) = x + 2$ and g is the function defined by $g(x) = x - 2$. Now observe what happens when we "apply" these two functions in succession.

First, we apply f, then g, to the number 2; that is,

$$2 \xrightarrow{f} f(2) \xrightarrow{g} g[f(2)]$$

and since $f(2) = 4$, we have $g[f(2)] = g(4) = 4 - 2 = 2$.

In general, $g[f(x)]$ is the result obtained when we first "apply" f to an element x and then "apply" g to the result. The function defined by $y = g[f(x)]$ is called a *composition* of f by g. Thus, for $f(x) = x + 2$ and $g(x) = x - 2$, we have $f[g(x)] = f(x - 2) = (x - 2) + 2 = x$ and $g[f(x)] = g(x + 2) = (x + 2) - 2 = x$. Here $f[g(x)] = g[f(x)] = x$.

Let us consider a second example, where $f(x) = 2x - 5$ and $g(x) = (x + 5)/2$; then

$$f[g(x)] = f\left(\frac{x+5}{2}\right) = 2\left(\frac{x+5}{2}\right) - 5 = x + 5 - 5 = x$$

and

$$g[f(x)] = g(2x - 5) = \frac{(2x - 5) + 5}{2} = \frac{2x}{2} = x$$

Once again we have a situation in which $f[g(x)] = g[f(x)] = x$.

It is not true in general that $f[g(x)] = g[f(x)]$. For example, if $f(x) = x^2$ and $g(x) = x + 1$, we have

$$f[g(x)] = f(x + 1) = (x + 1)^2 = x^2 + 2x + 1$$

and

$$g[f(x)] = g(x^2) = x^2 + 1$$

Since $x^2 + 2x + 1 \neq x^2 + 1$, $f[g(x)] \neq g[f(x)]$ for these two functions f and g.

The functions f and g defined in the first two examples above are called "invertible functions." These examples can be generalized as follows.

DEFINITION

Let f and g be two functions so related that $f[g(x)] = x$ for every element x in the domain of g, and $[g(f(x)] = x$ for every element x in the domain of f. Then f and g are said to be invertible, and each is said to be the inverse of the other. In symbols, we write $g = f^{-1}$ or $f = g^{-1}$.

Thus, in the first example, we write $f^{-1}(x) = x - 2$ for $f(x) = x + 2$, or we can write $g^{-1}(x) = x + 2$ for $g(x) = x - 2$. Similarly, in the second example, we can write $f^{-1}(x) = (x + 5)/2$ when $f(x) = 2x - 5$ or we write $g^{-1}(x) = 2x - 5$ when $g(x) = (x + 5)/2$. In general,

$$f[f^{-1}(x)] = x \quad \text{and} \quad f^{-1}[f(x)] = x$$

EXAMPLES

In each of the following examples show that f and g are invertible and, in fact, are inverses of each other.

1. $f(x) = x^3$ and $g(x) = \sqrt[3]{x}$

   SOLUTION. Let us examine $f[g(x)]$ and $g[f(x)]$.

   $$f[g(x)] = f(\sqrt[3]{x}) = (\sqrt[3]{x})^3 = x \quad \text{for all numbers in the domain of } g$$

   and

   $$g[f(x)] = g(x^3) = \sqrt[3]{x^3} = x \quad \text{for all numbers } x \text{ in the domain of } f.$$

   Since $f[g(x)] = g[f(x)] = x$, it follows that $f = g^{-1}$ and $g = f^{-1}$.

2. $f(x) = 5x + 9$ and $g(x) = (x - 9)/5$

   SOLUTION

   $$f[g(x)] = f\left(\frac{x-9}{5}\right) = 5\left(\frac{x-9}{5}\right) + 9 = x - 9 + 9 = x$$

   for all numbers x in the domain of g

and

$$g[f(x)] = g(5x + 9) = \frac{5x + 9 - 9}{5} = x$$

for all numbers x in the domain of f

Hence, $f = g^{-1}$ and $g = f^{-1}$.

2.1 Existence of Inverse Functions

Consider the two functions

$$f = \{(1, 2), (3, 1), (4, 3)\} \text{ and } g = \{(1, 2), (3, 2), (4, 5)\}$$

Notice that for each member of the domain of f, there is one and only one corresponding member of the range. Similarly, for each member of the range of f there is one and only one corresponding member of the domain of f, that is,

$$1 \longleftrightarrow 2$$
$$3 \longleftrightarrow 1$$
$$4 \longleftrightarrow 3$$

Whereas, for g, 2 in the range corresponds to more than one member of the domain, that is,

We say that f is a "one-to-one" function, whereas g is not one to one.

In general, a function f is one to one if each member of the range of the function corresponds to one and only one member of the domain; that is, if $x_1 \neq x_2$, then $f(x_1) \neq f(x_2)$.

We can use the graph of f to determine whether a function f is one to one. For example, the function f defined by $f(x) = x^3$ is one to one since for each y there is one and only one x; that is, each of all possible horizontal lines intersects the curve no more than once (Figure 1).

Figure 1

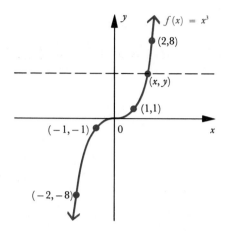

However, the function g defined by $g(x) = x^2$ is not one to one, since any horizontal line above the x axis intersects the curve twice (Figure 2).

Figure 2

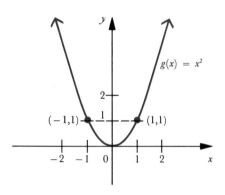

The following property characterizes the existence of the inverse function.

PROPERTY

If a function f is one to one, then f has an inverse, and, conversely, if f has an inverse, then f is one to one.

Thus, we can examine the graph of the function to determine whether or not a function has an inverse. For example, $f(x) = x^3$ has an inverse, whereas $g(x) = x^2$ has no inverse.

If $f(x) = 3x + 2$, then we can find the value of f^{-1} at a given number y by solving the equation $y = f(x)$, (that is, $y = 3x + 2$) for x. Thus, we have $x = \frac{1}{3}y - \frac{2}{3}$; or, in other words, $f^{-1}(y) = \frac{1}{3}y - \frac{2}{3}$. The letter that we use to denote a number in the domain of the inverse function is of no importance whatsoever, so this last equation can be written as $f^{-1}(u) = \frac{1}{3}u - \frac{2}{3}$ or $f^{-1}(t) = \frac{1}{3}t - \frac{2}{3}$, or even $f^{-1}(x) = \frac{1}{3}x - \frac{2}{3}$, and it will still define the same function f^{-1}. Now we will verify that $f[f^{-1}(x)] =$

$f^{-1}[f(x)] = x$ as follows:

$$f[f^{-1}(x)] = f(\tfrac{1}{3}x - \tfrac{2}{3}) = 3(\tfrac{1}{3}x - \tfrac{2}{3}) + 2 = x - 2 + 2 = x$$

and

$$f^{-1}[f(x)] = f^{-1}(3x + 2) = \tfrac{1}{3}(3x + 2) - \tfrac{2}{3} = x + \tfrac{2}{3} - \tfrac{2}{3} = x$$

Geometrically, we should note that the graph of f^{-1} is obtained from the graph of f by reflecting ("flipping") the graph across the line $y = x$ (Figure 3).

Figure 3

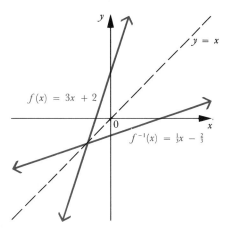

EXAMPLES

Given the function f, find f^{-1} and graph f^{-1} and f on the same coordinate system.

1 $f(x) = 3x - 4$

SOLUTION. f is a one-to-one function (Figure 4); hence, f^{-1} exists. If we let $y = 3x - 4$, then $x = (y + 4)/3$ results from solving for x in terms of y; therefore, $f^{-1}(y) = (y + 4)/3$, or $f^{-1}(x) = (x + 4)/3$. This can be verified as follows:

$$f[f^{-1}(x)] = f\left(\frac{x+4}{3}\right) = 3\left(\frac{x+4}{3}\right) - 4 = x$$

and

$$f^{-1}[f(x)] = f^{-1}(3x - 4) = \frac{3x - 4 + 4}{3} = x$$

After graphing $f(x) = 3x - 4$ and $f^{-1}(x) = (x + 4)/3$ on the same coordinate system, we can observe that the graph of the inverse function

f^{-1} can be obtained by reflecting the graph of f across the line $y = x$ (Figure 4).

Figure 4

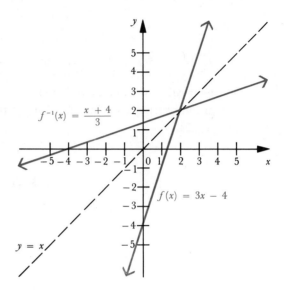

2 $f(x) = x^2, x \in [0, \infty)$

SOLUTION. The graph of $f(x) = x^2$, $x \in [0, \infty)$, indicates that f is a one-to-one function (Figure 5); hence, f^{-1} exists. If we let $y = x^2$, then, solving for x in terms of y, we obtain $x = \sqrt{y}$, since $x \in [0, \infty)$, so that $f^{-1}(y) = \sqrt{y}$, or $f^{-1}(x) = \sqrt{x}$. This can be verified as follows:

$$f(f^{-1}(x)) = f(\sqrt{x}) = (\sqrt{x})^2 = x$$

and

$$f^{-1}(f(x)) = f^{-1}(x^2) = \sqrt{x^2} = x \quad \text{for } x \in [0, \infty)$$

The graph of f^{-1} is a reflection of the graph of f across $y = x$ (Figure 5).

Figure 5

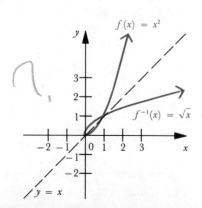

3 $f(x) = x^3 + 1$

SOLUTION. The graph of $f(x) = x^3 + 1$ indicates that f is a one-to-one function (Figure 6); hence, f^{-1} exists. If we let $y = x^3 + 1$, then $x^3 = y - 1$ or $x = \sqrt[3]{y - 1}$, so that $f^{-1}(y) = \sqrt[3]{y - 1}$ or $f^{-1}(x) = \sqrt[3]{x - 1}$. This can be verified as follows:

$$f(f^{-1}(x)) = f(\sqrt[3]{x - 1}) = (\sqrt[3]{x - 1})^3 + 1 = x$$

and

$$f^{-1}(f(x)) = f^{-1}(x^3 + 1) = \sqrt[3]{x^3 + 1 - 1} = \sqrt[3]{x^3} = x$$

The graph of f^{-1} is a reflection of the graph of f across $y = x$ (Figure 6).

Figure 6

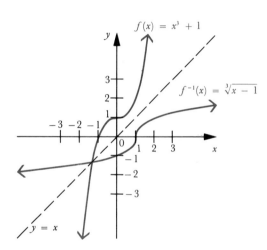

4 $f(x) = -2/x, \; x \in (-\infty, 0)$

Figure 7

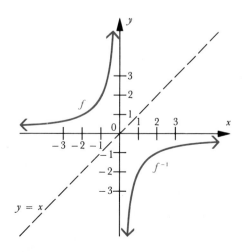

SOLUTION. The graph of $f(x) = -2/x$ indicates that f^{-1} exists (Figure 7). Let $y = -2/x$; then $x = -2/y$. This implies that $f^{-1}(y) = -2/y$ or $f^{-1}(x) = -2/x$. The graph of f^{-1} can be obtained by a reflection of the graph of f across $y = x$ (Figure 7).

5 Prove that every increasing function has an inverse.

PROOF. Assume that $y = f(x)$ is an increasing function. If x_1 and x_2 are different members of the domain of f and $x_1 < x_2$, then, since f is increasing, $f(x_1) < f(x_2)$ (Figure 8). This means that no two different ordered pairs of the function f have the same second members. In other words, each member of the range is the image of one and only one member of the domain; that is, f is one to one, from which we can conclude by the property that f^{-1} exists.

Figure 8

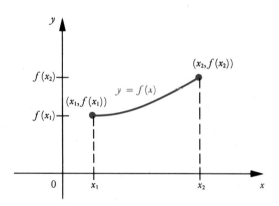

PROBLEM SET 1

1 Assume that $f(x) = x^2$ and $g(x) = 2x + 1$.
   a) Form $f[g(x)]$.
   b) Form $g[f(x)]$.
   c) Does $f[g(x)]$ equal $g[f(x)]$?

2 Verify that $g = f^{-1}$ for each of the following pairs of functions by using the composition of the functions.
   a) $f = \{(5, -3), (6, -2), (3, 4)\}$, $g = \{(-3, 5), (-2, 6), (4, 3)\}$
   b) $f(x) = 2x - 3$ and $g(x) = \dfrac{x + 3}{2}$
   c) $f(x) = 2x - 1$ and $g(x) = \dfrac{x + 1}{2}$
   d) $f(x) = \dfrac{1}{x}$ and $g(x) = \dfrac{1}{x}$, $x \neq 0$

3 Graph each pair of functions in Problem 2 on the same coordinate system to observe that the graph of f^{-1} is a reflection of the graph of f across the line $y = x$.

4 Find the inverse function f^{-1} of each of the following functions. Verify that $f(f^{-1}(x)) = x$ and $f^{-1}(f(x)) = x$.
   a) $f(x) = 1 - 3x$
   b) $f(x) = x^3 - 1$
   c) $f(x) = \dfrac{3}{x}$
   d) $f(x) = \tfrac{2}{3}x + \tfrac{1}{2}$

5 Let $f(x) = x^3 + 4$.
   a) Is f one to one?
   b) Does f^{-1} exist? If so, find it.
   c) If f^{-1} exists, find $f^{-1}(4)$, $f^{-1}(0)$, and $f^{-1}(-4)$.

6 Let $f(x) = -3x + 7$.
   a) Is f a decreasing function?
   b) Does f^{-1} exist? If so, find it.
   c) If f^{-1} exists, find $f^{-1}(0)$ and $f^{-1}(1)$ and sketch the graph of f^{-1}.

7 Let $f(x) = x^3 + x$.
   a) Is f an increasing function?
   b) Does f^{-1} exist?

8 Prove that every decreasing function has an inverse.

9 Examine the graphs of each of the following functions to determine whether or not f^{-1} exists. If f^{-1} exists, find it and graph f^{-1} on the same coordinate system as f.
   a) $f(x) = \dfrac{5}{x}$
   b) $f(x) = -2x + 5$
   c) $f(x) = |2 - x|$
   d) $f(x) = 1 + \dfrac{7x}{2}$
   e) $f(x) = -2x^3 + 1$
   f) $f(x) = \sqrt{2 - x}$
   g) $f(x) = -2x + 3$
   h) $f(x) = |x| - x$

3 Exponential Functions and Their Properties

Mathematical applications frequently require the use of functional relationships in which the variable occurs as an exponent such as in $f(x) = 2^x$ and $f(x) = (\tfrac{1}{3})^x$. These two functions are examples of *exponential functions*.

DEFINITION

If b is a positive number, then the function $f(x) = b^x$, $b \neq 1$, defines an *exponential function* with *base b*.

As usual, we take the domain of the function to be as much as possible of the real numbers. We know from Chapter 1 that this includes the positive and negative integers and, indeed, all the rational numbers. We here assume that b^x may be adequately defined for all irrational numbers as well (Example 1 displays an intuitive reason for this). Hence, the domain of this exponential function is the entire set of real numbers. We claim that the range is the set of positive real numbers, and Example 1 will justify this claim.

EXAMPLES

1   Find the domain and the range and sketch the graph of each of the following exponential functions.

a) $f(x) = 3^x$

SOLUTION. $f(x) = 3^x$ has base 3. The domain of $f(x) = 3^x$, as with all exponential functions, is the set of the real numbers R. The range of f is the set of positive real numbers, since the graph of f never touches or goes below the x axis. Using the table in Figure 1 we can locate some specific points and graph the function.

Figure 1

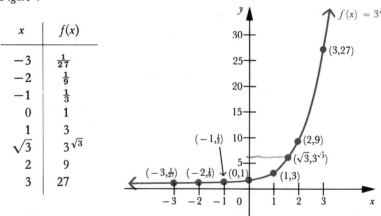

In doing so, we are assuming that all exponential functions are continuous over the entire real line (see Chapter 3, Section 3). Thus, we are assuming that a point such as $(\sqrt{3}, 3^{\sqrt{3}})$ is on the graph of $f(x) = 3^x$, although we shall defer a precise definition of the expression $3^{\sqrt{3}}$ to a later course in calculus. Notice that the graph of $f(x) = 3^x$ goes up to the right as x gets larger, so that $f(x)$ is an *increasing function* and hence has an inverse.

b) $f(x) = (\frac{1}{4})^x$

SOLUTION. $f(x) = (\frac{1}{4})^x$ has base R. The domain is the set of real numbers R, and the range is the set of positive real numbers. The graph of $f(x) = (\frac{1}{4})^x$ goes down to the right as x gets larger, so that the function $f(x) = (\frac{1}{4})^x$ is a *decreasing function* (Figure 2) and, consequently, has an inverse.

Figure 2

x	$f(x)$
-2	16
-1	4
0	1
1	$\frac{1}{4}$
2	$\frac{1}{16}$

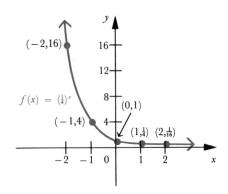

2. Let f be an exponential function with base b. Show that $f(u + v) = f(u) \cdot f(v)$ for any real number u and v.

   SOLUTION. Since f is an exponential function with base b, $f(x) = b^x$ and $f(u + v) = b^{u+v}$. By the rules of exponents, $b^{u+v} = b^u b^v$, so that $f(u + v) = b^{u+v} = b^u \cdot b^v = f(u) \cdot f(v)$.

3. If the graph of an exponential function contains the point $(3, 64)$, what is the base?

   SOLUTION. Since the function f is exponential, it is of the form $f(x) = b^x$. $(3, 64)$ on the graph of f implies that $64 = b^3$, so that $4^3 = b^3$, or $b = 4$ is the base.

4. For what value of x does $11^x = (121)^{2/3}$ hold?

   SOLUTION. We can solve this equation if we can express the number $(121)^{2/3}$ as a power of 11. Since

   $$(121)^{2/3} = (11^2)^{2/3} = 11^{4/3} \quad \text{(why?)}$$

   we have

   $$11^x = (121)^{2/3} = (11)^{4/3}$$

   Therefore, since $y = 11^x$ is one to one, the exponents are equal, so that $x = \frac{4}{3}$.

5. Solve $2^{2x+2} - 9(2^x) + 2 = 0$.

   SOLUTION. $2^{2x+2} - 9(2^x) + 2 = 0$ can be written as $(2^x)^2 2^2 - 9(2^x) + 2 = 0$. If we let $y = 2^x$, we get $4y^2 - 9y + 2 = 0$, that is, $(4y - 1)(y - 2) = 0$, so that $y = \frac{1}{4}$ or $y = 2$. Therefore, $2^x = \frac{1}{4}$ or $2^x = 2$. Hence, $x = -2$ or $x = 1$.

6. Suppose that the percentage of television viewers who remember what product is regularly advertised during the daily 6:30 P.M. news program

is given as $y = 60 - 20(3^{-x})$, where x is the number of times they have seen the program.

a) Calculate the percentages that correspond to $x = -1, 1, 2, 3, 4,$ and 5.
b) Use the result of part a to plot the graph of this modified exponential function.

SOLUTION

a) The percentages of television viewers are computed as follows:

If $x = -1$, then $y = 60 - 20(3^1) = 0$
If $x = 1$, then $y = 60 - 20(3^{-1}) = \frac{160}{3}$
If $x = 2$, then $y = 60 - 20(3^{-2}) = \frac{520}{9}$
If $x = 3$, then $y = 60 - 20(3^{-3}) = \frac{1,600}{27}$
If $x = 4$, then $y = 60 - 20(3^{-4}) = \frac{4,840}{81}$
If $x = 5$, then $y = 60 - 20(3^{-5}) = \frac{14,560}{243}$

b) Plotting these points we have the graph of the modified exponential function (Figure 3).

Figure 3

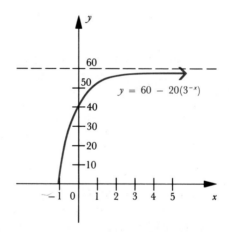

PROBLEM SET 2

1 Indicate the domain and range. Graph the function. Is the function increasing or decreasing?

a) $f(x) = 2^x$
b) $f(x) = 1^x$
c) $f(x) = 3^{x+1}$
d) $f(x) = -2^x$
e) $f(x) = -(\frac{1}{3})^x$
f) $f(x) = 2^{-x}$
g) $f(x) = (\frac{1}{5})^{-x}$
h) $f(x) = (0.1)^x$
i) $f(x) = 5(3^x)$
j) $f(x) = -(4)^x$

2 Use the graph of $f(x) = 3^x$ (Figure 1) to approximate each of the following numbers.

a) $3^{1/2}$
b) $3^{-0.16}$
c) $3^{3.14}$
d) $3^{0.12}$
e) $3^{1.41}$
f) $3^{\sqrt{2}}$

3   What is the base of the exponential function $f(x) = b^x$, whose graph contains the following point?
    a) (2, 9)   b) (2, 16)
    c) (3, 27)   d) (5, 3125)
    e) (0, 1)   f) $(\frac{1}{2}, \sqrt{10})$

4   For what values of b is the exponential function $f(x) = b^x$, $b > 0$, $b \neq 1$, an increasing function? A decreasing function?

5   Is an exponential function even, odd, or neither? Justify your answer by using the definition of an even function and the definition of an odd function.

6   Let $f(x) = 2^x$ and $g(x) = -x$. Find
    a) $f[g(x)]$ b) $g[f(x)]$
    c) $g[f(3)]$ d) $f[g(-1)]$
    e) $f[g(2)]$ f) $g[f(0)]$

7   Let $f(x) = 2^x - 2^{-x}$ and $g(x) = 2^x + 2^{-x}$. Find
    a) $f(x) + g(x)$ b) $f(x) - g(x)$
    c) $f(x) \cdot g(x)$ d) $f(x)/g(x)$
    e) $[f(x)]^2 - [g(x)]^2$ f) $f(x^2) - g(x^2)$

8   Solve each of the following equations.
    a) $3^{x-1} = 1$ b) $3^{x-5} = 27$
    c) $2^{2x} = 32$ d) $4^{(x+3)/2} = 2$
    e) $5^{4x} = 125$ f) $3 = (\frac{1}{27})^x$
    g) $3^{2x-1} = \frac{1}{81}$ h) $3^{x+1} = 9^x \cdot 27^{x+1}$
    i) $2^{2x+2} - 2^{x+2} = -1$ j) $3^{2x+1} - 4(3^x) + 1 = 0$

9   A department store's annual profit from the sales of a certain toy is given by the equation $y = 8{,}000 + 30{,}000(2^{-x})$, where y is in dollars and x denotes the number of years the toy has been on the market.
    a) Calculate the store's annual profit for $x = 1, 2, 3, 5, 10$, and 15.
    b) Use the results of part a to plot the graph of this modified exponential function.

10 A survey by a tire company showed that the proportion of tires still usable after having been driven for x miles is given by $y = 2^{-0.003x}$.
    a) Calculate y for $x = 1{,}000, 2{,}000, 3{,}000, 6{,}000, 9{,}000$, and $20{,}000$.
    b) Use the results of part a to plot the graph of this exponential function.

4 Logarithmic Functions and Their Properties

We have seen that the exponential function $f(x) = b^x$, $b > 0$, $b \neq 1$, is either an increasing function or a decreasing function (Problem Set 2, Problem 4). Since an increasing or decreasing function is one to one, such a function has an inverse. The inverse of an exponential function is called a *logarithmic function*.

DEFINITION

The *logarithmic function* with base b, denoted by $\log_b x$, is defined to be the inverse $f^{-1}(x)$ of the exponential function $f(x) = b^x$. Thus, $y = \log_b x$, where $b > 0$, $b \neq 1$, is equivalent to $x = b^y$.

Note that the domain of $x = b^y$ is the set of real numbers R, and the range is the set of positive real numbers, so that the domain of the function defined by $y = f(x) = \log_b x$ is the set of positive real numbers and the range is the set of real numbers R. For example,

$$\log_2 8 = 3 \quad \text{is equivalent to} \quad 2^3 = 8$$
$$\log_5 \tfrac{1}{25} = -2 \quad \text{is equivalent to} \quad 5^{-2} = \tfrac{1}{25}$$
$$\log_{100} 1{,}000 = \tfrac{3}{2} \quad \text{is equivalent to} \quad 100^{3/2} = 1{,}000$$

Since, in general, the graph of the inverse, f^{-1}, is the reflection of the graph of f across the line $y = x$, we can graph a logarithmic function by first graphing the corresponding exponential function and reflecting it across the line $y = x$. For example, the graph of $y = f(x) = \log_3 x$ can be obtained from the graph of $y = f(x) = 3^x$ as shown in Figure 1.

Figure 1

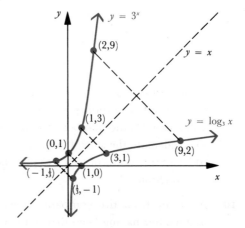

From the graph of $y = f(x) = \log_3 x$, we see that the function is increasing and continuous. Also, the domain of $f(x) = \log_3 x$ is the set of positive numbers and its range is the set of real numbers R. Note that the domain of the exponential function $f(x) = 3^x$ is the set of real numbers R and its range is the set of positive numbers.

A more direct way to graph a logarithmic function $y = \log_b x$ is to use the equivalent equation $x = b^y$ to locate some points on the graph. Thus, to graph $y = \log_3 x$, we can use the equivalent equation $x = 3^y$.

SECTION 4 LOGARITHMIC FUNCTIONS AND THEIR PROPERTIES

Hence, for $y = 0$, $x = 3^0 = 1$; for $y = 1$, $x = 3^1 = 3$; and, for $y = 2$, $x = 3^2 = 9$. One should note that we are reversing (actually, inversing) the usual technique in finding points on the graph. Here we have selected a value of $y = f(x)$ first and then determined the corresponding value of x (Figure 2).

Figure 2

$x = 3^y$	y
$\frac{1}{9}$	-2
$\frac{1}{3}$	-1
1	0
3	1
9	2
27	3

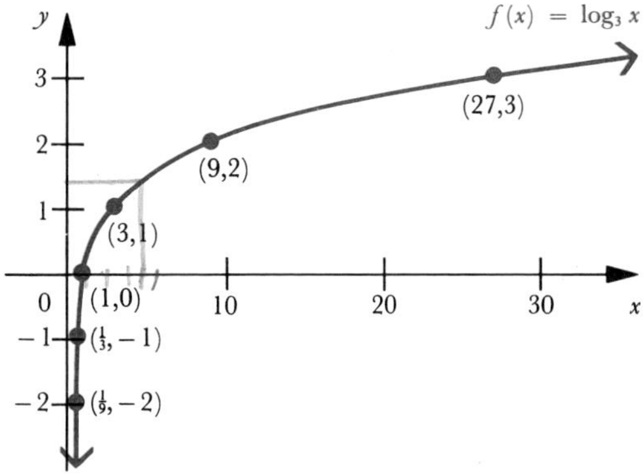

$f(x) = \log_3 x$

Points shown on graph: $(27, 3)$, $(9, 2)$, $(3, 1)$, $(1, 0)$, $(\frac{1}{3}, -1)$, $(\frac{1}{9}, -2)$

EXAMPLES

Find the domain and the range of each of the following functions, and sketch the graph.

1 $f(x) = \log_2 x$

SOLUTION. The domain of f is the set of positive real numbers, and the range is the set of real numbers R. Notice that $f(x) = \log_2 x$ is equivalent to $2^{f(x)} = x$; hence, the table in Figure 3 can be determined by the equivalent exponential equation. The graph of $f(x) = \log_2 x$ goes up to the right as x gets larger (Figure 3); hence, we say that $f(x) = \log_2 x$ is an increasing, continuous function.

Figure 3

$x = 2^{f(x)}$	$f(x)$
$\frac{1}{2}$	-1
1	0
2	1
4	2
8	3

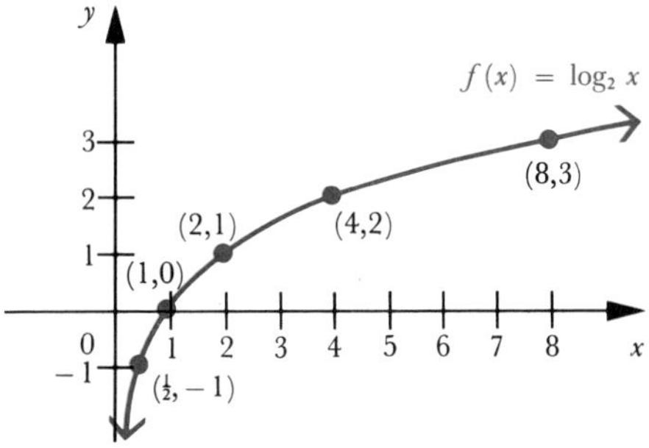

$f(x) = \log_2 x$

Points shown on graph: $(8, 3)$, $(4, 2)$, $(2, 1)$, $(1, 0)$, $(\frac{1}{2}, -1)$

2 $f(x) = \log_{1/4} x$

SOLUTION. The domain of f is the set of positive real numbers, and the range is the set of real numbers R. $f(x) = \log_{1/4} x$ is equivalent to

$(\frac{1}{4})^{f(x)} = x$. The table is determined by using the exponential equation $x = (\frac{1}{4})^{f(x)}$. The graph of $f(x) = \log_{1/4} x$ goes down to the right as x gets larger. Therefore, the function f is a decreasing, continuous function (Figure 4).

Figure 4

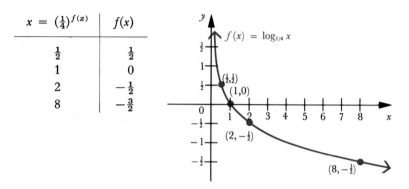

3   Find the domain of the function $f(x) = \log_4 (2 - 3x)$.

SOLUTION.   Since the domain of logarithmic functions is the set of positive real numbers, then $2 - 3x > 0$ for $f(x) = \log_4 (2 - 3x)$. Solving $2 - 3x > 0$, we have $x < \frac{2}{3}$. Hence, the domain of $f(x) = \log_4 (2 - 3x)$ is $\{x \mid x < \frac{2}{3}\} = (-\infty, \frac{2}{3})$.

PROBLEM SET 3

1   Write each of the following exponential statements as an equivalent logarithmic statement.
   a) $5^3 = 125$
   b) $4^{-2} = \frac{1}{16}$
   c) $\sqrt[5]{32} = 2$
   d) $(\frac{1}{3})^{-2} = 9$
   e) $\sqrt{9} = 3$
   f) $\frac{1}{6^2} = \frac{1}{36}$
   g) $x^3 = a$
   h) $7^0 = 1$
   i) $\pi^t = z$
   j) $w^{1/t} = \frac{1}{z}$

2   Write each of the following logarithmic statements as an equivalent exponential statement.
   a) $\log_9 81 = 2$
   b) $\log_{10} 0.0001 = -4$
   c) $\log_{1/3} 9 = -2$
   d) $\log_{10} \frac{1}{10} = -1$
   e) $\log_{\sqrt{16}} 2 = \frac{1}{2}$
   f) $\log_{36} 216 = \frac{3}{2}$
   g) $\log_x 2 = 4$
   h) $\log_t \frac{1}{z} = \frac{1}{w}$
   i) $\log_x 1 = 0$
   j) $\log_9 \frac{1}{3} = -\frac{1}{2}$

3   Graph the following functions and indicate the domain and range. Is the function increasing or decreasing?
   a) $f(x) = -\log_2 x$
   b) $f(x) = \log_5 x$
   c) $f(x) = \log_{1/2} x$
   d) $f(x) = \log_4 x$
   e) $f(x) = \log_6 x$

4   Use the graph of $f(x) = \log_3 x$ (Figure 2) to approximate each of the following numbers.
   a) $\log_3 4$
   b) $\log_3 5$
   c) $\log_3 9$
   d) $\log_3 3.5$
   e) $\log_3 1.3$
   f) $\log_3 0.2$
   g) $\log_3 0.7$

5   Graph $f(x) = \log_{10} x$ then approximate each of the following values.
   a) $\log_{10} 0.05$
   b) $\log_{10} 1.2$
   c) $\log_{10} 1.75$
   d) $\log_{10} 5.2$
   e) $\log_{10} 0.1$

6   Graph $f(x) = \log_b x$, where $b = 2, 3, 4, 5$ on the same coordinate system. How do the graphs compare?

7   Use the graph of Figure 1 to explain the symmetry of $f(x) = \log_3 x$ and $f(x) = 3^x$ with respect to the line $y = x$.

8   How do the graphs of logarithmic functions with bases less than 1 compare with the graphs of functions with bases greater than 1?

9   Find $f[f^{-1}(x)]$ and $f^{-1}[f(x)]$ if $f(x) = \log_3 x$. Simplify the results by using the fact that $f(x) = \log_3 x$ is equivalent to $x = 3^{f(x)}$.

10 Graph $y = \log_{10} x$ then find the values of x when
   a) $y = 0$
   b) $y < 0$
   c) $y > 0$

11 Find the domain of the following functions.
   a) $y = \log(1 - 2x)$
   b) $y = \log(2x + 1)$
   c) $y = \log x^2$
   d) $y = \log |x|$
   e) $y = \log |x + 1|$

12 Graph $y = \log_{10} x$, show that if $x_1 < x_2$ then $\log_{10} x_1 < \log_{10} x_2$; that is $y = \log_{10} x$ is an increasing function.

5 Properties of Logarithms

Since $y = \log_b x$ is equivalent to $b^y = x$, the properties of logarithms can be derived from the properties of exponents.

PROPERTIES OF LOGARITHMS

Suppose that M, N, and b, $b \neq 1$, are positive real numbers and that r is any real number. Then

(i) $b^{\log_b x} = x$

(v) $\log_b b^x = x$

(ii) $\log_b MN = \log_b M + \log_b N$

(iii) $\log_b N^r = r \log_b N$

(iv) $\log_b (M/N) = \log_b M - \log_b N$

PROOF OF i. $b^{\log_b x} = x$. Let $u = \log_b x$; then, by the definition on page 296, $b^u = x$, so that by substituting $\log_b x$ for u, we have

$$b^{\log_b x} = x$$

PROOF OF ii. $M = b^{\log_b M}$ and $N = b^{\log_b N}$ by Property i. Hence,

$$MN = b^{\log_b M} \cdot b^{\log_b N}$$

and, applying the properties of exponents, we have

$$MN = b^{\log_b M + \log_b N}$$

so that

$$\log_b MN = \log_b M + \log_b N$$

PROOF OF iii.

$$N = b^{\log_b N}$$

so that

$$N^r = (b^{\log_b N})^r$$

and, by the properties of exponents,

$$N^r = b^{r \log_b N}$$

so that $\log_b N^r = \log_b b^{r \log_b N}$

then $\log_b N^r = r \log_b N$

PROOF OF iv. Since M/N can be written as MN^{-1}, we have

$$\log_b (M/N) = \log_b MN^{-1} = \log_b M + \log_b N^{-1}$$

so that

$$\log_b (M/N) = \log_b M + (-1) \log_b N$$

or

$$\log_b (M/N) = \log_b M - \log_b N$$

In order to solve logarithmic equations, we will make use of the following property: $\log_b M = \log_b N$ if and only if $M = N$. This result follows from the fact that the logarithm function is a one-to-one function.

EXAMPLES

1. Evaluate $\log_b b$, $\log_b 1$, and $\log_b b^p$.

    SOLUTION. If $\log_b b = t$, then $b^t = b$, so that $t = 1$. Hence, $\log_b b = 1$. Also, $\log_b 1 = t$ implies that $b^t = 1$, so that $t = 0$, that is, $\log_b 1 = 0$. Finally, $\log_b b^p = p \log_b b = p \cdot 1 = p$.

2. Let $\log_b 2 = 0.35$, $\log_b 3 = 0.55$, and $\log_b 5 = 0.82$. Use the properties of logarithms to find each of the following values.

    a) $\log_b \frac{2}{3}$
    b) $\dfrac{\log_b 2}{\log_b 3}$
    c) $\log_b 2^3$
    d) $(\log_b 2)^3$
    e) $\log_b 24$
    f) $\log_b \sqrt{\frac{2}{3}}$
    g) $\log_b \dfrac{60}{b}$
    h) $\log_b 0.6$

    SOLUTIONS

    a) $\log_b \frac{2}{3} = \log_b 2 - \log_b 3 = 0.35 - 0.55 = -0.20$
    b) $\dfrac{\log_b 2}{\log_b 3} = \dfrac{0.35}{0.55} = \dfrac{7}{11} = 0.64$
    c) $\log_b 2^3 = 3 \log_b 2 = 3(0.35) = 1.05$
    d) $(\log_b 2)^3 = (0.35)^3 = 0.043$
    e) $\log_b 24 = \log_b (2^3 \cdot 3) = \log_b 2^3 + \log_b 3$
       $= 3 \log_b 2 + \log_b 3 = 3(0.35) + 0.55 = 1.60$
    f) $\log_b (\frac{2}{3})^{1/2} = \frac{1}{2} \log_b \frac{2}{3} = \frac{1}{2}(\log_b 2 - \log_b 3)$
       $= \frac{1}{2}(0.35 - 0.55) = -0.10$
    g) $\log_b \dfrac{60}{b} = \log_b 60 - \log_b b$
       $= \log_b (2^2 \cdot 3 \cdot 5) - \log_b b$
       $= 2 \log_b 2 + \log_b 3 + \log_b 5 - \log_b b$
       $= 2(0.35) + 0.55 + 0.82 - 1$
       $= 0.70 + 0.55 + 0.82 - 1 = 1.07$

h) $\log_b (0.6) = \log_b \frac{3}{5} = \log_b 3 - \log_b 5$
$= 0.35 - 0.82 = -0.47$

3 Solve for x if $\log_3 (x + 1) + \log_3 (x + 3) = 1$.

SOLUTION

$$\log_3 (x + 1) + \log_3 (x + 3) = \log_3 (x + 1)(x + 3) = 1 \quad \text{(why?)}$$

Hence,

$$x^2 + 4x + 3 = 3 \quad \text{(why?)}$$

which is equivalent to $x^2 + 4x = 0$. The solution of $x^2 + 4x = 0$ is $\{0, -4\}$; however, -4 does not satisfy the original equation (why?), so that $x = 0$ is the only solution. Hence, the solution set is $\{0\}$.

4 Solve for x if $\log_4 (x + 3) - \log_4 x = 1$.

SOLUTION

$$\log_4 (x + 3) - \log_4 x = \log_4 \frac{x + 3}{x} = 1$$

Therefore,

$$\frac{x + 3}{x} = 4^1 = 4$$

from which it follows that $x = 1$. Hence, the solution set is $\{1\}$.

5 Solve for x if $5^{\log_5 x} = 3$.

SOLUTION. Using the Property, part i on page 300 that is, $b^{\log_b x} = x$, we have $x = 3$. Hence, the solution set is $\{3\}$.

PROBLEM SET 4

1 Simplify each of the following expressions.

a) $\dfrac{\log_2 \sqrt{\frac{1}{8}} - \log_2 16^{1/3}}{\log_8 64 \cdot \log_8 8^{1/6}}$

b) $\dfrac{\log_7 7^{4 \cdot 2} + \log_7 7^{-0 \cdot 6}}{\log_{25} 5^{3/2} + \log_{25} 125}$

c) $\dfrac{\log_3 \sqrt{243}\sqrt{81\sqrt[3]{3}}}{\log_2 \sqrt[4]{64} + \log_2 4^{-10}}$

d) $\dfrac{\log_\pi \pi^2 + \log_\pi \pi^3}{\log_4 16 + \log_4 32}$

2 Determine the value of x in each of the following equations.
a) $6^{\log_6 5} + 7^{\log_7 6} = 3^{\log_3 x}$
b) $\log_7 7^4 = x$
c) $9^{\log_x 7} = 7$
d) $\log_3 3^x = 4$
e) $\log_x 3^4 = 4$

3. Prove that $\log_b (xyz) = \log_b x + \log_b y + \log_b z$, where b, x, y, and z are positive, with $b \neq 1$. (*Hint:* Use Property ii, page 300.)

4. Use $\log_{10} 2 = 0.3010$ and $\log_{10} 3 = 0.4771$ to find
   a) $\log_{10} 4$
   b) $\log_{10} 5$
   c) $\log_{10} 18$
   d) $\log_{10} 60$
   e) $\log_{10} 1{,}000$
   f) $\log_{10} 0.5$
   g) $\log_{10} \sqrt[5]{2}$
   h) $\log_{10} \tfrac{1}{3}$
   i) $\log_{10} 3{,}000$
   j) $\log_{10} \tfrac{18}{60}$

5. Use $x_1 = 10{,}000$, $x_2 = 10$, $b = 10$ and $p = 3$ to show the following statements are false.
   a) $\log_b \dfrac{x_1}{x_2} = \dfrac{\log_b x_1}{\log_b x_2}$
   b) $\dfrac{\log_b x_1}{\log_b x_2} = \log_b x_1 - \log_b x_2$
   c) $\log_b x_1 \cdot \log_b x_2 = \log_b x_1 + \log_b x_2$
   d) $\log_b x_1 x_2 = \log_b x_1 \cdot \log_b x_2$
   e) $\log_b x_1^p = (\log_b x_1)^p$
   f) $(\log_b x_1)^p = p \log_b x_1$

6. Solve each of the following equations.
   a) $\log_{10} (x + 1) = 1$
   b) $\log_5 (2x - 7) = 0$
   c) $\log_4 (x + \tfrac{1}{2}) = 3$
   d) $\log_7 (2x - 3) = 2$
   e) $\log_3 (3x + 1) = 2$
   f) $\log_3 (x^2 - 2x) = 1$
   g) $\log_{10} (x^2 + 21x) = 2$
   h) $\log_3 (x + 1)(x + 2) = 1$

7. Solve each of the following equations.
   a) $\log_2 (x^2 - 9) - \log_2 (x + 3) = 2$
   b) $\log_{10} (x + 1) - \log_{10} x = 1$
   c) $\log_3 (x + 24) + \log_3 x = 4$
   d) $\log_4 x + \log_4 (6x + 11) = 1$
   e) $\log_3 x + \log_3 (x - 6) = \log_3 7$
   f) $\log_5 \sqrt{3x + 4} - \log_5 \sqrt{x} = 0$
   g) $\dfrac{\log_{10} (7x - 12)}{\log_{10} x} = 2$

8. Use the properties of logarithms to simplify each of the following expressions.
   a) $\log_{10} (a^2 - ab) - \log_{10} (2a - 2b)$
   b) $3 \log_{10} \dfrac{a^2 b}{c^2} + 2 \log_{10} \dfrac{bc^2}{a^4} + 2 \log_{10} \dfrac{abc}{2}$
   c) $\log_{10} \dfrac{3x^2 - x}{3} - \log_{10} (3x - 1)$
   d) $\log_{10} \left(\dfrac{1}{4} - \dfrac{1}{x^2} \right) - \log_{10} \left(\dfrac{1}{2} - \dfrac{1}{x} \right)$
   e) $\log_{10} \left(a + \dfrac{a}{b} \right) - \log_{10} \left(c + \dfrac{c}{b} \right)$

9 Let $\log_b 2 = A$, $\log_b 3 = B$, and $\log_b 5 = C$. Express $\log_b (0.006)$ in terms of A, B, and C.

6 Common Logarithms

The two logarithmic bases used most often for purposes of computation are base 10 and base e, where e, an irrational number, is approximately equal to 2.718. Logarithms with base 10 are called *common logarithms*. By convention, we usually do not write the base 10 when using logarithmic notation, and $\log x$ is the abbreviated way of writing $\log_{10} x$. Logarithms with base e are called *natural logarithms*, and the natural logarithm, $\log_e x$, is usually written $\ln x$.

6.1 Logarithms—Base 10

For certain values of x it is easy to determine $\log x$. For example, $\log 10 = 1$ and $\log 100 = \log 10^2 = 2$. (Why?) In general, $\log 10^n = n$. (Why?)

For values of x not expressible as powers of 10, other methods are required to determine the value of $\log x$. Let us consider two examples to see how far we could get in computing $\log x$.

Suppose that $x = 5{,}340$. Using scientific notation, we can represent x as $5.34 \cdot 10^3$, so that

$$\begin{aligned} \log 5{,}340 &= \log (5.34 \cdot 10^3) \\ &= \log 5.34 + \log 10^3 \\ &= \log 5.34 + 3 \qquad \text{(why?)} \end{aligned}$$

Hence, determining $\log 5{,}340$ has been reduced to finding $\log 5.34$.

Suppose that $x = 0.000234$. Then

$$\begin{aligned} \log 0.000234 &= \log (2.34 \cdot 10^{-4}) \\ &= \log 2.34 + \log 10^{-4} \qquad \text{(why?)} \\ &= \log 2.34 + (-4) \end{aligned}$$

Here the problem is reduced to finding $\log 2.34$.

We will now generalize the procedure suggested by the two examples. For any positive number x that can be represented in scientific notation as $x = s \cdot 10^n$, where $1 \le s < 10$ and n is an integer, we have $\log x = \log (s \cdot 10^n) = \log s + \log 10^n = \log s + n$, so that $\log x = \log s + n$. This latter form is called the *standard form* of $\log x$, where $\log s$ is called the *mantissa* of $\log x$ and n is called the *characteristic* of $\log x$. Notice that since $y = \log x$ is an increasing function (see Problem Set 3, Problem 12) and $1 \le s < 10$, it follows that $\log 1 \le \log s < \log 10$; that is,

$0 \leq \log s < 1$. In other words, the mantissa is always a number between 0 and 1, possibly equal to 0.

Hence, the task of determining the value of $\log x$ is reduced to determining $\log s$, where s is always between 1 and 10. However, the approximate values of $\log s$ can be determined from the *common log tables* (Table I in Appendix A).

EXAMPLES

In each of the following problems express the number in scientific notation and then determine the common logarithm. Indicate the characteristic and the mantissa. Express the given numbers as a power of 10.

1 53,900

SOLUTION

$$53{,}900 = (5.39)(10^4)$$

so that

$$\log 53{,}900 = \log 5.39 + 4$$

From Table I, we find that $\log 5.39 = 0.7316$, so that

$$\log 53{,}900 = 0.7316 + 4 = 4.7316$$

Hence,

$$10^{4.7316} = 53{,}900$$

The mantissa is 0.7316 and the characteristic is 4.

2 0.0035

SOLUTION

$$0.0035 = (3.5)(10^{-3})$$

so that

$$\log 0.0035 = \log 3.5 - 3$$

From Table I, we find that $\log 3.5 = 0.5441$, so that

$$\log 0.0035 = 0.5441 - 3 = -2.4559$$

Hence,

$$10^{-2.4559} = 0.0035$$

The mantissa is 0.5441 and the characteristic is -3. Note that although $\log 3.5 = -2.4559$, we say that the mantissa of $\log 3.5$ is 0.5441 and *not* 0.4559 or even -0.4559. This is because in applications, the equivalent form $0.5441 - 3$ is much more convenient to work with.

3   28.4

SOLUTION

$$28.4 = (2.84)(10^1)$$

so that

$$\begin{aligned}\log 28.4 &= \log 2.84 + 1 \\ &= 0.4533 + 1 \\ &= 1.4533\end{aligned}$$

Hence,

$$10^{1.4533} = 28.4$$

The mantissa is 0.4533 and the characteristic is 1.

Now we will "reverse" the process of the preceding examples. That is, given a number r, determine the value of x such that $\log x = r$. This number is called the *antilogarithm* of r and is sometimes written as $x = \text{antilog } r$.

As was the case for finding values of the logarithm, it is easy to determine x for some values of r, but not for others. For example, if $\log x = -2$, $x = 0.01$ (why?); if $\log x = 5$, $x = 100{,}000$ (why?); however, if $\log x = 4.4969$, the value of x is not so easy to determine.

The antilog of 4.4969, or the solution of the equation $\log x = 4.4969$, can be determined by reversing the process of determining the logarithm. First, we write $\log x = 4.4969$ in standard form, that is, as the sum of a number between 0 and 1 and an integer:

$$\begin{aligned}\log x &= 4.4969 \\ &= 0.4969 + 4\end{aligned}$$

Second, we use Table I to find a value s such that $\log s = 0.4969$. Here $s = 3.14$. Hence,

$$\begin{aligned}\log x &= 4.4969 \\ &= 0.4969 + 4 \\ &= \log 3.14 + 4 = \log 3.14 + \log 10^4 \\ &= \log 3.14(10^4) = \log 31{,}400\end{aligned}$$

so that

$$x = 31{,}400$$

EXAMPLES

1. Solve $\log x = 2.7210$.

   SOLUTION

   $$\log x = 0.7210 + 2$$
   $$= \log s + 2$$

   Using Table I, we find that $\log 5.26 = 0.7210$, so that

   $$\log x = \log 5.26 + 2 = \log 5.26 + \log 10^2$$
   $$= \log 5.26(10^2) = \log 526$$

   so that

   $$x = 526$$

2. Solve $\log x = 0.5105 + (-3)$.

   SOLUTION

   $$\log x = 0.5105 + (-3)$$
   $$= \log s + (-3)$$

   Using Table I, we find that $\log 3.24 = 0.5105$, so that

   $$\log x = \log 3.24 + (-3) = \log 3.24 + \log 10^{-3}$$
   $$= \log 3.24(10^{-3}) = \log 0.00324$$

   so that

   $$x = 0.00324$$

3. Find $10^{-2.0804}$.

   SOLUTION. Let $x = 10^{-2.0804}$, so that

   $$\log x = -2.0804$$

   Since the mantissa must be positive, we have

   $$\log x = (-2.0804 + 3) - 3 \quad \text{or} \quad \log x = 0.9196 - 3$$

Using Table I, we have log 8.31 = 0.9196, so that

$$\log x = \log 8.31 - 3 = \log 8.31 + \log 10^{-3}$$
$$= \log 8.31(10^{-3}) = \log 0.00831$$

Hence,

$$x = 0.00831$$

4 Find antilog (-2.0804).

SOLUTION. Note that antilog x is the inverse function of log x. Hence, antilog $x = 10^x$, so that

$$\text{antilog}\,(-2.0804) = 10^{-2.0804}$$
$$= 0.00831 \quad \text{from Example 3, above.}$$

6.2 Interpolation

The logarithms and antilogarithms which we computed in Section 6.1 were special in the sense that we were able to find the necessary numbers in Table I. However, this will not always be the case. Suppose, for example, that we wanted to find log 1.234 or antilog 0.2217. We would not be able to find log 1.234 or the mantissa 0.2217 in Table I. This problem can be resolved by using an approximation method called *linear interpolation*.

For example, to determine log 1.234 by using linear interpolation, we proceed as follows. From Table I

$$\log 1.24 = 0.0934 \quad \text{and} \quad \log 1.23 = 0.0899$$

Note that 1.234 lies between 1.23 and 1.24. Let us examine that portion of the graph of $y = \log x$, where $1.23 < x < 1.24$ (Figure 1).

Figure 1

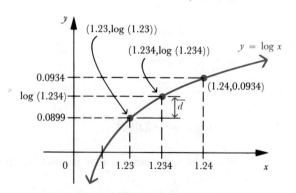

Now, log 1.234 is the length of the ordinate associated with the abscissa 1.234, so that log 1.234 = 0.0899 + $\bar{d}$, where $\bar{d}$ is the "distance" between log 1.234 and 0.0899. $\bar{d}$ is the number which we will approximate. First, we "replace" the arc of the log curve with a line segment (Figure 2).

Figure 2

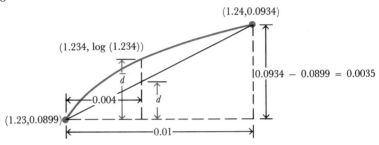

Next, we assume that $\bar{d}$ is approximately the same as d in Figure 2. Finally, d can be determined by using the proportionality of the sides of the similar right triangles which have been formed. Thus,

$$\frac{d}{0.0035} = \frac{0.004}{0.01}$$

Hence,

$$d = 0.0014$$

Since log 1.234 = log 1.23 + $\bar{d}$, and we approximate $\bar{d}$ by d, we have

$$\log 1.234 = \log 1.23 + d$$

or

$$\log 1.234 = 0.0899 + 0.0014$$
$$= 0.0913$$

The same process can be used to find antilogs. For example, we can find x, so that log x = 0.2217, by linear interpolation. From Table I, we find

$$\log 1.67 = 0.2227 \quad \text{and} \quad \log 1.66 = 0.2201$$

Note that the given number 0.2217 lies between the two numbers 0.2201 and 0.2227. As before, we examine the graph (Figure 3) of $y = \log x$ for $1.66 < x < 1.67$.

Figure 3

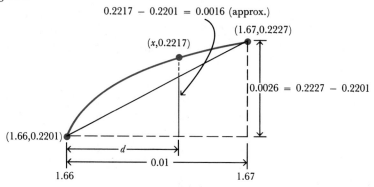

Hence,

$$\frac{d}{0.01} = \frac{0.0016}{0.0026}$$

That is, d is approximately 0.006, so that

$$x = 1.66 + 0.006$$
$$= 1.666$$

Essentially, then, linear interpolation is an approximation method which replaces an arc of a curve with a straight-line segment. The accuracy of this method of approximation depends on the "straightness" of the curve between endpoints.

Now we will present the two examples given above in a manner that simplifies the mechanics involved in linear interpolation.

EXAMPLES

1 Find log 1.234.

SOLUTION

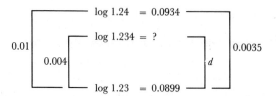

Now,

$$\frac{0.004}{0.01} = \frac{d}{0.0035}$$

so that

$$d = 0.0014$$

Thus,

$$\begin{aligned}\log 1.234 &= 0.0899 + d \\ &= 0.0899 + 0.0014 \\ &= 0.0913\end{aligned}$$

2 Solve $\log x = 0.2217$.

SOLUTION

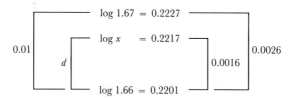

Now,

$$\frac{d}{0.01} = \frac{0.0016}{0.0026}$$

so that

$$d = 0.006$$

Hence,

$$\begin{aligned}x &= 1.66 + d \\ &= 1.66 + 0.006 \\ &= 1.666\end{aligned}$$

6.3 Computation with Logarithms

Now that the methods for finding common logarithms and antilogarithms have been introduced, we shall investigate the application of the properties of logarithms for computational purposes. These properties will be restated here for reference.

i $\log MN = \log M + \log N$
ii $\log (M/N) = \log M - \log N$
iii $\log N^r = r \log N$

EXAMPLES

1. Use logarithms to find $(53.7)(0.83)$.

    SOLUTION

    Let $x = (53.7)(0.83)$; then

    $$\begin{aligned}\log x &= \log (53.7)(0.83) \\ &= \log 53.7 + \log 0.83 \\ &= (\log 5.37 + 1) + [\log 8.3 + (-1)] \\ &= 0.7300 + 1 + 0.9191 + (-1) \\ &= 1.6491\end{aligned}$$

    Hence,

    $$\begin{aligned}x &= \text{antilog } 1.6491 \\ &= (\text{antilog } 0.6491) \cdot 10^1 = 44.6\end{aligned}$$

2. Use logarithms to find $0.837/0.00238$.

    SOLUTION

    Let $x = 0.837/0.00238$. Then

    $$\begin{aligned}\log x &= \log \frac{0.837}{0.00238} \\ &= [\log 8.37 + (-1)] - [\log 2.38 + (-3)] \\ &= (0.9227 - 1) - (0.3766 - 3) \\ &= 2.5461\end{aligned}$$

    Hence,

    $$\begin{aligned}x &= \text{antilog } 2.5461 \\ &= (\text{antilog } 0.5461) \cdot 10^2 \\ &= (3.517)10^2 = 351.7\end{aligned}$$

3. Use logarithms to find

    $$\frac{(289)(3.47)}{0.0987}$$

    SOLUTION. Let

    $$x = \frac{(289)(3.47)}{0.0987}$$

    Then

    $$\log x = \log 289 + \log 3.47 - \log 0.0987$$

Here, in order to simplify the use of logarithms for computations, we consider the following scheme:

$$\log 289 = 2.4609$$
$$+\log 3.47 = 0.5403$$
$$= 3.0012$$

$$-\log 0.0987 = -(0.9943 - 2)$$
$$\log x = 2.0069 + 2 = 4.0069$$

Hence,

$$x = 10{,}160$$

The above scheme for presenting a solution of a computational problem by logarithm removes many of the tedious steps used in the preceding examples.

4 Use logarithms to find $\sqrt[5]{17}$.

SOLUTION. Let $x = \sqrt[5]{17}$, so that

$$\log x = \log 17^{1/5}$$
$$= \tfrac{1}{5} \log 17$$
$$= \tfrac{1}{5}(1.2304)$$
$$= 0.2461 \quad \text{(approximately)}$$

Hence,

$$x = \text{antilog } 0.2461 = 1.762$$

5 Use logarithms to find

$$\frac{(134)^5 (0.35)^8}{(49)^3}$$

SOLUTION. Let

$$x = \frac{(134)^5 (0.35)^8}{(49)^3}$$

Then

$$\log x = 5 \log 134 + 8 \log 0.35 - 3 \log 49 \quad \text{(why?)}$$
$$5 \log 134 = 5(2.1271) = 10.6355$$
$$8 \log 0.35 = 8(0.5441 - 1) = 4.3528 - 8$$

and

$$3 \log 49 = 3(1.6902) = 5.0706$$

so that

$$\log x = 10.6355 + (4.3528 - 8) - 5.0706$$

or

$$x = \text{antilog } 1.9177 = 82.74$$

6 Use logarithms to find the value of $3.456/11.58$.

SOLUTION. Let

$$x = \frac{3.456}{11.58}$$

so that

$$\log x = \log 3.456 - \log 11.58$$
$$\log 3.456 = 0.5386$$
$$\log 11.58 = 1.0637$$

If we substract these logarithms, we obtain -0.5251. However, the mantissas given in Table I are all positive, so the antilog of -0.5251 cannot be found directly from the table. We can overcome this difficulty by writing $\log 3.456$ as $10.5386 - 10$ and then proceeding as before. Thus,

$$\begin{aligned} \log 3.456 &= 10.5386 - 10 \\ -\log 11.58 &= -1.0637 \\ \hline \log x &= 9.4749 - 10 \end{aligned}$$

or

$$\log x = 0.4749 + (-1)$$

Therefore,

$$x = \text{antilog } [0.4749 + (-1)] = 0.2985$$

7 Use logarithms to find $\sqrt[3]{83.5/99.6}$.

SOLUTION. Let $x = \sqrt[3]{83.5/99.6}$. Then

$$\log x = \tfrac{1}{3}(\log 83.5 - \log 99.6)$$

and so

$$\begin{array}{rll} \log 83.5 = & 1.9217 = & 11.9217 - 10 \\ -\log 99.6 = & -1.9983 = & -1.9983 \\ \hline \log \dfrac{83.5}{99.6} = & & 9.9234 - 10 \end{array}$$

and so

$$\log \sqrt[3]{\dfrac{83.5}{99.6}} = \dfrac{1}{3}\log\dfrac{83.5}{99.6} = \dfrac{1}{3}(9.9234 - 10) = 3.3078 - \dfrac{10}{3} = 0.3078 - \dfrac{1}{3}$$

However, we cannot yet find the antilog, since this is not yet in the form of a decimal fraction plus or minus an integer. This difficulty can be overcome by writing $\log 83.5/99.6$ as $29.9234 - 30$, so that

$$\dfrac{1}{3}\log\dfrac{83.5}{99.6} = \dfrac{1}{3}(29.9234 - 30) = 9.9745 - 10$$

Thus,

$$\log x = \log \sqrt[3]{\dfrac{83.5}{99.6}} = 0.9745 + (-1)$$

so that

$$x = \text{antilog}\,[0.9745 + (-1)] = 0.943$$

8 Use logarithms to solve $5^x = 7$.

SOLUTION. Since $5^x = 7$, we have $\log 5^x = \log 7$, so that

$$x \log 5 = \log 7$$

Hence,

$$x = \dfrac{\log 7}{\log 5} = \dfrac{0.8451}{0.6990} = 1.209$$

Having the common logarithm tables in hand, it is not very difficult to compute a logarithm with any base b, where $b > 0$ and $b \neq 1$. It is simply a matter of expressing the given logarithms in terms of common logarithms, and then using the table to compute them. This procedure is generalized by the following property.

PROPERTY: CHANGING BASES OF LOGARITHMS

$$\log_b x = \frac{\log_a x}{\log_a b}$$

where a and b are positive real numbers different from 1.

PROOF. We know that $b^{\log_b x} = x$ (why?), so that $\log_a b^{\log_b x} = \log_a x$. Hence, $\log_b x \cdot \log_a b = \log_a x$ (why?), from which it follows that

$$\log_b x = \frac{\log_a x}{\log_a b}$$

EXAMPLES

1 Express $\ln x = \log_e x$ in terms of common logarithms.

SOLUTION

$$\ln x = \log_e x = \frac{\log x}{\log e}$$

2 Express $\log_2 10$ in terms of common logarithms.

SOLUTION

$$\log_2 10 = \frac{\log_{10} 10}{\log_{10} 2} = \frac{1}{\log_{10} 2}$$

3 Show that $\log_a b \cdot \log_b a = 1$.

SOLUTION

$$\log_a b = \frac{\log_b b}{\log_b a} = \frac{1}{\log_b a}$$

Hence,

$$\log_a b \cdot \log_b a = \frac{1}{\log_b a} \cdot \log_b a = 1$$

4 Use logarithms to solve $e^{-3t} = 0.5$ for t.

SOLUTION. Since $e^{-3t} = 0.5$,

$$\log e^{-3t} = \log 0.5 \quad \text{or} \quad -3t \log e = \log 0.5$$

Hence,
$$t = \frac{\log 0.5}{-3 \log e}$$
$$= 0.23 \quad \text{(approximately)}$$

5 If P dollars represents the amount invested at an annual interest rate r, then the amount A_n accumulated in n years when interest is compounded t times a year is given by

$$A_n = P\left(1 + \frac{r}{t}\right)^{nt}$$

Suppose that $1,000 is placed at a yearly interest rate of 6 percent compounded every 4 months.
a) How much money is accumulated after 4 years?
b) In how many years would the money double at this rate?

SOLUTION. Here, $P = 1,000$, $r = 0.06$, and $t = 3$, so that

$$A_n = 1,000\left(1 + \frac{0.06}{3}\right)^{3n} \quad \text{or} \quad A_n = 1,000(1.02)^{3n}$$

a) If $n = 4$, $A_4 = 1,000(1.02)^{12}$. Now we can use logarithms to get

$$\log A_4 = \log [1,000(1.02)^{12}]$$
$$= \log 1,000 + \log (1.02)^{12}$$
$$= 3 + 12 \log 1.02$$
$$= 3 + 12(0.0086)$$
$$= 3.1032$$

Finally, we determine the antilogarithm:

$$\log A_4 = 0.1032 + 3$$
$$= \log s + 3$$
$$= \log 1.268 + 3$$

so that

$$A_4 = \$1,268 \quad \text{(approximately)}$$

b) We must solve

$$2,000 = 1,000(1.02)^{3n} \quad \text{or} \quad (1.02)^{3n} = 2$$

so that

$$\log (1.02)^{3n} = \log 2 \quad \text{or} \quad 3n \log 1.02 = \log 2$$

Hence,

$$n = \frac{\log 2}{3 \log 1.02} = 11.7 \text{ years} \quad \text{(approximately)}$$

PROBLEM SET 5

1. Express each of the following numbers in scientific notation and then compute the common logarithm. Interpolate if necessary. Indicate the mantissa and characteristic.
   - a) 0.015
   - b) 1547
   - c) 0.002304
   - d) 795.6
   - e) 33.33
   - f) 17
   - g) 0.00000137
   - h) 1370000
   - i) 5171
   - j) 0.00035
   - k) 17.53
   - l) 0.0607

2. Solve for the antilogarithm. Interpolate if necessary.
   - a) $\log x = 3.1452$
   - b) $\log x = -1.5050$
   - c) $\log x = 0.15$
   - d) $\log x = 2.4969$
   - e) $\log x = -9.5031$

3. Find the value of 10^x if
   - a) $x = 2.1038$
   - b) $x = 0.5428$
   - c) $x = -1.5031$
   - d) $x = -4.4510$
   - e) $x = 3.7297$
   - f) $x = 2.3511$

4. Graph $y = \sqrt{x}$. Use linear interpolation, together with the fact that $\sqrt{1} = 1$ and $\sqrt{4} = 2$, to approximate $\sqrt{2}$. Square your result to see how "close" your approximation is to $\sqrt{2}$.

5. Use common logarithms to compute each of the following values.
   - a) $(45.6)(0.357)$
   - b) $(0.00356)(0.786)$
   - c) $\dfrac{83.4}{20.7}$
   - d) $\dfrac{0.901}{1.03}$
   - e) $\sqrt[3]{99}$
   - f) $\sqrt[7]{0.035}$
   - g) $\dfrac{(3.87)^2(1.326)}{\sqrt{4.379}}$
   - h) $\dfrac{\sqrt{0.957}}{\sqrt[5]{32.46}}$
   - i) $\sqrt{\sqrt[3]{69.83}}$
   - j) $\sqrt{\sqrt{\sqrt{7}}}$
   - k) $\sqrt{\dfrac{(3.887)^3(47.32)^2}{(52.37 - 73.4)^2}}$
   - l) $\dfrac{(45.07)(0.5689)(2.346)}{(8.379)(100.7)(0.0034)}$

6. Use common logarithms to compute each of the following values. (Use $e = 2.72$.)
   - a) $\log_2 5$
   - b) $\log_{1/10} 10$
   - c) $\ln 3 = \log_e 3$
   - d) $\log_5 \tfrac{1}{3}$
   - e) $\log_{1/3} \tfrac{4}{3}$

7 Use logarithms to solve each of the following equations.
   a) $2^x = 7$
   b) $e^x = 5$
   c) $17^{2x-1} = 4^{-x}$
   d) $e^{-4/x} = 0.6$
   e) $3^{5x-1} = 0.32$

8 Suppose that $1,500.00 is put into a savings plan which yields a yearly interest rate of $5\frac{1}{2}$ percent. State how much money is accumulated after 5 years, and in how many years the money would triple if the money is compounded
   a) annually
   b) semiannually
   c) quarterly
   d) monthly

9 Suppose that $500.00 is placed at a yearly interest rate of 2 percent compounded annually.
   a) How much money is accumulated after 7 years?
   b) In how many years would the money double at this rate?

10 Find the amount and the interest if $4,000.00 placed in a bank at an annual rate of 2 percent is compounded semiannually for 12 years.

11 When Gus was born, $500 was placed to his credit in an account that pays 6 percent compounded quarterly. What will there be to his credit on his twentieth birthday?

REVIEW PROBLEM SET

1 Let $f(x) = 1 - 5x$ and $g(x) = 7x + 3$.
   a) Find f^{-1} and g^{-1}.
   b) Show that $f(f^{-1}(x)) = x$.
   c) Show that $g^{-1}(g(x)) = x$.

2 Examine the graphs of the given functions to determine whether or not the function f has an inverse f^{-1}. If f^{-1} exists, find it and graph f^{-1} on the same coordinate system as one containing the graph of the given function f.
   a) $f(x) = -\frac{3}{2}x + 1$
   b) $f(x) = -5|x|$
   c) $f(x) = -2x^3 + 3$
   d) $f = \{(x, y) \mid y = \frac{1}{x-1}\}$
   e) $f = \{(x, y) \mid y = 5\}$
   f) $f = \{(x, y) \mid y = \frac{7}{x}\}$

3 Sketch the graph of each of the following functions, and find the domain and the range. Indicate whether the function is increasing or decreasing and find the inverse function.
   a) $f(x) = 2^x$
   b) $f(x) = 7^x$
   c) $f(x) = 3(2^x)$
   d) $f(x) = -2(3^x)$
   e) $f(x) = 4^x$
   f) $f(x) = 2^{1/x}$, $x \neq 0$

4. For what values of x are each of the following equations true?
   a) $5^{x+2} = 625$
   b) $2^{3x+6} = 32$
   c) $3^{2x+1} = (27)^{x-1}$
   d) $3^{4x} = (27)^{3x-4}$
   e) $(1.2)^{2x+1} = (1.44)^{1-x}$
   f) $6^{3x+7} = (216)^{3-x}$

5. During its first 5 years of operation, a company grossed $80,000, $140,000, $200,000, $375,000, and $600,000.
   a) Plot these values at $x = 1, 2, 3, 4,$ and 5, draw a smooth curve which represents the company's gross earnings, and use it to predict the amount the company should gross during the seventh year.
   b) Calculate the values of $y = 50,000(2^{0.5x})$ for $x = 1, 2, 3, 4,$ and 5, and compare these figures with the company's actual gross earnings during the first 5 years by plotting the corresponding points on the figure of part a.

6. Express each of the following equations in exponential form.
   a) $\log_2 32 = 5$
   b) $\log_{27} \frac{1}{9} = -\frac{2}{3}$
   c) $\log_8 32 = \frac{5}{3}$
   d) $\log_{1/32} \frac{1}{16} = \frac{4}{5}$
   e) $\log_{27} 9 = \frac{2}{3}$
   f) $\log_{13} 1 = 0$

7. Find the value of each of the following logarithms.
   a) $\log_3 9$
   b) $\log_{100} 0.0001$
   c) $\log_6 1$
   d) $\log_4 8$
   e) $\log_9 \frac{1}{3}$
   f) $\log_5 0.04$
   g) $\log_x \sqrt[3]{x}$
   h) $\log_x x^3$
   i) $\log_{x^3} x^9$

8. Find the value of x in each of the following logarithm equations.
   a) $\log_3 x = 2$
   b) $\log_4 16 = x$
   c) $\log_{\sqrt{2}} x = -3$
   d) $\log_x 16 = -\frac{4}{3}$
   e) $\log_x 3 = \frac{1}{5}$
   f) $\log_x 81 = 4$
   g) $\log_{\sqrt{3}} 9\sqrt{3} = x$
   h) $\log_{10} x = -2$

9. Express each of the following logarithms as a sum, difference, or multiple of logarithms.
   a) $\log_2 xy^5$
   b) $\log_2 (3^6 \cdot 4^7)$
   c) $\log_a b^x c^y$
   d) $\log_a \frac{x^2 y^2}{z^3}$
   e) $\log_e \sqrt[7]{5^3 \cdot 5^6}$
   f) $\log_5 (2^6 \cdot 3^7 \cdot 5^2)$

10. a) If $\log_b x = 3$, find $\log_{1/b} x$.
    b) If $\log_b x = 3$, find $\log_b 1/x$.
    c) Show that $\log_{1/b} x = \log_b 1/x$.

11. Let $\log_a 2 = 0.69, \log_a 3 = 1.10, \log_a 5 = 1.62,$ and $\log_a 7 = 1.94$. Find each of the following values.
    a) $\log_a (3^5 \cdot 3^7)$
    b) $\log_a \frac{2^4}{3^4}$

c) $\log_a \sqrt[5]{5^3 \cdot 7^4}$
d) $\log_a \sqrt[3]{\frac{3}{2}}$
e) $\log_a \sqrt[3]{16}$
f) $\log_a \frac{60}{a}$
g) $\sqrt[3]{\log_a 16}$
h) $\log_a \frac{25}{27}$

12 Use common logarithms and the tables to compute each of the following values.
  a) $\log 55.6$
  b) $\log 844$
  c) $\log 0.534$
  d) $\log_3 5$
  e) $\log_5 \frac{1}{7}$
  f) $\log_{1/5} \frac{4}{5}$
  g) $\log 61.30$
  h) $\log (5.31)(0.917)$
  i) $\log \sqrt[6]{0.00983}$
  j) $\log \frac{0.00424}{(76)(1.16)}$

13 Solve each of the following equations.
  a) $\log_5 (x^2 - 4) = 0$
  b) $\log_5 (2x - 1) + \log_5 (2x + 1) = 1$
  c) $\log_{1/2} (4x^2 - 1) - \log_{1/2} (2x + 1) = 1$

14 Combine each of the following expressions into a single term.
  a) $\log \frac{11}{5} + \log \frac{14}{3} - \log \frac{22}{15}$
  b) $\log \frac{6}{7} - \log \frac{27}{4} + \log \frac{21}{16}$
  c) $\log x^3 - \log \frac{2}{x^4} + \log x^3 + \log \frac{2}{x}$

15 Sketch the graph of each of the following functions, and find the domain and range. Indicate whether the function is increasing or decreasing and determine the inverse.
  a) $g(x) = \log_{1/2} x$
  b) $g(x) = \log_2 (x + 1)$
  c) $f(x) = \log_3 (x + 2)$

16 Let $f(x) = \log_{16} x$. Find the domain and the range of f. Also find each of the following:
  a) $f(256)$
  b) $f(64)$
  c) $f(32)$
  d) $f(\sqrt[5]{2})$

17 JoAnne borrowed $4,000 from the teachers' credit union at an annual interest rate of 8 percent compounded quarterly. How much does she owe at the end of 3 years?

— 18 approximate by logs to 3 digits $\sqrt{5}$

CHAPTER 7

Sequences, Mathematical Induction, and the Binomial Theorem

CHAPTER 7

Sequences, Mathematical Induction, and the Binomial Theorem

1 Introduction

One purpose in this chapter is to introduce a basic kind of function called a sequence. In addition, finite sums and progressions will be considered. Other topics covered in this chapter include mathematical induction, combinations, permutations, the binomial theorem, and probability.

2 Sequences

We will begin by studying functions whose domains are the set of positive integers. Such functions are called *sequences*. For example, $f(n) = 2/n$, where n is a positive integer, is a sequence. The graph of f consists of discrete points (Figure 1). Notice that in graphing f, the points that are displayed in Figure 1 are not to be connected.

Quite often subscript notation is used to describe sequences. For example, $f(n) = 2/n$ can be written in the form $a_n = 2/n$, so that

Figure 1

$a_1 = f(1) = 2$
$a_2 = f(2) = 1$
$a_3 = f(3) = \frac{2}{3}$
$a_4 = f(4) = \frac{1}{2}$
$a_5 = f(5) = \frac{2}{5}$

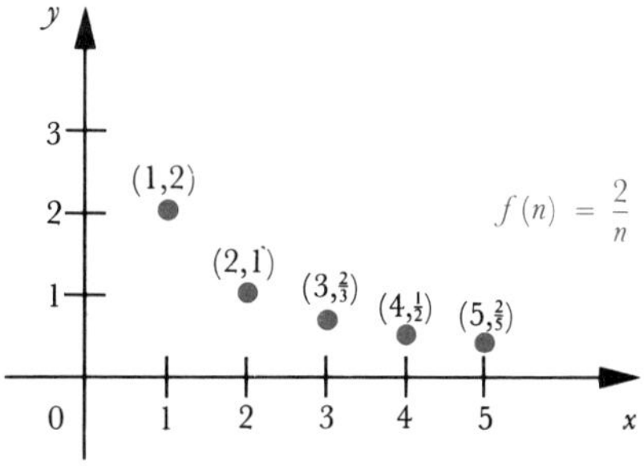

The subscript represents the domain value.

EXAMPLES

For each of the following sequences, write the first five terms, plot them, and describe the range.

1    $a_n = (-1)^n$

SOLUTION. The first five terms of the sequence $a_n = (-1)^n$ are $a_1 = -1$, $a_2 = 1$, $a_3 = -1$, $a_4 = 1$, and $a_5 = -1$. The range is the set $\{-1, 1\}$ (Figure 2).

Figure 2

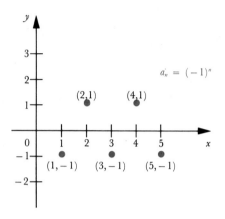

2    $a_n = 3 - \dfrac{1}{n}$

SOLUTION. The first five terms of the sequence $a_n = 3 - (1/n)$ are $a_1 = 2$, $a_2 = \frac{5}{2}$, $a_3 = \frac{8}{3}$, $a_4 = \frac{11}{4}$, and $a_5 = \frac{14}{5}$. The range is a subset of the rational numbers (Figure 3).

Figure 3

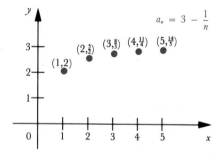

2.1 Summation Notation

At this point we shall be concerned with the sums of many terms of a sequence, and so a notation called *sigma notation* is introduced to facilitate writing these sums. This notation involves the use of the symbol Σ, the capital sigma of the Greek alphabet, which corresponds to the word sum. Some examples of the sigma notation are

$$\sum_{k=1}^{4} k^2 = 1^2 + 2^2 + 3^2 + 4^2$$

$$\sum_{k=0}^{3} (4k+1) = (4(0)+1) + (4(1)+1) + (4(2)+1) + (4(3)+1)$$

and

$$\sum_{k=1}^{5} \frac{1}{k} = 1 + \tfrac{1}{2} + \tfrac{1}{3} + \tfrac{1}{4} + \tfrac{1}{5}$$

In general, we write $\sum_{k=1}^{n} a_k$ as an abbreviated way of writing the finite sum $a_1 + a_2 + \cdots + a_n$, that is,

$$\sum_{k=1}^{n} a_k = a_1 + a_2 + \cdots + a_n$$

Here the symbol a_k represents the numbers $a_1, a_2, \ldots,$ to a_n in turn; we say that k runs from 1 to n. In more formal terminology, using the function concept, we would say the sequence is a function with domain $\{1, 2, \ldots, n\}$, that k is being used as a variable (the way we usually use x), and that a_k is the value of the function at k, for any k between 1 and n. For example, if S_n is defined by the equation $S_n = \sum_{k=1}^{n} (2k)$ for any positive integer n, then

$$S_3 = \sum_{k=1}^{3} (2k) = 2 \cdot 1 + 2 \cdot 2 + 2 \cdot 3 = 2 + 4 + 6 = 12$$

and

$$S_4 = \sum_{k=1}^{4} (2k) = 2 \cdot 1 + 2 \cdot 2 + 2 \cdot 3 + 2 \cdot 4 = 2 + 4 + 6 + 8 = 20$$

EXAMPLES

In Examples 1–3, evaluate each sum.

1 $\sum_{k=1}^{3} (4k^2 - 3k)$

SOLUTION. $a_k = 4k^2 - 3k$. To find the indicated sum, we substitute the integers 1, 2, and 3 for k in succession and then add the resulting numbers. Thus,

$$\sum_{k=1}^{3} (4k^2 - 3k) = [4(1)^2 - 3(1)] + [4(2)^2 - 3(2)] + [4(3)^2 - 3(3)]$$
$$= 1 + 10 + 27 = 38$$

2 $\sum_{i=3}^{6} i(i-2)$

SOLUTION. $a_i = i(i-2)$. Notice here that the index of summation begins with 3. To find the indicated sum, we substitute the integers 3, 4, 5, and 6 in succession and then add the resulting numbers. Thus,

$$\sum_{i=3}^{6} i(i-2) = [3(3-2)] + [4(4-2)] + [5(5-2)] + [6(6-2)]$$
$$= 3 + 8 + 15 + 24 = 50$$

3 $\sum_{k=2}^{5} \frac{k-1}{k+1}$

SOLUTION. Here $a_k = (k-1)/(k+1)$, so that

$$\sum_{k=2}^{5} \frac{k-1}{k+1} = \left(\frac{2-1}{2+1}\right) + \left(\frac{3-1}{3+1}\right) + \left(\frac{4-1}{4+1}\right) + \left(\frac{5-1}{5+1}\right)$$
$$= \frac{1}{3} + \frac{2}{4} + \frac{3}{5} + \frac{4}{6} = \frac{21}{10}$$

4 Write the following finite sum in sigma notation:

$$1 + \tfrac{1}{2} + \tfrac{1}{4} + \tfrac{1}{8} + \tfrac{1}{16}$$

SOLUTION

$$1 + \tfrac{1}{2} + \tfrac{1}{4} + \tfrac{1}{8} + \tfrac{1}{16} = (\tfrac{1}{2})^0 + (\tfrac{1}{2})^1 + (\tfrac{1}{2})^2 + (\tfrac{1}{2})^3 + (\tfrac{1}{2})^4$$
$$= \sum_{k=0}^{4} (\tfrac{1}{2})^k$$

or, equivalently,

$$1 + \tfrac{1}{2} + \tfrac{1}{4} + \tfrac{1}{8} + \tfrac{1}{16} = \sum_{k=1}^{5} (\tfrac{1}{2})^{k-1} \quad \text{(Why?)}$$

PROBLEM SET 1

1 Find the first five terms of each of the following sequence functions.

a) $f(n) = \dfrac{n(n+2)}{2}$ \qquad b) $a_n = \dfrac{n+4}{n}$

c) $f(n) = \dfrac{n(n-3)}{2}$ \qquad d) $f(n) = \dfrac{3}{n(n+1)}$

e) $a_n = (-1)^n + 3$ \qquad f) $f(n) = \dfrac{n^2 - 2}{2}$

2 Plot the five points of each of the sequence functions determined in Problem 1.

3 Find the numerical values of each of the following finite sums.

a) $\sum_{k=1}^{5} k$ \qquad b) $\sum_{k=0}^{4} \dfrac{2^k}{(k+1)}$

c) $\sum_{i=1}^{10} 2i(i-1)$
d) $\sum_{k=0}^{4} 3^{2k}$

e) $\sum_{k=2}^{5} 2^{k-2}$
f) $\sum_{i=2}^{6} \frac{1}{i}$

g) $\sum_{k=1}^{3} (2k+1)$
h) $\sum_{k=1}^{5} (3k^2 - 5k + 1)$

i) $\sum_{i=1}^{4} \frac{i}{i+1}$
j) $\sum_{k=1}^{4} k^k$

k) $\sum_{k=1}^{100} 5$
l) $\sum_{i=3}^{7} (i+2)$

m) $\sum_{k=1}^{5} \frac{1}{k(k+1)}$
n) $\sum_{k=1}^{4} \frac{3}{k}$

o) $\sum_{j=3}^{10} (3j-1)$

4 Express each of the following finite sums in sigma notation.
a) $1 + 4 + 7 + 10 + 13$
b) $\frac{1}{2} + \frac{1}{4} + \frac{1}{8} + \frac{1}{16} + \frac{1}{32}$
c) $\frac{3}{5} + \frac{9}{25} + \frac{27}{125} + \frac{81}{625}$
d) $\frac{1}{6} + \frac{2}{11} + \frac{3}{16} + \frac{4}{21}$

5 Determine whether each of the following statements is true or false. Give the reason.

a) $\sum_{k=0}^{100} k^3 = \sum_{k=1}^{100} k^3$ True $0^3 = 0$

b) $\sum_{k=1}^{100} 2 = 200$

c) $\sum_{k=0}^{100} (k+2) = \left(\sum_{k=0}^{100} k\right) + 2$ — $\sum_{k=0}^{100} k + \sum_{k=0}^{100} 2$

d) $\sum_{k=0}^{99} (k+1)^2 = \sum_{k=1}^{100} k^2$

e) $\sum_{k=0}^{100} k^2 = \left(\sum_{k=0}^{100} k\right)^2$

3 Progressions

Consider the following two examples of successions of numbers that follow specific patterns:

$1, 3, 5, \ldots$
$2, 4, 8, 16, \ldots$

In the first example, we are dealing with the positive odd integers so that the next terms follow the pattern 7, 9, 11, 13, and so on. In the second example, the numbers can be written as 2^1, 2^2, 2^3, and 2^4, so that the next terms follow the pattern $2^5 = 32$, $2^6 = 64$, $2^7 = 128$, and so on. Each of the above examples follows a specific pattern and we say they constitute *progressions*. The first is an example of an *arithmetic progression* and the second is an example of a *geometric progression*.

3.1 Arithmetic Progressions

Suppose that we borrow p dollars from a bank at the simple interest rate r. We notice that

the amount owed at the end of the first year is $p + (p \cdot r)$
the amount owed at the end of the second year is $p + 2(p \cdot r)$
the amount owed at the end of the third year is $p + 3(p \cdot r)$

and

the amount owed at the end of the nth year is expressed by
$s = p + n(p \cdot r)$

This equation forms a linear function (Figure 1).

Observe that each one of the amounts is obtained from the preceding amount by adding $p \cdot r$. This is an example of arithmetic progression.

Figure 1

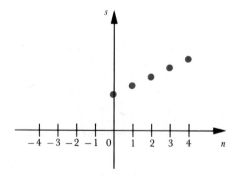

In general, if we begin with a number a_1 and repeatedly add the same number d, then the numbers a_1, $a_1 + d$, $a_1 + 2d$, $a_1 + 3d$, ..., $a_1 + (n-1)d$, ... are said to constitute an *arithmetic progression*. a_1 is called the *first term* of the arithmetic progression, d is called the *common difference*, and $a_1 + (n-1)d$ is called the *n*th *term*. We usually denote the nth term by a_n.

For example, 2, -1, -4, -7, ... is an arithmetic progression with the first term 2 and common difference -3. The tenth term, a_{10}, of this sequence is

$$2 + (10 - 1)(-3) = -25$$

Now, let us investigate the sum of the first n terms of an arithmetic progression. For convenience, we will let S_n represent the sum of the first n terms. For example, if $a_n = 1 + 2n$, then

$$S_1 = a_1 = 3$$
$$S_2 = a_1 + a_2 = 3 + 5 = 8$$
$$S_3 = a_1 + a_2 + a_3 = 3 + 5 + 7 = 15$$
$$\vdots$$
$$S_n = a_1 + a_2 + \cdots + a_n = 3 + 5 + \cdots + (1 + 2n)$$
$$= \sum_{k=1}^{n} [3 + (k-1)2]$$

In general,

$$S_n = a_1 + (a_1 + d) + (a_1 + 2d) + (a_1 + 3d) + \cdots$$
$$+ [a_1 + (n-3)d] + [a_1 + (n-2)d] + [a_1 + (n-1)d]$$
$$= \sum_{k=1}^{n} [a_1 + (k-1)d]$$

for the arithmetic progression $a_n = a_1 + (n-1)d$. Notice that the sum of the first term and last term is the same as the sum of the second term and the next to last term, and so on. That is,

$$(a_1) + [a_1 + (n-1)d] = a_1 + a_n$$
$$(a_1 + d) + [a_1 + (n-2)d]$$
$$= (a_1) + [a_1 + (n-1)d] = a_1 + a_n$$
$$(a_1 + 2d) + [a_1 + (n-3)d]$$
$$= (a_1) + [a_1 + (n-1)d] = a_1 + a_n$$

and so on. Since there are $n/2$ such pairs of terms, we have

$$S_n = \sum_{k=1}^{n} [a_1 + (k-1)d] = \frac{n}{2}(a_1 + a_n)$$

EXAMPLES

1   Find the twentieth term and the sum of the first 20 terms of an arithmetic progression whose first term is 2 and whose common difference is 4.

SOLUTION. $a_1 = 2$, $d = 4$, and $n = 20$. Hence,

$$a_{20} = a_1 + (20-1)d$$
$$= 2 + (19)(4) = 78$$

Furthermore, by the formula for S_n,

$$S_{20} = \frac{20}{2}(a_1 + a_{20})$$

$$= \frac{20}{2}(2 + 78)$$

$$= 800$$

2 For an arithmetic progression, the sum of the first 10 terms is 351 and the tenth term is 51. Find the first term and the common difference.

SOLUTION. From the formula for S_n, we have

$$S_{10} = \frac{10}{2}(a_1 + a_{10})$$

That is,

$$351 = \frac{10}{2}(a_1 + 51)$$

or, equivalently,

$$351 = 5a_1 + 255$$

so that

$$5a_1 = 96 \quad \text{or} \quad a_1 = 19.2$$

Also, $a_{10} = a_1 + 9d$, so that $51 = 19.2 + 9d$ or, equivalently,

$$9d = 31.8 \quad \text{so that} \quad d = \frac{31.8}{9} = \frac{53}{15}$$

3 How many terms are there in the arithmetic progression for which $a_1 = 3$, $d = 5$, and $S_n = 255$?

SOLUTION. From the formula for S_n, we have

$$255 = \frac{n}{2}(3 + a_n)$$

so that

$$510 = n(3 + a_n)$$

From the formula for a_n, we have

$$a_n = 3 + (n - 1)5$$
$$= 5n - 2$$

Hence, $510 = n(3 + 5n - 2)$; that is,

$$5n^2 + n - 510 = 0 \quad \text{or} \quad (5n + 51)(n - 10) = 0$$

so that

$$n = 10 \quad \text{or} \quad -\tfrac{51}{10}$$

Hence, $n = 10$ is the number of terms, since n must be a positive integer.

4 If an object falls vertically 16 feet during the first second, 48 feet during the second second, 80 feet during the third second, and so on, how far will it fall during the tenth second? During the first 10 seconds?

SOLUTION. The distances 16, 48, and 80 have a common difference of 32. Hence, they form an arithmetic progression with $a_1 = 16$ and $d = 32$, so that

$$a_n = 16 + (n - 1)32$$

For $n = 10$, we have

$$a_{10} = 16 + (10 - 1)32$$

or

$$a_{10} = 304$$

Thus, the object falls 304 feet during the tenth second. From the formula for S_n, with $n = 10$, $a_1 = 16$ and $a_{10} = 304$, we have

$$S_{10} = \tfrac{10}{2}(16 + 304)$$
$$= 5(320)$$
$$= 1{,}600$$

Thus, the object falls 1,600 feet during the first 10 seconds.

3.2 Geometric Progressions

If we borrow p dollars at an interest rate r compounded annually,

the amount A_1 owed at the end of the first year is $p + p \cdot r = p(1+r)$
the amount A_2 owed at the end of the second year is $p(1+r)^2$
the amount A_3 owed at the end of the third year is $p(1+r)^3$

and

the amount owed at the end of n years is expressed by $A_n = p(1+r)^n$

This is an example of a modified exponential function. Notice that each amount is obtained from the preceding amount by multiplying by $1 + r$, which makes a geometric progression.

In general, if we begin with a number a_1 and repeatedly multiply by the same number r, then the numbers $a_1, a_1 r, a_1 r^2, \ldots, a_1 r^{n-1}, \ldots$ are said to constitute a *geometric progression*. In this geometric progression a_1 is called the *first term*, r is called the *common ratio*, and the nth *term* a_n is given by $a_n = a_1 r^{n-1}$.

For example, $1, \frac{1}{2}, \frac{1}{4}, \frac{1}{8}, \ldots$ is a geometric progression in which $a_1 = 1$, $r = \frac{1}{2}$, and $a_n = (\frac{1}{2})^{n-1}$.

Just as we did with arithmetic progressions, we will derive a formula for S_n, the sum of the first n terms of a geometric progression. For example, for $a_n = 3(2)^{n-1}$

$$S_1 = a_1 = 3$$
$$S_2 = a_1 + a_2 = 3 + 6 = 9$$
$$S_3 = a_1 + a_2 + a_3 = 3 + 6 + 12 = 21$$
$$\vdots$$
$$S_n = a_1 + a_2 + \cdots + a_n = 3(2^n - 1)$$

In general, consider

$$S_n = \sum_{i=1}^{n} a_1 r^{i-1} = a_1 + a_1 r + a_1 r^2 + \cdots + a_1 r^{n-1}$$

for $a_n = a_1 r^{n-1}$

First, multiply both sides of that equation by r to get

$$rS_n = a_1 r + a_1 r^2 + a_1 r^3 + \cdots + a_1 r^n$$

Next, subtracting rS_n from S_n, we get

$$S_n - rS_n = a_1 - a_1 r^n$$

so that

$$(1-r)S_n = a_1 - a_1 r^n$$

or

$$S_n = \sum_{i=1}^{n} a_1 r^{i-1} = \frac{a_1 - a_1 r^n}{1 - r} = \frac{a_1(1 - r^n)}{1 - r} \quad \text{for } r \neq 1$$

$$S_n = \overbrace{a_1 + a_1 + \cdots + a_1}^{n \text{ terms}} = na_1 \quad \text{for } r = 1$$

EXAMPLES

1 Find the tenth term and the sum of the first 10 terms of the geometric progression whose first term is $\frac{1}{2}$ and whose common ratio is 2.

SOLUTION. Here we have $a = \frac{1}{2}$ and $r = 2$. Using the formula for a_n, we have

$$a_{10} = ar^{10-1} = \tfrac{1}{2}(2^9) = 256$$

Also, from the formula for S_n,

$$S_{10} = \frac{\tfrac{1}{2}(1 - 2^{10})}{1 - 2} = \frac{\tfrac{1}{2}(-1{,}023)}{-1} = 511.5$$

2 The sum of the first five terms of a geometric progression is $2\tfrac{7}{27}$, and the common ratio is $-\tfrac{1}{3}$. Find the terms of the geometric progression.

SOLUTION. From the formula for S_n, we have

$$2\tfrac{7}{27} = \frac{a[1 - (-\tfrac{1}{3})^5]}{1 - (-\tfrac{1}{3})}$$

so that

$$2\tfrac{7}{27} = \left(\frac{\tfrac{244}{243}}{\tfrac{4}{3}}\right) a$$

Hence, $a = 3$, and the terms of the geometric progression are

$$3,\ 3(-\tfrac{1}{3}),\ 3(-\tfrac{1}{3})^2,\ 3(-\tfrac{1}{3})^3,\ 3(-\tfrac{1}{3})^4, \ldots \quad \text{or} \quad 3,\ -1,\ \tfrac{1}{3},\ -\tfrac{1}{9},\ \tfrac{1}{27}, \ldots$$

3 The number of bacteria in a certain culture triples every 4 hours. If there were x number of bacteria present initially in a particular culture, how many are present 20 hours later?

SOLUTION. Here $a_1 = x$ and $r = 3$, so that the formula for a_n is

$$a_n = x \cdot 3^{n-1}$$

where a_n represents the number of bacteria after n four-hour intervals.

Since the culture triples every 4 hours, this will happen 5 times in 20 hours, so that $n = 5$.

Thus, $a_5 = x \cdot 3^{5-1}$ or $a_5 = 81x$.

3.3 Geometric Series

If a_n determines a sequence and $S_n = \sum_{k=1}^{n} a_k = a_1 + a_2 + \cdots + a_n$, then the sequence determined by S_n is called an *infinite series* and it is usually written $\sum_{k=1}^{\infty} a_k$. We will be concerned with determining the "sums" of geometric series.

Geometric series are series of the form $\sum_{k=1}^{\infty} ar^{k-1}$, where a is a constant and $|r| < 1$. For example,

$$\sum_{k=1}^{\infty} (\tfrac{1}{2})^{k-1} = 1 + \tfrac{1}{2} + \tfrac{1}{4} + \tfrac{1}{8} + \cdots + (\tfrac{1}{2})^n + \cdots$$

is a geometric series in which $a = 1$ and $r = \tfrac{1}{2}$.

The notion of an "infinite sum" is defined as follows. Suppose that $\sum_{k=1}^{\infty} a_k$ is an infinite series. First, *partial sums* are formed:

$$s_1 = a_1$$
$$s_2 = a_1 + a_2$$
$$s_3 = a_1 + a_2 + a_3$$
$$\cdots\cdots\cdots\cdots\cdots\cdots\cdots\cdots\cdots\cdots\cdots\cdots\cdots\cdots$$
$$s_n = a_1 + a_2 + \cdots + a_n = \sum_{k=1}^{n} a_k$$

Then the sum S of $\sum_{k=1}^{\infty} a_k$, written $S = \sum_{k=1}^{\infty} a_k$, is defined to be the "limit value" that s_n approaches as "n approaches infinity," if the limit value is finite.

Let us use this notion to determine the formula for the sum of a geometric series

$$\sum_{k=1}^{\infty} ar^{k-1}, \qquad |r| < 1$$

Here $a_k = ar^{k-1}$, so that

$$s_1 = a$$
$$s_2 = a + ar$$
$$s_3 = a + ar + ar^2$$
$$\cdots\cdots\cdots\cdots\cdots\cdots\cdots\cdots\cdots\cdots\cdots\cdots\cdots\cdots$$

(1) $\qquad s_n = a + ar + ar^2 + ar^3 + \cdots + ar^{n-1}$

Upon multiplying both sides of Eq. (1) by r, we get

(2) $\qquad rs_n = ar + ar^2 + ar^3 + \cdots + ar^n$

Next subtract Eq. (2) from Eq. (1) to get

$$s_n - rs_n = a - ar^n$$

so that

$$(1 - r)s_n = a - ar^n$$

Hence,

$$s_n = \frac{a - ar^n}{1 - r} \qquad \text{where } |r| < 1$$

However, this latter equation can be written as

$$s_n = \frac{a}{1 - r} - \frac{ar^n}{1 - r}$$

Intuitively, we can see that as n becomes increasingly large (as n approaches infinity) r^n approaches 0. [Consider what happens for example to the values of $(\tfrac{1}{2})^n$ as n becomes larger and larger by examining the graph of $f(n) = (\tfrac{1}{2})^n$.]

Consequently, s_n approaches $a/(1 - r)$ as n approaches infinity so that

$$\sum_{k=1}^{\infty} ar^{k-1} = \frac{a}{1 - r} \qquad \text{where } |r| < 1$$

For example,

$$\sum_{k=1}^{\infty} (\tfrac{1}{2})^{k-1} = \frac{1}{1 - \tfrac{1}{2}} = 2$$

The next example shows that geometric series have an interesting application in connection with the repeating decimals that were introduced in Chapter 1.

EXAMPLE

Use geometric series to find the rational numbers which correspond to the following decimals.
a) $0.3\overline{33}$ \qquad\qquad\qquad\qquad\qquad b) $0.24\overline{2424}$

SOLUTION

a) From the expression $0.33\overline{3}$, we obtain the geometric series

$$0.33\overline{3} = \frac{3}{10} + \frac{3}{100} + \frac{3}{1,000} + \frac{3}{10,000} + \cdots + \frac{3}{10^n} + \cdots$$

$$= \tfrac{3}{10} + \tfrac{3}{10}(\tfrac{1}{10}) + \tfrac{3}{10}(\tfrac{1}{10})^2 + \tfrac{3}{10}(\tfrac{1}{10})^3 + \cdots$$

$$+ \tfrac{3}{10}(\tfrac{1}{10})^{n-1} + \cdots$$

$$= \sum_{k=1}^{\infty} \tfrac{3}{10}(\tfrac{1}{10})^{k-1}$$

Here $a = \tfrac{3}{10}$ and $r = \tfrac{1}{10}$, so that the sum is given by

$$0.33\overline{3} = \frac{\tfrac{3}{10}}{1 - \tfrac{1}{10}} = \tfrac{1}{3}$$

b) $0.24\overline{2424}$ can be written as

$$0.24\overline{2424} = \frac{24}{100} + \frac{24}{10,000} + \frac{24}{1,000,000} + \cdots$$

$$= \sum_{k=1}^{\infty} (\tfrac{24}{100})(\tfrac{1}{100})^{k-1}$$

We have a geometric series in which $a = \tfrac{24}{100}$ and $r = \tfrac{1}{100}$. Hence,

$$0.24\overline{2424} = \frac{\tfrac{24}{100}}{1 - \tfrac{1}{100}} = \frac{24}{99} = \frac{8}{33}$$

PROBLEM SET 2

1 Determine which of the following sequences are arithmetic progressions, and find the common difference d and S_{10} for each such progression.
 a) 2, 5, 8, 11, ...
 b) 3, 5, 7, 9, ...
 c) 7, 12, 17, 22, ...
 d) $11a + 7b, 7a + 2b, 3a - 3b, \ldots$ (a and b are constants)
 e) 67, 54, 41, 28, ...
 f) $9a^2, 16a^2, 23a^2, 30a^2, \ldots$
 g) 5.7, 6.9, 8.1, 9.3, ...
 h) 1.4, 4.5, 7.6, 10.7, ...

2 a) Find the tenth and fifteenth terms of the arithmetic progression $-13, -6, 1, 8, \ldots$.

 b) Find the twelfth and thirty-fifth terms of the arithmetic progression $19, 17, 15, 13, \ldots$.

 c) Find the sixth and ninth terms of the arithmetic progression $a + 24b, 4a + 20b, 7a + 16b, \ldots$, where a and b are constants.

 d) Find the third and sixteenth terms of the arithmetic progression $7a^2 - 4b, 2a^2 + 7b, -3a^2 + 18b, \ldots$, where a and b are constants.

3 Find S_7 for the arithmetic progression $6, 3b + 1, 6b - 4, \ldots$, where b is a constant.

4 In each of the following parts, certain elements of an arithmetic progression are given. Find the indicated unknowns.
   a) $a_1 = 6; d = 3; a_{10}; S_{10}$
   b) $a_1 = 38; d = -2; n = 25; S_n$
   c) $a_1 = 17; S_{18} = 2{,}310; d; a_{18}$
   d) $d = 3; S_{25} = 400; a_1; a_{25}$
   e) $a_1 = 27; a_n = 48; S_n = 1{,}200; n; d$

5 Determine which of the following sequences are geometric progressions and give the value of the common ratio and S_5 for each such progression.
   a) $2, 6, 18, \ldots$
   b) $1, \frac{1}{5}, \frac{1}{25}, \ldots$
   c) $1, -2, 4, \ldots$
   d) $\frac{4}{9}, \frac{1}{6}, \frac{1}{16}, \ldots$
   e) $81, 54, 36, \ldots$
   f) $147, -21, 3, \ldots$
   g) $9, -6, 4, \ldots$
   h) $64, -32, 16, \ldots$

6 Find the indicated term of each of the following geometric progressions.
   a) The tenth term of $-4, 2, -1, \frac{1}{2}, \ldots$
   b) The eighth term of $\frac{1}{8}, \frac{1}{4}, \frac{1}{2}, \ldots$
   c) The fifth term of $32, 16, 8, \ldots$
   d) The eleventh term of $1, 1.03, (1.03)^2, \ldots$
   e) The nth term of $1, 1 + a, (1 + a)^2, \ldots$
   f) The twelfth term of $10^{-5}, 10^{-7}, 10^{-9}, \ldots$

7 a) Find the sixth and tenth terms and the sum of the first 10 terms of the geometric progression $6, 12, 24, 48, \ldots$.
   b) Find the sixth and eighth terms and the sum of the first eight terms of the geometric progression $2, 6, 18, \ldots$.
   c) Find the fifth term of the geometric progression $3, 6, 12, \ldots$.

8 Find the indicated element in each of the following geometric progressions with the given elements.
   a) $a_1 = 2; n = 3; S_n = 26; r$
   b) $r = 2; n = 5; a_n = -48; a_1$ and S_n
   c) $a_1 = 3; a_n = 192; n = 7; r$
   d) $a_6 = 3; a_9 = -81; r$ and a_1
   e) $a_5 = \frac{1}{8}; r = -\frac{1}{2}; a_8$ and S_8
   f) $a_1 = 1; r = (1.03)^{-1}; S_9$
   g) $a_1 = \frac{1}{16}; r = 2; a_n = 32; n$ and S_n
   h) $a_1 = \frac{243}{256}; a_n = 3; S_n = \frac{939}{256}; r$ and n
   i) $a_1 = 250; r = \frac{3}{5}; a_n = 32\frac{2}{5}; n$ and S_n

9 a) Write the first three terms of a geometric progression in which the fourth term is 2 and the seventh term is 54.
   b) Write the first four terms of a geometric progression in which the fifth term is $\frac{1}{7}$ and the seventh term is $\frac{4}{343}$.

10 Find the sum of each of the following geometric series.
   a) $\frac{1}{3} + \frac{1}{9} + \frac{1}{27} + \frac{1}{81} + \cdots + (\frac{1}{3})^n + \cdots$
   b) $\frac{2}{3} + \frac{4}{9} + \frac{8}{27} + \cdots + (\frac{2}{3})^n + \cdots$
   c) $\frac{1}{5} + \frac{1}{25} + \frac{1}{125} + \cdots + (\frac{1}{5})^n + \cdots$
   d) $\frac{9}{8} + \frac{9}{64} + \frac{9}{512} + \cdots + 9(\frac{1}{8})^n + \cdots$

11 Use geometric series to find the rational number that is represented by each of the following decimal numbers.
   a) $0.32\overline{32}$
   b) $0.04\overline{9999}$
   c) $0.4\overline{6464}$
   d) $0.072\overline{072072}$
   e) $3.561\overline{561561}$
   f) $32.421842\overline{184218}$

12 A man invests $500 at the beginning of each year for 10 years at 5 percent simple interest. How much will he have to his credit at the end of the tenth year?

13 A car costs $5,000 and depreciates 25 percent of the original cost during the first year, 21 percent during the second year, 17 percent during the third year, and so on, for 6 years. What is the car worth when it is 6 years old?

14 If a person saves 1 cent on the first day of June, 2 cents on the second day, 4 cents on the third day, and so on, find the total amount saved during the month.

15 A man invests $500 at the beginning of each year for 10 years at 5 percent compounded annually. How much will he have to his credit at the end of the tenth year?

4 Mathematical Induction

In Section 3 we established formulas for finding the sum of the first n terms of an arithmetic progression and of a geometric progression. In this section we shall consider another method of proof applicable to such problems that has a broader application than those methods used in Section 3. In fact, it is the keystone of that branch of mathematics called *number theory*.

Consider the following sums of consecutive odd positive integers.

$$\sum_{i=1}^{1} (2i - 1) = 1 = 1 = 1^2$$

$$\sum_{i=1}^{2} (2i - 1) = 1 + 3 = 4 = 2^2$$

$$\sum_{i=1}^{3} (2i - 1) = 1 + 3 + 5 = 9 = 3^2$$

$$\sum_{i=1}^{4} (2i - 1) = 1 + 3 + 5 + 7 = 16 = 4^2$$

$$\sum_{i=1}^{5} (2i - 1) = 1 + 3 + 5 + 7 + 9 = 25 = 5^2$$

We see from the above that in each case the indicated sum is equal to the square of the number of consecutive odd integers being added. Can we conclude that this is always the case? That is, is it true that

$$\sum_{i=1}^{n} (2i - 1) = 1 + 3 + 5 + \cdots + (2n - 1) = n^2$$

for all n, where n is a positive integer? In order to establish the sequential proof of the assertion in this equation we need a method of proof using the *principle of mathematical induction*.

PRINCIPLE OF MATHEMATICAL INDUCTION

Let $A_1, A_2, A_3, \ldots$ represent a sequence of assertions. That is, for each positive integer n, there is a corresponding assertion A_n. If the following two conditions hold:

i A_1 is true

ii For each fixed positive integer k, if A_k is true, then A_{k+1} is true; then it follows that every assertion $A_1, A_2, A_3, \ldots$ is true. That is, A_n is true for all positive integers n.

For example, let us apply the principle of mathematical induction to prove the assertion A_n, where

$$A_n: 1 + 3 + 5 + \cdots + (2n - 1) = n^2$$

For example,

$A_1: 1 = 1^2$
$A_2: 1 + 3 = 2^2$
$A_3: 1 + 3 + 5 = 3^2$

and

$A_4: 1 + 3 + 5 + 7 = 4^2$

In order to show that A_n is true for all positive integers n we need to verify the following conditions:

i A_1 is true.

ii If A_k is true, then A_{k+1} is also true, where k is a fixed positive integer.

Condition i is obviously true in this example since A_1 is the assertion which states $1 = 1^2$.

To verify condition ii we must show that A_k implies A_{k+1}. That is, we must show that if A_k is assumed to be true, then A_{k+1} must also be true. If A_k is true, we have

$$\sum_{i=1}^{k}(2i - 1) = 1 + 3 + 5 + \cdots + (2k - 1) = k^2$$

Adding $(2k + 1)$ to both sides of the equation, we have

$$[1 + 3 + 5 + \cdots + (2k - 1)] + (2k + 1) = k^2 + (2k + 1)$$
$$= k^2 + 2k + 1$$
$$= (k + 1)^2$$

But this result is precisely the assertion A_{k+1}. Hence, we have proved condition ii for the example. Thus by the principle of mathematical induction we can conclude that A_n is true for any positive integer. Therefore $\sum_{i=1}^{n}(2i - 1) = 1 + 3 + 5 + \cdots + (2n - 1) = n^2$ for any positive integer n.

Let us consider other examples using the principle of mathematical induction.

EXAMPLES

1 Use mathematical induction to prove that for any positive integer n,

$$\sum_{i=1}^{n} 2i = 2 + 4 + 6 + \cdots + 2n = n(n + 1)$$

PROOF. Let A_n be the assertion $2 + 4 + 6 + \cdots + 2n = n(n + 1)$. Using the principle of mathematical induction, we have

i A_1 becomes $2 = 1(1 + 1)$, which is true.

ii Assume that A_k is true. That is, assume that for any fixed integer k

$$\sum_{i=1}^{k} 2i = 2 + 4 + 6 + \cdots + 2k = k(k + 1)$$

and prove that as a result A_{k+1} is true where A_{k+1} is the assertion

$$\sum_{i=1}^{k+1} 2i = 2 + 4 + 6 + \cdots + 2k + (2k + 2) = (k + 1)(k + 2)$$

We can do so by adding $(2k + 2)$ to both sides of A_k. Thus,

$$2 + 4 + 6 + \cdots + 2k + (2k + 2) = k(k + 1) + (2k + 2)$$
$$= k(k + 1) + 2(k + 1) = (k + 1)(k + 2)$$

Hence, A_{k+1} is true if A_k is true and we have proved condition ii. Thus by the principle of mathematical induction we can conclude that A_n is true for any positive integer n. That is, $\sum_{i=1}^{n} 2i = 2 + 4 + 6 + \cdots + 2n = n(n + 1)$ for any positive integer n.

2 Use mathematical induction to prove that for any positive integer n,

$$\sum_{i=1}^{n} i^2 = 1^2 + 2^2 + 3^2 + \cdots + n^2 = \tfrac{1}{6}n(n + 1)(2n + 1)$$

PROOF.   Let A_n be the assertion

$$\sum_{i=1}^{n} i^2 = 1^2 + 2^2 + 3^2 + \cdots + n^2 = \tfrac{1}{6}n(n + 1)(2n + 1)$$

Using the principle of mathematical induction, we have

i A_1: $1^2 = \tfrac{1}{6}(1)(1 + 1)(2 \cdot 1 + 1)$, or $1 = 1$ which is true.

ii Assume that A_k is true. That is, assume that for any fixed integer k,

$$1^2 + 2^2 + 3^2 + \cdots + k^2 = \tfrac{1}{6}k(k + 1)(2k + 1)$$

and prove that A_{k+1} is true, where A_{k+1} is the assertion

$$1^2 + 2^2 + 3^2 + \cdots + k^2 + (k + 1)^2 = \tfrac{1}{6}(k + 1)(k + 2)(2k + 3)$$

Adding $(k + 1)^2$ to both sides of A_k we have

$$\begin{aligned}
1^2 + 2^2 + 3^2 + \cdots + k^2 + (k + 1)^2 &= \left(\tfrac{1}{6}k(k + 1)(2k + 1)\right) + \left((k + 1)^2\right) \\
&= (k + 1)[\tfrac{1}{6}k(2k + 1) + (k + 1)] \\
&= (k + 1)\frac{2k^2 + k + 6k + 6}{6} \\
&= (k + 1)\frac{2k^2 + 7k + 6}{6} \\
&= \tfrac{1}{6}(k + 1)(k + 2)(2k + 3)
\end{aligned}$$

Since this last equation is A_{k+1}, then A_{k+1} is true if A_k is true. Hence, we have proved condition ii. Thus, by the principle of mathematical induction, we can conclude that A_n is true for any positive integer n. That is,

$$\sum_{i=1}^{n} i^2 = 1^2 + 2^2 + 3^2 + \cdots + n^2 = \tfrac{1}{6}n(n + 1)(2n + 1)$$

3. Use mathematical induction to verify the formula for the sum of the first n terms of an arithmetic progression. That is, prove that

$$S_n = \sum_{i=1}^{n} [a_1 + (i-1)d] = \frac{n}{2}(a_1 + a_n)$$

PROOF. Let A_n be the assertion

$$S_n = \sum_{i=1}^{n} [a_1 + (i-1)d] = a_1 + (a_1 + d) + (a_1 + 2d) + \cdots$$
$$+ [a_1 + (n-1)d]$$
$$= \frac{n}{2}(a_1 + a_n)$$

i A_1 becomes $S_1 = \frac{1}{2}(a_1 + a_1) = 2a_1/2 = a_1$, which is true.

ii If A_k is true, that is, if

$$a_1 + (a_1 + d) + (a_1 + 2d) + \cdots + [a_1 + (k-1)d] = \frac{k}{2}(a_1 + a_k)$$

then we must show that, as a consequence, A_{k+1} is true, that is, that

$$a_1 + (a_1 + d) + (a_1 + 2d) + \cdots + (a_1 + kd) = \frac{k+1}{2}(a_1 + a_{k+1})$$

By adding $(a_1 + kd)$ to both sides of assertion A_k, we have

$$a_1 + (a_1 + d) + (a_1 + 2d) + \cdots + [a_1 + (k-1)d] + (a_1 + kd)$$
$$= \frac{k}{2}(a_1 + a_k) + (a_1 + kd)$$
$$= \frac{k}{2}(a_1 + a_k) + \frac{2(a_1 + kd)}{2}$$
$$= \frac{ka_1 + ka_k + 2a_1 + 2kd}{2}$$
$$= \frac{(ka_1 + a_1) + (a_1 + ka_k + 2kd)}{2}$$
$$= \frac{(k+1)a_1 + (a_1 + kd + ka_k + kd)}{2}$$
$$= \frac{(k+1)a_1 + (a_{k+1} + ka_{k+1})}{2}$$
$$= \frac{(k+1)a_1 + (k+1)a_{k+1}}{2}$$
$$= \frac{k+1}{2}(a_1 + a_{k+1})$$

Hence A_{k+1} is true if A_k is true. Thus by the principle of mathematical induction we can conclude that A_n is true for any positive integer n.

4 Prove that $x + y$ is a factor of $x^{2n-1} + y^{2n-1}$ for any positive integer n.

PROOF. Using the principle of mathematical induction,

i A_1 becomes the assertion that $x^{2(1)-1} + y^{2(1)-1} = x + y$ is divisible by $x + y$, which is obviously true.

ii If A_k is true, that is, if $x^{2k-1} + y^{2k-1}$ is divisible by $x + y$, then we must prove that the assertion A_{k+1} is true; that is, we must prove that

$$x^{2(k+1)-1} + y^{2(k+1)-1} = x^{2k+1} + y^{2k+1}$$

is divisible by $x + y$.

If we add and subtract $x^2 y^{2k-1}$ to $x^{2k+1} + y^{2k+1}$, we have

$$\begin{aligned} x^{2k+1} + y^{2k+1} &= x^{2k+1} + x^2 y^{2k-1} - x^2 y^{2k-1} + y^{2k+1} \\ &= x^2(x^{2k-1} + y^{2k-1}) + y^{2k-1}(y^2 - x^2) \\ &= x^2(x^{2k-1} + y^{2k-1}) + y^{2k-1}(y - x)(y + x) \end{aligned}$$

If A_k is true, then each term of this last expression is divisible by $x + y$, so that $x^{2k+1} + y^{2k+1}$ is also divisible by $x + y$. Hence, by the principle of mathematical induction, we can conclude that for any positive integer n, $x^{2n-1} + y^{2n-1}$ is divisible by $x + y$.

PROBLEM SET 3

Use the principle of mathematical induction to prove each of the following assertions for all positive integers.

1 $\sum_{i=1}^{n} i = 1 + 2 + 3 + 4 + \cdots + n = \dfrac{n(n + 1)}{2}$

2 $\sum_{i=1}^{n} (4i - 1) = 3 + 7 + 11 + \cdots + (4n - 1) = n(2n + 1)$

3 $\sum_{i=1}^{n} 6i = 6 + 12 + 18 + \cdots + 6n = 3n(n + 1)$

4 $\sum_{i=1}^{n} (4i - 3) = 1 + 5 + 9 + \cdots + (4n - 3) = n(2n - 1)$

5 $\sum_{i=1}^{n} 4^i = 4 + 4^2 + 4^3 + \cdots + 4^n = \dfrac{4(4^n - 1)}{3}$

6 $\sum_{i=1}^{n} i 2^{i-1} = 1 + 2 \cdot 2 + 3 \cdot 2^2 + 4 \cdot 2^3 + \cdots + n 2^{n-1} = 1 + (n - 1)2^n$

7 $\sum_{i=1}^{n} i 3^{i-1} = 1 + 2 \cdot 3 + 3 \cdot 3^2 + 4 \cdot 3^3 + \cdots + n 3^{n-1} = \dfrac{1 + (2n - 1)3^n}{4}$

8. $\displaystyle\sum_{i=1}^{n} 2i(2i+2) = 2\cdot 4 + 4\cdot 6 + 6\cdot 8 + \cdots + 2n(2n+2)$
$$= \frac{4n}{3}(n+1)(n+2)$$

9. $\displaystyle\sum_{i=1}^{n} (3i-1)(3i+2) = 2\cdot 5 + 5\cdot 8 + 8\cdot 11 + \cdots$
$$+ (3n-1)(3n+2) = n(3n^2 + 6n + 1)$$

10. $\displaystyle\sum_{i=1}^{n} (3i-2)(3i+3) = 1\cdot 6 + 4\cdot 9 + 7\cdot 12 + \cdots$
$$+ (3n-2)(3n+3) = 3n(n^2 + 2n - 1)$$

11. $\displaystyle\sum_{i=1}^{n} \frac{1}{i(i+1)} = \frac{1}{1\cdot 2} + \frac{1}{2\cdot 3} + \frac{1}{3\cdot 4} + \cdots + \frac{1}{n(n+1)} = \frac{n}{n+1}$

12. $\displaystyle\sum_{i=1}^{n} \frac{1}{(2i-1)(2i+1)} = \frac{1}{1\cdot 3} + \frac{1}{3\cdot 5} + \frac{1}{5\cdot 7} + \cdots$
$$+ \frac{1}{(2n-1)(2n+1)} = \frac{n}{2n+1}$$

13. $\displaystyle\sum_{i=1}^{n} (2i-1)^3 = 1^3 + 3^3 + 5^3 + \cdots + (2n-1)^3 = n^2(2n^2 - 1)$

14. $x - y$ is a factor of $x^n - y^n$.

15. $x + y$ is a factor of $x^{2n} - y^{2n}$.

16. The sum of the first n terms of a geometric progression $a_n = a_1 r^{n-1}$ is given by the formula
$$S_n = \sum_{k=1}^{n} a_1 r^{k-1} = \frac{a_1(1 - r^n)}{1 - r}$$

5 Combinations and Permutations

In many practical applications of mathematics it is often necessary to determine the number of different ways in which the elements of a set can be arranged to form subsets of a given size. Subsets that are formed without regard to the order in which the elements are presented are called *combinations*, whereas those subsets that are formed where order is significant are called *permutations*. Finding the number of three-man committees that can be formed from ten men is an example of finding the number of combinations. However, to find the number of different four-digit license plates that can be made from the set of digits $\{1, 2, 3, 4, 5, 6, 7, 8, 9\}$ is an example of permutations, since order is obviously important. Our purpose in this section is to determine formulas for finding the number of different combinations or permutations that can be formed from a given set.

5.1 Combinations

A set of k elements chosen from a given set of n different elements, without regard to the order in which they are chosen or arranged, is called a *combination* of n different elements taken k at a time. The notation C_k^n, also written $\binom{n}{k}$, is used to denote the number of combinations of n elements taken k at a time.

For example, the number of two-element combinations that can be formed from the set $\{a, b, c\}$ is 6. The combinations are $\{a, b\}$, $\{a, c\}$, $\{a, d\}$, $\{b, c\}$, $\{b, d\}$, and $\{c, d\}$. We write $C_2^4 = \binom{4}{2} = 6$.

Also the number of three-man committees that can be formed from a group of five can be determined as follows. Assume that the five are represented as a, b, c, d, and e; then the possible three-man committees are

$\{a, b, c\}$ $\quad$ $\{a, d, e\}$
$\{a, b, d\}$ $\quad$ $\{b, c, d\}$
$\{a, b, e\}$ $\quad$ $\{b, c, e\}$
$\{a, c, d\}$ $\quad$ $\{b, d, e\}$
$\{a, c, e\}$ $\quad$ $\{c, d, e\}$

Consequently, there are 10 possible committees and we write

$$C_3^5 = \binom{5}{3} = 10$$

The formula for C_k^n is conveniently expressed using *factorial* notation. We shall first introduce this notation before continuing with the discussion of combinations. The symbol $n!$ (read "n factorial" or "factorial n") is defined by

$$n! = 1 \cdot 2 \cdot 3 \cdots (n-1)n$$

or

$$n! = n(n-1)(n-2) \cdots 2 \cdot 1$$

Thus,

$$4! = 4 \cdot 3 \cdot 2 \cdot 1 = 24 \quad \text{and} \quad 6! = 6 \cdot 5 \cdot 4 \cdot 3 \cdot 2 \cdot 1 = 720$$

Notice that

$$(n-1)! = (n-1)(n-2)(n-3) \cdots 4 \cdot 3 \cdot 2 \cdot 1$$

and, by multiplying both sides of this equation by n, we have the following recursive relationship:

$$n! = n[(n-1)!]$$

Setting $n = 1$ in the above relationship, we have

$$1! = 1[(1-1)!] \quad \text{or} \quad 1! = 1 \cdot 0!$$

Therefore, we shall define

$$0! = 1$$

Expressions involving factorial notation may be simplified as follows:

$$\frac{7!}{5!} = \frac{7 \cdot 6 \cdot 5 \cdot 4 \cdot 3 \cdot 2 \cdot 1}{5 \cdot 4 \cdot 3 \cdot 2 \cdot 1} = 7 \cdot 6 = 42$$

$$\frac{8!}{3! \cdot 5!} = \frac{8 \cdot 7 \cdot 6 \cdot 5 \cdot 4 \cdot 3 \cdot 2 \cdot 1}{(3 \cdot 2 \cdot 1)(5 \cdot 4 \cdot 3 \cdot 2 \cdot 1)} = 8 \cdot 7 = 56$$

EXAMPLES

1   Write in expanded form and simplify.

a) $\dfrac{7!}{5! \cdot 3!}$ b) $\dfrac{3!}{2!4!}$ c) $\dfrac{3! + 5!}{4! - 6!}$ d) $\dfrac{(n+1)!}{(n-1)!}$

SOLUTION

a) $\dfrac{7!}{5! \cdot 3!} = \dfrac{7 \cdot 6 \cdot 5!}{5! \cdot 3 \cdot 2 \cdot 1} = 7$

b) $\dfrac{3!}{2!4!} = \dfrac{3 \cdot 2!}{2!4!} = \dfrac{3}{4 \cdot 3 \cdot 2} = \dfrac{1}{8}$

c) $\dfrac{3! + 5!}{4! - 6!} = \dfrac{3!(1 + 5 \cdot 4)}{3!(4 - 4 \cdot 5 \cdot 6)} = \dfrac{1 + 20}{4 - 120} = -\dfrac{21}{116}$

d) $\dfrac{(n+1)!}{(n-1)!} = \dfrac{(n+1)(n)[(n-1)!]}{(n-1)!} = (n+1)n = n^2 + n$

Using the factorial notation, the formula for $\binom{n}{k}$ can be expressed as follows. If $0 \le k \le n$, then the combination C_k^n or $\binom{n}{k}$ is defined by

$$\binom{n}{k} = \frac{n!}{k!(n-k)!}$$

Thus, the number of combinations of four letters a, b, c, and d taken two at a time is

$$\binom{4}{2} = \frac{4!}{2!(4-2)!} = \frac{4!}{2!2!} = 6$$

Also, the number of five-man committees that can be formed from a group of six Democrats is

$$\binom{6}{5} = \frac{6!}{5!(6-5)!} = 6$$

2 Evaluate the following expressions.

a) $\binom{8}{5}$ b) $\binom{10}{8}$

SOLUTION

a) $\binom{8}{5} = \frac{8!}{5!(8-5)!} = \frac{8!}{5!3!} = \frac{8 \cdot 7 \cdot 6 \cdot 5!}{5!3!} = \frac{8 \cdot 7 \cdot 6}{3 \cdot 2 \cdot 1} = 56$

b) $\binom{10}{8} = \frac{10!}{8!(10-8)!} = \frac{10!}{8!2!} = \frac{10 \cdot 9 \cdot 8!}{8!2!} = \frac{10 \cdot 9}{2 \cdot 1} = 45$

3 Find the number of combinations of 20 things taken 17 at a time.

SOLUTION. The number of combinations of 20 things taken 17 at a time is

$$\binom{20}{17} = \frac{20!}{17!(20-17)!} = \frac{20 \cdot 19 \cdot 18 \cdot 17!}{17!3!} = \frac{20 \cdot 19 \cdot 18}{3 \cdot 2 \cdot 1} = 1{,}140$$

4 How many different hands consisting of 5 cards could be dealt from a deck of 52 cards?

SOLUTION. The number of different hands is

$$\binom{52}{5} = \frac{52!}{5!(52-5)!} = \frac{52!}{5!47!} = \frac{52 \cdot 51 \cdot 50 \cdot 49 \cdot 48 \cdot 47!}{5!47!}$$
$$= 2{,}598{,}960$$

5 From a group of 20 employees, 4 are to be selected to work on a special project. In how many different ways can the 4 employees be selected?

SOLUTION. Here we wish to know how many 4-element combinations can be formed from a set of 20 elements.

$$\binom{20}{4} = \frac{20!}{4!(20-4)!} = 4{,}845$$

6 Show that

$$\binom{n}{k} = \frac{n(n-1)(n-2)\cdots(n-k+1)}{k!}$$

SOLUTION

$$\binom{n}{k} = \frac{n!}{k!(n-k)!}$$
$$= \frac{n(n-1)(n-2)\cdots(n-k+1)[(n-k)!]}{k!(n-k)!}$$
$$= \frac{n(n-1)(n-2)\cdots(n-k+1)}{k!}$$

5.2 Permutations

A set of k elements chosen from a given set of n different elements, in which the order they are arranged or chosen is significant, is called a *permutation* of n different elements taken k at a time.

For example, the set of digits $\{7, 8, 9\}$ may be arranged as a 3 digit number without repetitions in the following six ways, each of which is a permutation of the set

    789   879   978

    798   897   987

The number of permutations may be calculated beforehand by noting that the first place can be filled in any one of three different ways; after it is filled, the second place can be filled in any one of two different ways by the remaining two digits. Then the third place is filled by the remaining digit. We see that the number of permutations can be written $3 \cdot 2 \cdot 1 = 3! = 6$.

In general, the *number of permutations* of n elements taken k at a time, written P_k^n, is defined to be

$$P_k^n = n(n-1)(n-2)(n-3)\cdots(n-k+1)$$

For example, $P_3^5 = 5 \cdot 4 \cdot 3 = 60$ and $P_2^4 = 4 \cdot 3 = 12$.

In particular, if $k = n$, we obtain the number of permutations of n different things taken n at a time as

$$P_n^n = n(n-1)(n-2)\cdots(3)(2)(1) = n!$$

EXAMPLES

1   Find P_1^4, P_2^5, P_3^6, P_5^8, and P_4^4.

SOLUTION

$$P_1^4 = 4$$
$$P_2^5 = 5 \cdot 4 = 20$$
$$P_3^6 = 6 \cdot 5 \cdot 4 = 120$$
$$P_5^8 = 8 \cdot 7 \cdot 6 \cdot 5 \cdot 4 = 6,720$$
$$P_4^4 = 4 \cdot 3 \cdot 2 \cdot 1 = 24$$

2 How many three-digit integers can be formed from the set of digits $\{1, 2, 3, 4, 5\}$, if no repetition of digits is allowed?

SOLUTION. We have five different digits to be taken three at a time. Therefore, the required number is $P_3^5 = 5 \cdot 4 \cdot 3 = 60$.

3 In how many ways can five students be seated in a row of five desks?

SOLUTION. We have five different students to be seated in five different desks. This can be computed as $P_5^5 = 5! = 5 \cdot 4 \cdot 3 \cdot 2 \cdot 1 = 120$.

An alternate expression for P_k^n can be obtained by observing that

$$P_k^n = n(n-1)(n-2) \cdots (n-k+1)$$
$$= \frac{n(n-1)(n-2) \cdots (n-k+1)(n-k)!}{(n-k)!}$$

or

$$P_k^n = \frac{n!}{(n-k)!}$$

4 Find the number of permutations of six elements taken three at a time.

SOLUTION. Using the formula

$$P_k^n = \frac{n!}{(n-k)!}$$

we have

$$P_3^6 = \frac{6!}{3!} = \frac{6 \cdot 5 \cdot 4 \cdot 3!}{3!} = 120$$

5 Find n if $P_2^n = 110$.

SOLUTION

$$P_2^n = \frac{n!}{(n-2)!} = n(n-1)$$

so that

$$n(n-1) = 110 \quad \text{or} \quad n^2 - n - 110 = 0$$

$(n-11)(n+10) = 0$. Then $n = 11$. Note that the solution $n = -10$ is discarded here. (why?)

6 In how many ways can four differently colored marbles be arranged in a row of 4?

SOLUTION. The number of arrangements of four different marbles in a row is expressed by

$$P_4^4 = 4! = 24$$

7 It is required to seat five men and four women in a row of 9 so that the women occupy the even places. How many such arrangements are possible?

SOLUTION. The men may be seated in P_5^5 ways, the women in P_4^4 ways. Each arrangement of men is associated with each arrangement of women. Hence, the required number of arrangements is given by the product $P_5^5 \cdot P_4^4 = 5! \cdot 4! = 2,880$.

PROBLEM SET 4

1 Write in expanded form and simplify.
  a) $\dfrac{4!}{6!}$
  b) $\dfrac{3! \cdot 8!}{4! \cdot 7!}$
  c) $\dfrac{1}{4!} + \dfrac{1}{3!}$
  d) $\dfrac{2!}{4! - 3!}$
  e) $\dfrac{0}{0!}$
  f) $\dfrac{4! \cdot 6!}{8! - 5!}$
  g) $\dfrac{(n-2)!}{(n+1)!}$
  h) $\dfrac{(n+1)!}{(n-3)!}$
  i) $\dfrac{(n+k)!}{(n+k-2)!}$

2 Simplify each of the following expressions.
  a) $\binom{5}{0}$
  b) $\binom{5}{1}$
  c) $\binom{5}{2}$
  d) $\binom{n}{0}$
  e) $\binom{n}{1}$
  f) $\binom{n}{n}$
  g) $\binom{n}{2}$
  h) $\binom{12}{0}$
  i) $\binom{12}{12}$

3 Find n in each of the following cases.
  a) $\binom{n}{3} = \binom{50}{47}$
  b) $\binom{n}{7} = \binom{n}{5}$

4 Show that $\binom{n}{k} = \binom{n}{n-k}$.

5. a) Compute $4! + 3!$
   b) Compute $(4 + 3)!$
   c) Does $4! + 3! = (4 + 3)!$?

6. How many different committees of three persons can be chosen from a group of five persons?

7. How many different hands of 13 cards can be dealt from a deck of 52 cards?

8. In how many different ways can a hand consisting of three aces and two cards that are not aces be selected from a standard bridge deck?

9. Evaluate each of the following expressions.
   a) P_2^7 b) P_4^6 c) P_5^8 d) P_3^4

10. Find n in each of the following cases.
    a) $P_2^{n+1} = 3P_2^{n-3}$ b) $7P_2^n = 12P_2^{n-2}$
    c) $P_5^n = 20P_3^n$ d) $P_3^{n+1} = P_4^n$

11. Show that each of the following equations is true.
    a) $P_3^n = n \cdot P_2^{n-1}$ b) $P_3^n - P_2^n = (n - 3)P_2^n$
    c) $k! C_k^n = P_k^n$

12. A president, a vice-president, a secretary, and a treasurer of an organization are selected from 30 members. In how many ways can this selection be made?

13. How many three-digit numbers can be formed from the digits 1, 3, 4, 7, 8, and 9 if no repetitions of digits are allowed?

14. There are 11 flags that are displayed together, one above another, on a flag pole. How many signals are possible if 6 of the flags are different colored square flags, 5 are different colored triangular flags, and that no similar shaped flags will be next to each other.

15. In how many ways may five boys and five girls sit in a row if the boys and girls are assigned to alternate seats?

16. In how many ways may 10 books be arranged on a shelf?

17. How many different 5 member committees can be formed from 9 people if each person can serve on more than one committee?

18. How many different license plates can be made that have three different letters and three different digits?

6 Binomial Theorem

In Chapter 1 we considered the special products $(a + b)^2$ and $(a + b)^3$.

In this section we shall develop a formula for the expansion of $(a + b)^n$, where n is any positive integer.

$$(a + b)^1 = a + b$$
$$(a + b)^2 = a^2 + 2ab + b^2$$
$$(a + b)^3 = a^3 + 3a^2b + 3ab^2 + b^3$$
$$(a + b)^4 = a^4 + 4a^3b + 6a^2b^2 + 4ab^3 + b^4$$
$$(a + b)^5 = a^5 + 5a^4b + 10a^3b^2 + 10a^2b^3 + 5ab^4 + b^5$$

The above pattern holds for the expansion of $(a + b)^n$, where n is a positive integer. Hence, in general

1. There are $n + 1$ terms. The "first term" is a^n and the "last term" is b^n.

2. The powers of a decrease by 1 and the powers of b increase by 1 for each term, the sum of the exponents of a and b is n for each term.

3. If the coefficient of any term is multiplied by the exponent on a and divided by 1 more than the exponent on b, the result is the coefficient of the next term.

One method of displaying the coefficients in the expansion of $(a + b)^n$, for $n = 1, 2, 3, \ldots$, is the following array of numbers, known as *Pascal's triangle*:

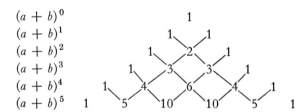

The coefficients in each line—except the first and last, which are always 1—can be found by adding the pair of coefficients from the preceding line as indicated by the V's. The pattern displayed by the terms in the expansion of $(a + b)^n$ is easier to express using the notation introduced in Section 5.1. Thus, we have

The "first term" is a^n, and the coefficient is $1 = \binom{n}{0}$

The "second term" contains $a^{n-1}b$, and the coefficient is $\dfrac{n}{1!} = \binom{n}{1}$

The "third term" contains $a^{n-2}b^2$, and the coefficient is

$$\frac{n(n-1)}{2!} = \binom{n}{2}$$

The "fourth term" contains $a^{n-3}b^3$, and the coefficient is

$$\frac{n(n-1)(n-2)}{3!} = \binom{n}{3}$$

and so forth.

The "$k+1$" term contains $a^{n-k}b^k$, and the coefficient is

$$\frac{n(n-1)(n-2)\cdots(n-k+1)}{k!} = \binom{n}{k}$$

The "nth term" contains b^n, and the coefficient is

$$\frac{n(n-1)(n-2)\cdots 1}{n!} = \binom{n}{n} = 1$$

These results are generalized in the *binomial theorem* (also known as the *binomial expansion*).

BINOMIAL THEOREM

Let a and b be real numbers and let n be a positive integer; then

$$(a+b)^n = \binom{n}{0}a^n + \binom{n}{1}a^{n-1}b + \cdots + \binom{n}{k}a^{n-k}b^k + \cdots$$

$$+ \binom{n}{n}b^n = \sum_{k=0}^{n}\binom{n}{k}a^{n-k}b^k$$

PROOF. Using the principle of mathematical induction, we have

i A_1 is the statement: $(a+b)^1 = (a^1 + b^1)$, which is obviously true.

ii   We must show that if A_n is true, then A_{n+1} is also true. (Here we are using n instead of k.)

To this end assume that A_n is true; that is, assume that

$$(a+b)^n = \binom{n}{0}a^n + \binom{n}{1}a^{n-1}b + \cdots + \binom{n}{k}a^{n-k}b^k + \cdots$$

$$+ \binom{n}{n}b^n$$

for n a positive integer. Multiplying both sides by $(a + b)$, we obtain

$$(a + b)^n(a + b) = (a + b)\left[\binom{n}{0}a^n + \binom{n}{1}a^{n-1}b + \cdots\right.$$
$$\left. + \binom{n}{k}a^{n-k}b^k + \cdots + \binom{n}{n}b^n\right]$$
$$= \binom{n}{0}(a^{n+1} + a^n b) + \binom{n}{1}(a^n b + a^{n-1}b^2) + \cdots$$
$$+ \binom{n}{k}(a^{n-k+1}b^k + a^{n-k}b^{k+1}) + \cdots$$
$$+ \binom{n}{n}(ab^n + b^{n+1})$$
$$= \binom{n}{0}a^{n+1} + \left[\binom{n}{0} + \binom{n}{1}\right]a^n b$$
$$+ \left[\binom{n}{1} + \binom{n}{2}\right]a^{n-1}b^2 + \cdots$$
$$+ \left[\binom{n}{k-1} + \binom{n}{k}\right]a^{n+1-k}b^k + \cdots$$
$$+ \binom{n}{n}b^{n+1}$$

But

$$\binom{n}{k-1} + \binom{n}{k} = \frac{n!}{(k-1)!(n-k+1)!} + \frac{n!}{k!(n-k)!}$$
$$= \frac{n!k + n!(n-k+1)}{k!(n-k+1)!} = \frac{n!(n+1)}{k!(n+1-k)!}$$
$$= \frac{(n+1)!}{k!(n+1-k)!} = \binom{n+1}{k}$$

so that

$$(a + b)^{n+1} = \binom{n+1}{0}a^{n+1} + \binom{n+1}{1}a^n b + \cdots$$
$$+ \binom{n+1}{k}a^{n+1-k}b^k + \cdots + \binom{n+1}{n+1}b^{n+1}$$

But the latter assertion is precisely A_{n+1}, and the proof is complete.

For example to expand $(x + 2y^2)^5$, we substitute x for a and $2y^2$ for b, so that

$$(x + 2y^2)^5 = [x + (2y^2)]^5 = \binom{5}{0}x^5 + \binom{5}{1}x^4(2y^2) + \binom{5}{2}x^3(2y^2)^2$$
$$+ \binom{5}{3}x^2(2y^2)^3 + \binom{5}{4}x(2y^2)^4 + \binom{5}{5}(2y^2)^5$$

$$= x^5 + \frac{5}{1!}x^4(2y^2) + \frac{5\cdot 4}{2!}x^3(2y^2)^2 + \frac{5\cdot 4\cdot 3}{3!}x^2(2y^2)^3$$
$$+ \frac{5\cdot 4\cdot 3\cdot 2}{4!}x(2y^2)^4 + \frac{5\cdot 4\cdot 3\cdot 2\cdot 1}{5!}(2y^2)^5$$
$$= x^5 + 10x^4y^2 + 40x^3y^4 + 80x^2y^6 + 80xy^8 + 32y^{10}$$

In order to write any particular term of the binomial expansion $(a + b)^n$, or to find the term where b has any particular exponent, observe that the $k + 1$ term (denoted by u_{k+1}) in the binomial expansion is given by

$$u_{k+1} = \binom{n}{k}a^{n-k}b^k = \frac{n!}{k!(n-k)!}a^{n-k}b^k$$
$$= \frac{n(n-1)(n-2)\cdots(n-k+1)}{k!}a^{n-k}b^k$$

For example, the sixth term u_6 of $(x^2 + 2y)^{12}$ is

$$\binom{12}{5}(x^2)^7(2y)^5 = \frac{12\cdot 11\cdot 10\cdot 9\cdot 8}{5\cdot 4\cdot 3\cdot 2\cdot 1}(x^2)^7(2y)^5 = 25{,}344x^{14}y^5$$

and the term involving y^4 in the expansion of $(x^2 + 2y)^{12}$ is

$$\binom{12}{4}(x^2)^8(2y)^4 = \frac{12\cdot 11\cdot 10\cdot 9}{4!}(x^2)^8(2y)^4 = 7{,}920x^{16}y^4$$

EXAMPLES

1 Expand $(x + y)^7$.

SOLUTION

By the binomial theorem

$$(x+y)^7 = \binom{7}{0}x^7 + \binom{7}{1}x^6y + \binom{7}{2}x^5y^2 + \binom{7}{3}x^4y^3 + \binom{7}{4}x^3y^4$$
$$+ \binom{7}{5}x^2y^5 + \binom{7}{6}xy^6 + \binom{7}{7}y^7$$
$$= x^7 + \frac{7}{1!}x^6y + \frac{7\cdot 6}{2!}x^5y^2 + \frac{7\cdot 6\cdot 5}{3!}x^4y^3 + \frac{7\cdot 6\cdot 5\cdot 4}{4!}x^3y^4$$
$$+ \frac{7\cdot 6\cdot 5\cdot 4\cdot 3}{5!}x^2y^5 + \frac{7\cdot 6\cdot 5\cdot 4\cdot 3\cdot 2}{6!}xy^6 + y^7$$
$$= x^7 + 7x^6y + 21x^5y^2 + 35x^4y^3 + 35x^3y^4$$
$$+ 21x^2y^5 + 7xy^6 + y^7$$

2 Expand $(x - 2)^4$.

SOLUTION

$$(x - 2)^4 = [x + (-2)]^4$$
$$= \binom{4}{0} x^4 + \binom{4}{1} x^3(-2) + \binom{4}{2} x^2(-2)^2$$
$$+ \binom{4}{3} x(-2)^3 + \binom{4}{4} (-2)^4$$
$$= x^4 + 4x^3(-2) + 6x^2(-2)^2 + 4x(-2)^3 + (-2)^4$$
$$= x^4 - 8x^3 + 24x^2 - 32x + 16$$

3   Find the eighth term of the expansion of $(x - y)^{12}$.

SOLUTION.   In this example, $a = x$, $b = -y$, $n = 12$, and $k + 1 = 8$ or $k = 7$. Substituting in the formula for u_{k+1}, the eighth term is

$$\binom{12}{7} x^5 (-y)^7 = -\binom{12}{7} x^5 y^7$$
$$= -\frac{12 \cdot 11 \cdot 10 \cdot 9 \cdot 8 \cdot 7 \cdot 6}{7 \cdot 6 \cdot 5 \cdot 4 \cdot 3 \cdot 2 \cdot 1} x^5 y^7$$
$$= -792 x^5 y^7$$

4   Find the term involving x^7 in the expansion of $(2 - x)^{12}$.

SOLUTION.   In this example, $a = 2$, $b = -x$, $n = 12$, and $k = 7$. (Why?) Substituting in the formula for u_{k+1}, the term involving x^7 is

$$\binom{12}{7} (2)^5 (-x)^7 = -\binom{12}{7} (2)^5 x^7$$
$$= -\frac{12 \cdot 11 \cdot 10 \cdot 9 \cdot 8 \cdot 7 \cdot 6}{7 \cdot 6 \cdot 5 \cdot 4 \cdot 3 \cdot 2 \cdot 1} 32 x^7$$
$$= -25{,}344 x^7$$

PROBLEM SET 5

1   Use the binomial theorem to expand each of the following expressions.
 a) $(2z + x)^8$
 b) $(x - 3)^8$
 c) $(y^2 - 2x)^5$
 d) $\left(\frac{1}{a} + \frac{x}{2}\right)^5$

2   Find the first four terms of the expansion.
 a) $(x^2 - 2a)^{10}$
 b) $\left(2a - \frac{1}{b}\right)^6$
 c) $\left(\sqrt{\frac{x}{2}} + 2y\right)^7$
 d) $\left(\frac{1}{a} + \frac{x}{2}\right)^{11}$
 e) $(a^{3/2} - 2x^2)^8$

3 Find the first five terms in the following expansions and simplify them.
   a) $(x + y)^{16}$ b) $(a^2 + b^2)^{12}$
   c) $(a - 2b^2)^{11}$ d) $(a + 2y^2)^8$
   e) $(x - 2y)^7$ f) $\left(1 - \dfrac{x}{y^2}\right)^8$
   g) $(a^3 - a^2)^9$ h) $\left(x + \dfrac{1}{2y}\right)^{15}$

4 Find the indicated term for each of the following expressions.
   a) $\left(\dfrac{x^2}{2} + a\right)^{15}$, fourth term   b) $(y^2 - 2z)^{10}$, sixth term
   c) $\left(2x^2 - \dfrac{a^2}{3}\right)^9$, seventh term   d) $(x + \sqrt{a})^{12}$, middle term
   e) $\left(a + \dfrac{x^3}{3}\right)^9$, term containing x^{12} f) $\left(2\sqrt{y} - \dfrac{x}{2}\right)^{10}$, term containing y^4

7 Probability

Probability is applied in a variety of practical situations in business, economics, and the social and physical sciences. The definition and the rules of probability will be presented here in terms of motivational problems rather than formally. However, before discussing probability, it is necessary to establish some basic terminology. We say that activities such as tossing a coin, reading the daily temperature on a thermometer, or drawing a card from a deck of cards are called *experiments*. With any kind of experiment there is associated a set of possible results called *outcomes*.

For example, when a coin is tossed, there are two possible outcomes. They are heads (denoted by H) or tails (denoted by T). If two different coins are tossed simultaneously, the possible outcomes are: (H, H), (H, T), (T, H), and (T, T). In each of the above examples these are the only possible outcomes. The set of all possible outcomes in an experiment is called a *sample space*.

For example, suppose that an urn contains a white ball (W), a red ball (R), and a green ball (G), and an experiment consists of drawing one ball from the urn and then another one without replacing the first ball that is drawn. To determine the sample space, we note that the first ball removed will be W, R, or G. If the first ball is W, the next ball will be either R or G. Hence, there are two possible outcomes if the first ball is W. Similarly, there are two possible outcomes if the first is either R or G. Hence, the sample space for this experiment is

$$\{(W, R), (W, G), (R, W), (R, G), (G, W), (G, R)\}$$

Each element of this set represents the order in which the two balls are drawn. Note that outcome (W, R) is different from (R, W).

In general, we call each subset of a sample space an *event*. Suppose we toss a die and observe the number. The sample space S is given by $S = \{1, 2, 3, 4, 5, 6\}$. The event "a number less than 3 showing" is the subset $\{1, 2\}$ of the sample space S.

EXAMPLE

A single die is rolled. Write the following events in set notation.
a) E_1: The number showing is odd.
b) E_2: The number showing is even.
c) E_3: The number showing is greater than three.
d) E_4: The number showing is less than seven.
e) E_5: The number showing is seven.

SOLUTION. In set notation, these events can be written as the following subsets of the sample space $\{1, 2, 3, 4, 5, 6\}$.
a) $E_1 = \{1, 3, 5\}$
b) $E_2 = \{2, 4, 6\}$
c) $E_3 = \{4, 5, 6\}$
d) $E_4 = \{1, 2, 3, 4, 5, 6\}$
e) $E_5 = \emptyset$

Notice that each of these sets is a subset of the sample space.

7.1 Probability of an Event

If a fair die (one that is perfectly symmetrical) is tossed in an unbiased manner (without trying to force a particular outcome), then any one face of the die has the same chance of showing topside as does any other face. (From here on, in all problems involving dice, we assume that this is the case.) In this case we are inclined to say, for example, that "The chances that the die will land with 1 up are 1 out of 6." Similarly, we might say, "There are 3 chances out of 6 that the number showing is even," or "The chances that the number showing will be greater than 4 are 2 out of 6."

Mathematical probability is, for the most part, based on the layman's interpretation of the word "probability," as illustrated in the preceding paragraph. Thus we define *the probability of an event* as follows:

DEFINITION

If an experiment with finite sample space S is conducted, then the probability that the actual outcome belongs to the event E is, (the number of elements in E)/(the number of elements in S). This ratio is also referred to as *the probability of the event E*.

Letting $n(E)$ and $n(S)$ denote the number of elements in E and S, respectively, and letting $p(E)$ denote the probability of E, we have

$$p(E) = \frac{n(E)}{n(S)}$$

Using this definition of probability we can deduce the following three properties.

If E is an event of a finite sample space S and $p(E)$ denotes the probability of E, we have

i $p(S) = 1$ since $p(S) = n(S)/n(S) = 1$.

ii $p(E) = 0$ if $E = \emptyset$, since $n(E) = 0$ suggests that $p(E) = n(E)/n(S) = 0/n(S) = 0$.

iii $0 \leq p(E) \leq 1$. Since $E \subseteq S$, it follows that $n(E) \leq n(S)$. (Why?)

Consequently, $p(E) = n(E)/n(S) \leq n(S)/n(S) = 1$. Also, since $n(E) \geq 0$, we have $0 \leq n(E)/n(S) = p(E)$. Hence, $0 \leq p(E) \leq 1$.

EXAMPLES

1 Consider the experiment of tossing a die with sample space $S = \{1, 2, 3, 4, 5, 6\}$. What is the probability of the event E, an even number showing?

SOLUTION. The event E is given by $E = \{2, 4, 6\}$. We see that $n(E) = 3$ and $n(S) = 6$, so that

$$p(E) = \frac{n(E)}{n(S)} = \frac{3}{6} = \frac{1}{2}$$

2 In drawing a card from a well-shuffled deck of 52 cards, find the probability of the card being an ace.

SOLUTION. Since there are four aces in the deck of 52 cards, $n(E) = 4$ and $n(S) = 52$, so that

$$p(E) = \frac{n(E)}{n(S)} = \frac{4}{52} = \frac{1}{13}$$

3 Suppose that an urn contains 3 red balls, 4 blue balls, and 2 green balls. Assume further that the balls are thoroughly mixed and that the experiment consists of selecting a ball out of the urn at random. (This means one ball is as likely to be selected as another.)
a) What is the probability of drawing a red ball?

b) What is the probability of drawing a blue ball?
c) What is the probability of drawing a green ball?

SOLUTION. We can set up a sample space by numbering the balls 1, 2, 3, 4, 5, 6, 7, 8, and 9. Let us assign 1, 2, and 3 to the red balls; 4, 5, 6 and 7 to the blue balls and 8 and 9 to the green balls. Then

$S = \{1, 2, 3, 4, 5, 6, 7, 8, 9\}$
Event E_1, drawing a red ball, is $E_1 = \{1, 2, 3\}$
Event E_2, drawing a blue ball, is $E_2 = \{4, 5, 6, 7\}$
Event E_3, drawing a green ball, is $E_3 = \{8, 9\}$

a) Here $n(S) = 9$ and $n(E_1) = 3$, so that the probability of drawing a red ball is

$$p(E_1) = \frac{n(E_1)}{n(S)} = \frac{3}{9} = \frac{1}{3}$$

b) Here $n(S) = 9$ and $n(E_2) = 4$, so that the probability of drawing a blue ball is

$$p(E_2) = \frac{n(E_2)}{n(S)} = \frac{4}{9}$$

c) Here $n(S) = 9$ and $n(E_3) = 2$, so that the probability of drawing a green ball is

$$p(E_3) = \frac{n(E_3)}{n(S)} = \frac{2}{9}$$

Note that in this example $p(E_1) + p(E_2) + p(E_3) = \frac{3}{9} + \frac{4}{9} + \frac{2}{9} = 1$. That is, the probability of drawing either a red, blue, or green ball in this example is 1, which means that the event of drawing a red, green, or blue ball cannot fail to happen. This example illustrates an interesting feature of problems in probability, which is that choosing the sample space correctly can be the most difficult part. If we had chosen the space to be $\{R, B, G\}$, standing for red, blue, and green, respectively, the correct result would have been impossible to find; on the other hand, by realizing that we could give numbers to the balls, we found the remaining work to be trivial.

4 What is the probability of drawing 2 hearts from a deck of 52 cards?

SOLUTION. The wording of this example omits a good deal of information. The reader should notice that the experiment being performed consists of drawing one card, and then another, from the deck. Thus the sample space (the set of all possible outcomes) consists of all possible pairs of 52 cards. The number of elements in the sample space S is the number of ways of drawing 2 cards from a deck of 52 cards, which is the number of combinations of 52 objects taken at 2 a time. The number of ways of drawing 2 cards from a deck of 52 cards is C_2^{52}, or $\binom{52}{2}$, and the number of outcomes in event E of drawing 2 hearts is C_2^{13}, or $\binom{13}{2}$. Thus,

$$n(E) = C_2^{13} = \binom{13}{2} = \frac{13!}{2!(13-2)!} = \frac{13 \cdot 12}{1 \cdot 2} = 78$$

and

$$n(S) = C_2^{52} = \binom{52}{2} = \frac{52!}{2!(52-2)!} = \frac{52 \cdot 51}{1 \cdot 2} = 1{,}326$$

so that the probability of drawing two hearts is

$$p(E) = \frac{n(E)}{n(S)} = \frac{78}{1326} = \frac{1}{17}$$

5 What is the probability of being dealt a five-card poker hand containing all spades?

SOLUTION. The number of elements in the sample space S is the number of ways of drawing a five-card poker hand (just a five-card hand) from a deck of 52 cards, which is C_5^{52}, or $\binom{52}{5}$, possible ways. The number of outcomes in event E of forming five-card hands containing all spades from 13 spades is C_5^{13}, or $\binom{13}{5}$. Thus,

$$n(S) = C_5^{52} = \binom{52}{5} = \frac{52!}{5!(52-5)!} = \frac{52!}{5! \cdot 47!}$$
$$= \frac{48 \cdot 49 \cdot 50 \cdot 51 \cdot 52}{1 \cdot 2 \cdot 3 \cdot 4 \cdot 5} = 2{,}598{,}960$$

and

$$n(E) = C_5^{13} = \binom{13}{5} = \frac{13!}{5!(13-5)!} = \frac{13!}{5! \cdot 8!}$$
$$= \frac{13 \cdot 12 \cdot 11 \cdot 10 \cdot 9}{1 \cdot 2 \cdot 3 \cdot 4 \cdot 5} = 1{,}287$$

so that the probability is

$$p(E) = \frac{n(E)}{n(S)} = \frac{1{,}287}{2{,}598{,}960}$$

7.2 Probability of More Than One Event

So far we have been concerned with problems dealing only with one event occurring in a single trial. Here we shall consider the problem of finding the probability of two or more events taking place. Two or more events are called *mutually exclusive* events if they have no elements in common. Thus, if E_1 and E_2 are two mutually exclusive events, then $E_1 \cap E_2 = \emptyset$. For example, in the experiment of selecting a student from a class, event E_1, a boy is selected, and E_2, a girl is selected, are mutually exclusive. Note that if two events are mutually exclusive they cannot occur simultaneously. This can be formalized as follows:

THEOREM 1

If E_1 and E_2 are two mutually exclusive events of a sample space S, the probability of event "E_1 or E_2" (denoted by $E_1 \cup E_2$) is $p(E_1 \cup E_2) = p(E_1) + p(E_2)$.

PROOF. E_1 and E_2 are mutually exclusive events, so that $E_1 \cap E_2 = \emptyset$ and $n(E_1 \cup E_2) = n(E_1) + n(E_2)$. Hence,

$$p(E_1 \cup E_2) = \frac{n(E_1 \cup E_2)}{n(S)} = \frac{n(E_1) + n(E_2)}{n(S)} = \frac{n(E_1)}{n(S)} + \frac{n(E_2)}{n(S)}$$

$$= p(E_1) + p(E_2)$$

EXAMPLES

1. If a bead is drawn from a container containing 3 red beads, 4 white beads, and 5 black beads, what is the probability that it is either a black bead or a white bead?

   SOLUTION. The probability of event E_1 drawing a black bead, is

   $$p(E_1) = \tfrac{5}{12}$$

   The probability of event E_2 drawing a white bead, is

   $$p(E_2) = \tfrac{4}{12}$$

   Since E_1 and E_2 are mutually exclusive events, the probability of drawing either a black bead or a white bead is

   $$p(E_1 \text{ or } E_2) = p(E_1 \cup E_2) = p(E_1) + p(E_2)$$
   $$= \tfrac{5}{12} + \tfrac{4}{12} = \tfrac{9}{12} = \tfrac{3}{4}$$

2. If a single card is drawn from a deck of 52 cards, what is the probability that it is a black ace or a red face card?

SOLUTION. Since there are two black aces, the probability of the event E_1 drawing a black ace, is $p(E_1) = \frac{2}{52}$. The probability of the event E_2 drawing a red face card, is $p(E_2) = \frac{6}{52}$ because there are 6 red face cards. Since E_1 and E_2 are mutually exclusive events, the probability of drawing either a black ace or a red face card is

$$p(E_1 \text{ or } E_2) = p(E_1 \cup E_2) = p(E_1) + p(E_2) = \tfrac{2}{52} + \tfrac{6}{52} = \tfrac{8}{52} = \tfrac{2}{13}$$

3 (Complement Event) Use Theorem 1 to prove that if E_1 and E_2 are two events of a sample space S that satisfy the conditions that $E_1 \cup E_2 = S$ and $E_1 \cap E_2 = \emptyset$, then $p(E_2) = 1 - p(E_1)$.

PROOF. E_2 in satisfying the above conditions is called the *complement* of E_1 and is denoted by $E_2 = E_1^c$. From Theorem 1 we have

$$p(E_1 \cup E_2) = p(E_1) + p(E_2)$$

so that

$$p(S) = p(E_1) + p(E_2)$$

Since $p(S) = 1$, it follows that

$$p(E_2) = 1 - p(E_1) \quad \text{or} \quad p(E_1^c) = 1 - p(E_1)$$

4 A ball is drawn at random from a box containing 8 red balls, 6 white balls, and 7 blue balls. Determine the probability that it is
a) red b) not red c) white
d) not white e) blue f) not blue

SOLUTION. Let B, R, and W denote the events of drawing a blue ball, a red ball, and a white ball, respectively. Let S be a sample space; then $n(S) = 21$, $n(B) = 7$, $n(R) = 8$, and $n(W) = 6$, so that

a) $p(R) = \dfrac{n(R)}{n(S)} = \dfrac{8}{21}$

b) $p(R^c) = 1 - p(R) = 1 - \tfrac{8}{21} = \tfrac{13}{21}$

c) $p(W) = \dfrac{n(W)}{n(S)} = \dfrac{6}{21} = \dfrac{2}{7}$

d) $p(W^c) = 1 - p(W) = 1 - \tfrac{2}{7} = \tfrac{5}{7}$

e) $p(B) = \dfrac{n(B)}{n(S)} = \dfrac{7}{21} = \dfrac{1}{3}$

f) $p(B^c) = 1 - p(B) = 1 - \tfrac{1}{3} = \tfrac{2}{3}$

Sometimes we want to know the probability that "E_1 or E_2" will occur, where E_1 and E_2 are *not* mutually exclusive. For example, we may ask, on the simultaneous toss of two dice, for the probability that at least one of the dice shows a 2. In this experiment there are six outcomes where the first die shows a 2; call this event E_1. There are also six outcomes where the second die shows a 2; call this event E_2. The general pattern of the sample space, together with E_1 and E_2, is shown in the following table.

	Outcomes of second die						
		1	2	3	4	5	6
Outcomes of first die	1	(1,1)	(1,2)	(1,3)	(1,4)	(1,5)	(1,6)
	2	(2,1)	(2,2)	(2,3)	(2,4)	(2,5)	(2,6)
	3	(3,1)	(3,2)	(3,3)	(3,4)	(3,5)	(3,6)
	4	(4,1)	(4,2)	(4,3)	(4,4)	(4,5)	(4,6)
	5	(5,1)	(5,2)	(5,3)	(5,4)	(5,5)	(5,6)
	6	(6,1)	(6,2)	(6,3)	(6,4)	(6,5)	(6,6)

We see from the above table that the two events intersect, and that intersection is the event $\{(2, 2)\}$.

The outcome (2, 2) occurs in both events E_1 and E_2. E_1 and E_2 are *not* mutually exclusive since $E_1 \cap E_2 \neq \emptyset$. Now the probability of at least one of the dice showing a 2 can be computed directly from the table as

$$p(E_1 \cup E_2) = \tfrac{11}{36}$$

It is also possible to use $p(E_1)$ and $p(E_2)$ to compute $p(E_1 \cup E_2)$ as

$$\begin{aligned} p(E_1 \cup E_2) &= p(E_1) + p(E_2) - p(E_1 \cap E_2) \\ &= p(\{2 \text{ on the first die}\}) \\ &\quad + p(\{2 \text{ on the second die}\}) - p(\{(2, 2)\}) \end{aligned}$$

The subtraction is necessary; otherwise the outcome (2, 2) would be counted twice, once in E_1 and once in E_2. Hence, from the table we get

$$p(E_1 \cup E_2) = \tfrac{6}{36} + \tfrac{6}{36} - \tfrac{1}{36} = \tfrac{11}{36}$$

This example illustrates the following theorem.

THEOREM 2

If E_1 and E_2 are any two events of a finite sample space S, then

$$p(E_1 \cup E_2) = p(E_1) + p(E_2) - p(E_1 \cap E_2)$$

PROOF. Suppose that E_1 and E_2 are two events.

$$p(E_1 \cup E_2) = \frac{n(E_1 \cup E_2)}{n(S)}$$

But $n(E_1 \cup E_2) = n(E_1) + n(E_2) - n(E_1 \cap E_2)$, since $n(E_1 \cap E_2)$ is added twice in $n(E_1) + n(E_2)$ if E_1 and E_2 are not mutually exclusive, or else E_1 and E_2 are mutually exclusive, in which case $n(E_1 \cap E_2) = 0$. Hence,

$$\begin{aligned} p(E_1 \cup E_2) &= \frac{n(E_1 \cup E_2)}{n(S)} = \frac{n(E_1) + n(E_2) - n(E_1 \cap E_2)}{n(S)} \\ &= \frac{n(E_1)}{n(S)} + \frac{n(E_2)}{n(S)} - \frac{n(E_1 \cap E_2)}{n(S)} \\ &= p(E_1) + p(E_2) - p(E_1 \cap E_2) \end{aligned}$$

EXAMPLES

1 A card is drawn from a deck of 52 cards. Use Theorem 2 to find the probability that the card is a king or a heart.

SOLUTION. Let the event E_1 be the set of kings, so that $p(E_1) = \frac{4}{52}$, and the event E_2 be the set of hearts, so that $p(E_2) = \frac{13}{52}$. Then $E_1 \cap E_2 = \{\text{king of hearts}\}$, so that $p(E_1 \cap E_2) = \frac{1}{52}$. Therefore,

$$\begin{aligned} p(E_1 \cup E_2) &= p(E_1) + p(E_2) - p(E_1 \cap E_2) \\ &= \tfrac{4}{52} + \tfrac{13}{52} - \tfrac{1}{52} \\ &= \tfrac{16}{52} = \tfrac{4}{13} \end{aligned}$$

2 If two cards are drawn from a deck of 52 cards, what is the probability that either both are black or both are queens?

SOLUTION. Let E_1 be the event that both cards are black. Since there are 26 black cards in a deck, $n(E_1) = C_2^{26} = \binom{26}{2}$. If E_2 is the event that both cards are queens, then $n(E_2) = C_2^4 = \binom{4}{2}$ because there are four queens in a deck. The deck contains 52 cards and the outcomes consist of sets of two cards. Hence the number of elements in the sample space S is $n(S) = C_2^{52} = \binom{52}{2}$. Since there is only one pair of black queens, $n(E_1 \cap E_2) = 1$. We have

$$p(E_1) = \frac{n(E_1)}{n(S)} = \frac{\binom{26}{2}}{\binom{52}{2}} = \frac{325}{1{,}326} \qquad p(E_2) = \frac{n(E_2)}{n(S)} = \frac{\binom{4}{2}}{\binom{52}{2}} = \frac{6}{1{,}326}$$

and

$$p(E_1 \cap E_2) = \frac{n(E_1 \cap E_2)}{n(S)} = \frac{1}{\binom{52}{2}} = \frac{1}{1{,}326}.$$

so that

$$p(E_1 \cup E_2) = p(E_1) + p(E_2) - p(E_1 \cap E_2)$$
$$= \tfrac{325}{1{,}326} + \tfrac{6}{1{,}326} - \tfrac{1}{1{,}326} = \tfrac{330}{1{,}326} = \tfrac{165}{663}$$

7.3 Independent and Dependent Events

Let us perform the experiment of tossing three coins one after the other. The sample space S is given by the set of triples $S = \{HHH, HHT, HTH, THH, HTT, THT, TTH, TTT\}$, where H represents a head and T represents a tail. Let E_1 be the event tails on the first coin, and E_2 be the event heads on the third coin. Then

$$E_1 = \{THH, THT, TTH, TTT\}$$
$$E_2 = \{HHH, HTH, TTH, THH\}$$

and

$$E_1 \cap E_2 = \{THH, TTH\}$$

Hence, $n(E_1) = 4$, $n(E_2) = 4$, $n(S) = 8$ and $n(E_1 \cap E_2) = 2$. Then the probability of the event tails on the first coin and heads on the third coin is

$$p(E_1 \text{ and } E_2) = p(E_1 \cap E_2) = \tfrac{2}{8} = \tfrac{1}{4}$$

Observe that

$$p(E_1 \cap E_2) = \tfrac{4}{8} \cdot \tfrac{4}{8} = \tfrac{16}{64} = \tfrac{1}{4}$$

The result of the above example often happens in situations in which the probability that the two events "E_1 and E_2" will occur together, namely,

$$p(E_1 \cap E_2) = p(E_1) \cdot p(E_2)$$

Let E_1 and E_2 be two events in a sample space S. They may have no relation to each other, that is, the occurrence of one event does not affect the other, or they may be so related that the occurrence of one event does affect the other. If they are not related, then we say that the probability of one (event E_1) will be *independent* of whether the second (event E_2) has occurred. If the occurrence of one event (E_1) affects the occurrence of the other event (E_2), then they are called *dependent events*. In general, we have the following property: If E_1 and E_2 are events in a sample space S, and if $p(E_1 \cap E_2) = p(E_1) \cdot p(E_2)$, then E_1 and E_2 are independent events. If two events E_1 and E_2 are not independent, then they are dependent events. The proof of this property is trivial, once independence has been carefully defined, but this is more than we wish to cover here.

EXAMPLES

1   Find the probability of obtaining two heads on tossing two coins in order. Use the notion of independent events.

SOLUTION. Since the occurrence of either of these events is not affected by the occurrence of the other, the events are independent. The probability of event E_1, heads on the first coin, is $\frac{1}{2}$. The probability of event E_2, heads on the second coin, is also $\frac{1}{2}$, so that

$$p(E_1 \cap E_2) = p(E_1) \cdot p(E_2) = \tfrac{1}{2} \cdot \tfrac{1}{2} = \tfrac{1}{4}$$

2   A box contains 4 red balls, 5 white balls, and 7 black balls. If a ball is drawn from the box and then replaced, and then a second ball is drawn, what is the probability that both balls are white?

SOLUTION. The probability of event E_1, drawing a white ball on the first drawing, is $\frac{5}{16}$, and the probability of event E_2, drawing a white ball on the second drawing, is also $\frac{5}{16}$ since the first ball is replaced. The probability of drawing a white ball on both attempts is

$$p(E_1 \cap E_2) = p(E_1) \cdot p(E_2) = (\tfrac{5}{16})(\tfrac{5}{16}) = \tfrac{25}{256}$$

since these events are independent.

3   In Example 2, if a ball is drawn, and without replacing it a second ball is drawn, what is the probability that both balls are white?

SOLUTION. The probability of event E_1, drawing a white ball on the first drawing, is $\frac{5}{16}$. However, assuming that the first drawing resulted in a white ball, there are only 4 white balls left in the remaining 15 balls. Hence, the probability of event E_2', drawing a white ball on the second drawing given that the first drawing was a white ball, is $\frac{4}{15}$, so that the probability of drawing two white balls is

$$p(E_1 \cap E_2') = p(E_1) \cdot p(E_2') = (\tfrac{5}{16})(\tfrac{4}{15}) = \tfrac{20}{240} = \tfrac{1}{12}$$

PROBLEM SET 6

1. Write sample spaces for the following experiments.
   a) A coin is tossed three times.
   b) Two coins are tossed once.
   c) Two coins are tossed twice.

2. A game consists of throwing a die and observing the numbers that show on the face.
   a) What is the sample space for this game?
   b) What is the probability that the face showing is even and less than 4?

3. A newly wed couple decided that they would like to have three children.
   a) Describe the sample space for the children in terms of sex and order of their birth, using set notations.
   b) Describe the event of having two girls and one boy and the first child being a girl, using set notations.
   c) What is the probability of the event of having two girls and one boy and the first child being a girl?

4. In an experiment, "a single die is thrown"
   a) What is the sample space of this experiment?
   b) What is the probability of making a point greater than one and less than six in the experiment?

5. Determine the probability p for each of the following events.
   a) An odd number appears in a single toss of a fair die.
   b) The sum seven appears in a single toss of a fair die.
   c) At least one tail appears in two tosses of a fair coin.
   d) A head appears in three successive tosses of a coin.
   e) An ace, ten of diamonds, or two of spades appears in drawing a single card from a well-shuffled ordinary deck of 52 cards.

6. A bag contains 5 white, 3 black, and 4 red balls. A ball is drawn at random from the bag. Determine the probability that it is
   a) red   b) not red   c) black   d) not black
   e) white   f) not white   g) red or black   h) red or white
   i) black or white   j) not red or black   k) not red or white

7. Two dice are cast. Let E_1 be the event that both dice show the same numeral, and E_2 be the event that the sum of numbers thrown is greater than 8. Find the following probabilities.
   a) $p(E_1)$
   b) $p(E_2)$
   c) $p(E_1^c)$
   d) $p(E_2^c)$
   e) $p(E_1 \cup E_2)$
   f) $p(E_1^c \cup E_2^c)$

8 Two dice are thrown simultaneously. What is the probability that
   a) The sum of the faces showing is less than 7?
   b) The sum of the faces showing is greater than 7?
   c) The sum of the faces showing is 7?

9 A coin is tossed three times. List the possible outcomes and find the probabilities of the following events.
   a) The first or the third toss is H (heads).
   b) At least one toss is H.
   c) At least two tosses are H.
   d) All three tosses are H.

10 Find the probability of three quarters falling all tails up when tossed simultaneously.

11 Find the probability of a three turning up in one toss of a die.

12 Find the probability of tossing a one or a three in one toss of a die.

13 What is the probability of obtaining a tail in a toss of a coin?

14 A dime and a quarter are tossed simultaneously. Find the probability that the dime will fall tails and the quarter will fall heads; that both coins fall heads or tails.

15 A bag contains 4 red balls and 2 blue balls; another contains 3 red balls and 5 black balls. If one is drawn from each bag, find the probability that
   a) both are red   b) both are blue   c) one is red and one is blue

16 In a college of 28,000 students, 280 students participate in sports. What is the probability that a specified student participates in sports?

17 If the probability that an American League baseball team A will win the World Series is $\frac{1}{7}$ and the probability that a National League baseball team B will win is $\frac{1}{6}$, what is the probability that team A or B will win the World Series?

18 If the probabilities that each of two sisters will be married at the age of 20 are 0.63 and 0.68, find the probability that both of the sisters will be married at the age of 20.

19 If the probability that a student will attend Wayne State University is $\frac{1}{4}$ and the probability that he will attend the University of Michigan is $\frac{1}{5}$, find the probability that he will attend Wayne or Michigan.

20 A ball is drawn and replaced, and then another is drawn from a bag that contains 6 red and 3 green. What is the probability that the first one will be red and the second will be green?

REVIEW PROBLEM SET

1. For each of the following sequences, write the first four terms, describe the range, and plot the four points.
   a) $f(n) = 1 + (-1)^n$
   b) $f(n) = 2^n$
   c) $f(n) = 5 - \dfrac{3}{n}$
   d) $a_n = \dfrac{2}{n+1}$
   e) $f(n) = \dfrac{n(4n+1)}{5}$

2. For each of the following arithmetic progressions, find the indicated term and the indicated sum.
   a) $4, 9, 14, \ldots$; ninth term and S_9
   b) $21, 19, 17, \ldots$; tenth term and S_{10}
   c) $42, 39, 36, \ldots$; eleventh term and S_{11}
   d) $0.3, 1.2, 2.1, \ldots$; fifteenth term and S_{15}
   e) $\tfrac{1}{6}, \tfrac{1}{3}, \tfrac{1}{2}, \ldots$; twenty-fourth term and S_{24}
   f) $\tfrac{1}{6}, \tfrac{1}{4}, \tfrac{1}{3}, \ldots$; thirtieth term and S_{30}
   g) $\tfrac{5}{8}, 1\tfrac{1}{8}, 1\tfrac{5}{8}, \ldots$; sixteenth term and S_{16}

3. Find the value of x so that each of the following will be arithmetic progressions.
   a) $2, 1 + 2x, 21 - 3x, \ldots$
   b) $2x, \tfrac{1}{2}x + 3, 3x - 10, \ldots$
   c) $3x, 2x + 1, x^2 - 4, \ldots$

4. For each of the following geometric progressions, find the indicated term and the indicated sum.
   a) $3, 12, 48, \ldots$; eighth term and S_8
   b) $16, 8, 4, \ldots$; ninth term and S_9
   c) $81, -27, 9, \ldots$; sixth term and S_6
   d) $\sqrt{2}, 2, 2\sqrt{2}, \ldots$; tenth term and S_{10}
   e) $3, -3\sqrt{2}, 6, \ldots$; eighteenth term and S_{18}
   f) $2, -2\sqrt{2}, 4, \ldots$; twentieth term and S_{20}
   g) $9, -3\sqrt{3}, 3, \ldots$; thirteenth term and S_{13}

5. Determine the value of x so that each of the following will be geometric progressions.
   a) $x - 6, x + 6, 2x + 2, \ldots$
   b) $\tfrac{1}{2}x, x + 2, 3x + 1, \ldots$
   c) $x - 7, x + 5, 8x - 5, \ldots$
   d) $x + 1, x + 2, x - 3, \ldots$

6. Find the numerical values of each of the following finite sums.
   a) $\sum_{k=1}^{5} k(2k - 1)$
   b) $\sum_{k=5}^{10} (2k - 1)^2$
   c) $\sum_{k=1}^{4} 2k^2(k - 3)$
   d) $\sum_{k=1}^{6} 3^{k+1}$

e) $\sum_{k=2}^{6} (k+1)(k+2)$ f) $\sum_{k=4}^{7} \frac{1}{k(k-3)}$

7 Evaluate each of the following expressions.

a) $\binom{15}{9}$ b) $\binom{6}{3}$ c) $\binom{n}{n-1}$

d) $\binom{14}{5}$ e) $\binom{7}{5}$ f) $\binom{8}{3}$

8 a) Show that

$$\binom{n}{k} + \binom{n}{k+1} = \binom{n+1}{k+1}$$

b) Let $a = b = 1$ in the expansion of $(a+b)^n$, and find the sum

$$\binom{n}{0} + \binom{n}{1} + \binom{n}{2} + \cdots + \binom{n}{n}$$

9 Use the binomial theorem to expand each of the following expressions.
   a) $(x+2y)^4$
   b) $(x-3y)^4$
   c) $(1+x)^5$
   d) $(2x+1)^5$
   e) $(1-2x)^6$
   f) $(a-b)^6$
   g) $(3x+y)^4$
   h) $\left(x - \frac{1}{x}\right)^8$
   i) $(3x+\sqrt{x})^5$
   j) $\left(3y + \frac{1}{3\sqrt{y}}\right)^6$
   k) $\left(2x + \frac{1}{y}\right)^3$
   l) $\left(x^3 - \frac{1}{\sqrt{x}}\right)^9$

10 Find the indicated term in each of the following binomial expansions.
    a) fifth term of $(x+y)^{10}$
    b) sixth term of $(x-y)^{11}$
    c) fifth term of $(2x+y)^{10}$
    d) sixth term of $(x-3y)^9$
    e) fourth term of $(3x+y)^{11}$
    f) third term of $(2x+y)^{20}$

11 How many different amounts of money can be formed from a penny, a nickel, a dime, a quarter, and a half-dollar?

12 How many different committees of 4 persons each can be chosen from a group of 10 persons?

13 Determine the probability for each of the following events.
    a) An even number appears in a single toss of a fair die.
    b) At least one head appears in two tosses of a fair coin.
    c) A king, ace, jack of clubs, or queen of diamonds appears in drawing a single card from a well-shuffled deck of 52 cards.
    d) The sum of 8 appears in a single toss of a pair of fair dice.

14 A ball is drawn at random from a box containing 2 red balls, 3 white balls, and 4 blue balls. Determine the probability that it is
  a) red b) white c) blue
  d) not red e) not white f) not blue

15 In how many ways can five differently colored marbles be arranged in a row of five?

16 In how many ways can 10 people be seated on a bench if only four seats are available?

17 How many different batting orders for a baseball team are possible
  a) If the pitcher is to bat last?
  b) If the first baseman is fourth, the pitcher is last, and the catcher is fifth?

18 Find the probability of 5 dimes falling all heads up when tossed simultaneously?

19 A ball is drawn at random from a box containing 10 red, 30 white, 20 blue, and 15 orange marbles. Find the probability that it is
  a) orange or red b) white
  c) not red or blue d) red or blue
  e) not blue f) white or blue
  g) not white or blue

20 If the probability that car A wins a race is $\frac{1}{6}$ and the probability that car B wins the race is $\frac{1}{5}$, what is the probability that one or the other of the cars will win?

21 If the probabilities that each of two roommates will graduate from college are 0.76 and 0.81, find the probability that both roommates will graduate from college.

22 A bag contains 6 white and 3 black balls. If 2 balls are drawn and the first is not replaced before the second is drawn, what is the probability that the first one will be white and the second one will be black?

CHAPTER 8

Linear Systems:
Matrices and Determinants

CHAPTER 8

Linear Systems: Matrices and Determinants

1 Introduction

In this chapter we will consider two procedures for solving a system of linear equations. The first part of the chapter will be devoted to the discussion of a general method for solving such systems known as the *row-reduction method*. It involves the use of a special device known as a *matrix*, and it is a generalization of the elimination method covered in Chapter 4, Section 3.2. Next, a function that associates a number with a matrix, called a *determinant*, will be used to solve systems by a method called *Cramer's rule*.

We will restrict our attention in this text to linear systems with two equations and two unknowns or with three equations and three unknowns, although it is possible to generalize the two methods to handle linear systems with more than three unknowns.

2 Matrices and Row Reduction

Consider the process of finding the solution of the following linear system by using the elimination method.

(A) $\begin{cases} 3x - y = 1 \\ x + 2y = 0 \end{cases}$

First, multiply the second equation by -3, then add the first equation to the second equation, placing the result in place of the original second equation to get

(B) $\begin{cases} 3x - y = 1 \\ -7y = 1 \end{cases}$

System (B) is equivalent to system (A), by which we mean that system (B) has the same solution as system (A). This fact shall not be proven here.

Next, multiply the second equation by $-\frac{1}{7}$ to get

(C) $\begin{cases} 3x - y = 1 \\ y = -\frac{1}{7} \end{cases}$

In system (C) replace equation one with the sum of equations one and two to get

(D) $\begin{cases} 3x = \frac{6}{7} \\ y = -\frac{1}{7} \end{cases}$

Finally, multiply the first equation by $\frac{1}{3}$ to get the solution

(E) $\begin{cases} x = \frac{2}{7} \\ y = -\frac{1}{7} \end{cases}$

In the elimination method there is little reason to continue writing the variables. All we need to do is to maintain a record of the coefficients of each of the variables. The standard device used for doing this is a *matrix*.

A *matrix* is a rectangular array of numbers. The *size* of a matrix is described by specifying the number of rows and columns. The numbers occurring in a matrix are called the *entries* of the matrix. For example, the matrix

$$A = \begin{bmatrix} 1 & 2 \\ 3 & -1 \end{bmatrix}$$

has two rows and two columns and we say that A is a 2×2 matrix and 1, 2, 3, and -1 are the entries of A. The matrix

$$B = \begin{bmatrix} 1 & 3 & 5 \\ -2 & 5 & 6 \end{bmatrix}$$

has two rows and three columns, so that B is a 2×3 matrix. B has entries 1, 3, 5, -2, 5, and 6. The matrix

$$C = \begin{bmatrix} 3 \\ 1 \\ 2 \end{bmatrix}$$

is a 3×1 matrix with entries 3, 1, and 2. Notice that when the size of a matrix is specified, the number of rows is given first and then the number of columns.

The entry of a matrix is identified by using double subscripts to indicate its row and its column position. For example, the entry of the second row and the second column of a matrix D is denoted by a_{22}; the entry of the first row and the fourth column of a matrix D is a_{14}. In general, the entry of the ith row and the jth column of matrix D is denoted by a_{ij}.

For example, if

$$A = \begin{bmatrix} 3 & 2 & -1 \\ 4 & 1 & 5 \end{bmatrix}$$

then $a_{11} = 3$, $a_{12} = 2$, $a_{13} = -1$, $a_{21} = 4$, $a_{22} = 1$, and $a_{23} = 5$.

Let us now see how matrix notation can be used as a sort of shorthand notation for solving the linear system (A) on page 377 by the elimination method.

We want to solve the following system.

(A) $\begin{cases} 3x - y = 1 \\ x + 2y = 0 \end{cases}$

The matrix form of (A) is written as

$$A = \begin{bmatrix} 3 & -1 & \vdots & 1 \\ 1 & 2 & \vdots & 0 \end{bmatrix}$$

A is called an *augmented matrix*. The augmented matrix can be used to "maintain a record" of the coefficients of the linear equations when the elimination method is applied to solve (A). Notice that an "augmented matrix" is just an ordinary matrix; the word comes from the idea of the "coefficient matrix augmented by the numbers from the right-hand sides of the equations." ($\downarrow$ is used to indicate that the systems are equivalent.)

System	Matrix form of system
(A) $\begin{cases} 3x - y = 1 \\ x + 2y = 0 \end{cases}$	$A = \begin{bmatrix} 3 & -1 & \vdots & 1 \\ 1 & 2 & \vdots & 0 \end{bmatrix}$
First, multiply the second equation by -3.	First, multiply the second row by -3.
(B) $\begin{cases} 3x - y = 1 \\ -3x - 6y = 0 \end{cases}$	$B = \begin{bmatrix} 3 & -1 & \vdots & 1 \\ -3 & -6 & \vdots & 0 \end{bmatrix}$
Next, replace equation two with the sum of equations one and two.	Next, replace row two with the sum of rows one and two.

(C) $\begin{cases} 3x - y = 1 \\ - 7y = 1 \end{cases}$ $C = \begin{bmatrix} 3 & -1 & \vdots & 1 \\ 0 & -7 & \vdots & 1 \end{bmatrix}$

$\Big\downarrow$ Multiply equation two by $-\frac{1}{7}$. $\Big\downarrow$ Multiply row two by $-\frac{1}{7}$.

(D) $\begin{cases} 3x - y = 1 \\ y = -\frac{1}{7} \end{cases}$ $D = \begin{bmatrix} 3 & -1 & \vdots & 1 \\ 0 & 1 & \vdots & -\frac{1}{7} \end{bmatrix}$

$\Big\downarrow$ Replace equation one with the sum of equations one and two. $\Big\downarrow$ Replace row one with the sum of rows one and two.

(E) $\begin{cases} 3x = \frac{6}{7} \\ y = -\frac{1}{7} \end{cases}$ $E = \begin{bmatrix} 3 & 0 & \vdots & \frac{6}{7} \\ 0 & 1 & \vdots & -\frac{1}{7} \end{bmatrix}$

$\Big\downarrow$ Multiply equation one by $\frac{1}{3}$. $\Big\downarrow$ Multiply row one by $\frac{1}{3}$.

(F) $\begin{cases} x = \frac{2}{7} \\ y = -\frac{1}{7} \end{cases}$ $F = \begin{bmatrix} 1 & 0 & \vdots & \frac{2}{7} \\ 0 & 1 & \vdots & -\frac{1}{7} \end{bmatrix}$

Hence, the solution of (A) is $\{(\frac{2}{7}, -\frac{1}{7})\}$.

Realizing that the numbers to the left of the dashed line represent the coefficients of x and y, we have $\{(\frac{2}{7}, -\frac{1}{7})\}$ as the solution set.

Consider the system of linear equations

(A) $\begin{cases} x_1 - 2x_2 + 3x_3 = -1 \\ 2x_1 - x_2 + 2x_3 = 2 \\ 3x_1 + x_2 + 2x_3 = 3 \end{cases}$

The matrix form of system (A) is written as

$$A = \begin{bmatrix} 1 & -2 & 3 & \vdots & -1 \\ 2 & -1 & 2 & \vdots & 2 \\ 3 & 1 & 2 & \vdots & 3 \end{bmatrix}$$

System

(A) $\begin{cases} x_1 - 2x_2 + 3x_3 = -1 \\ 2x_1 - x_2 + 2x_3 = 2 \\ 3x_1 + x_2 + 2x_3 = 3 \end{cases}$

Matrix form of system

$A = \begin{bmatrix} 1 & -2 & 3 & \vdots & -1 \\ 2 & -1 & 2 & \vdots & 2 \\ 3 & 1 & 2 & \vdots & 3 \end{bmatrix}$

First, replace the second equation by the sum of -2 times the first equation and the second equation.

First, replace the second row by the sum of -2 times the first row and the second row.

(B) $\begin{cases} x_1 - 2x_2 + 3x_3 = -1 \\ 3x_2 - 4x_3 = 4 \\ 3x_1 + x_2 + 2x_3 = 3 \end{cases}$

$B = \begin{bmatrix} 1 & -2 & 3 & | & -1 \\ 0 & 3 & -4 & | & 4 \\ 3 & 1 & 2 & | & 3 \end{bmatrix}$

Next, replace equation three with the sum of -3 times equation one and the third equation.

Next, replace row three with the sum of -3 times row one and row three.

(C) $\begin{cases} x_1 - 2x_2 + 3x_3 = -1 \\ 3x_2 - 4x_3 = 4 \\ 7x_2 - 7x_3 = 6 \end{cases}$

$C = \begin{bmatrix} 1 & -2 & 3 & | & -1 \\ 0 & 3 & -4 & | & 4 \\ 0 & 7 & -7 & | & 6 \end{bmatrix}$

Multiply equation two by $\frac{1}{3}$.

Multiply row two by $\frac{1}{3}$.

(D) $\begin{cases} x_1 - 2x_2 + 3x_3 = -1 \\ x_2 - \frac{4}{3}x_3 = \frac{4}{3} \\ 7x_2 - 7x_3 = 6 \end{cases}$

$D = \begin{bmatrix} 1 & -2 & 3 & | & -1 \\ 0 & 1 & -\frac{4}{3} & | & \frac{4}{3} \\ 0 & 7 & -7 & | & 6 \end{bmatrix}$

Now, replace the third equation with -7 times the second equation plus the third equation.

Now, replace the third row with -7 times the second row plus the third row.

(E) $\begin{cases} x_1 - 2x_2 + 3x_3 = -1 \\ x_2 - \frac{4}{3}x_3 = \frac{4}{3} \\ \frac{7}{3}x_3 = -\frac{10}{3} \end{cases}$

$E = \begin{bmatrix} 1 & -2 & 3 & | & -1 \\ 0 & 1 & -\frac{4}{3} & | & \frac{4}{3} \\ 0 & 0 & \frac{7}{3} & | & -\frac{10}{3} \end{bmatrix}$

Next replace equation one with 2 times equation two plus equation one.

Next replace row one with 2 times row two plus row one.

(F) $\begin{cases} x_1 + \frac{1}{3}x_3 = \frac{5}{3} \\ x_2 - \frac{4}{3}x_3 = \frac{4}{3} \\ \frac{7}{3}x_3 = -\frac{10}{3} \end{cases}$

$F = \begin{bmatrix} 1 & 0 & \frac{1}{3} & | & \frac{5}{3} \\ 0 & 1 & -\frac{4}{3} & | & \frac{4}{3} \\ 0 & 0 & \frac{7}{3} & | & -\frac{10}{3} \end{bmatrix}$

Multiply equation three by $\frac{3}{7}$.

Multiply row three by $\frac{3}{7}$.

(G) $\begin{cases} x_1 \phantom{- \tfrac{4}{3}x_3} + \tfrac{1}{3}x_3 = \tfrac{5}{3} \\ x_2 - \tfrac{4}{3}x_3 = \tfrac{4}{3} \\ x_3 = -\tfrac{10}{7} \end{cases}$
$G = \begin{bmatrix} 1 & 0 & \tfrac{1}{3} & \vdots & \tfrac{5}{3} \\ 0 & 1 & -\tfrac{4}{3} & \vdots & \tfrac{4}{3} \\ 0 & 0 & 1 & \vdots & -\tfrac{10}{7} \end{bmatrix}$

| Now replace equation one with the sum of $-\tfrac{1}{3}$ times equation three and equation one. | Now replace row one with the sum of $-\tfrac{1}{3}$ times row three and row one. |

(H) $\begin{cases} x_1 \phantom{- \tfrac{4}{3}x_3} = \tfrac{15}{7} \\ x_2 - \tfrac{4}{3}x_3 = \tfrac{4}{3} \\ x_3 = -\tfrac{10}{7} \end{cases}$
$H = \begin{bmatrix} 1 & 0 & 0 & \vdots & \tfrac{15}{7} \\ 0 & 1 & -\tfrac{4}{3} & \vdots & \tfrac{4}{3} \\ 0 & 0 & 1 & \vdots & -\tfrac{10}{7} \end{bmatrix}$

| Finally, replace equation two with the sum of $\tfrac{4}{3}$ times equation three and equation two. | Finally, replace row two with the sum of $\tfrac{4}{3}$ times row three and row two. |

(I) $\begin{cases} x_1 = \tfrac{15}{7} \\ x_2 = -\tfrac{4}{7} \\ x_3 = -\tfrac{10}{7} \end{cases}$
$I = \begin{bmatrix} 1 & 0 & 0 & \vdots & \tfrac{15}{7} \\ 0 & 1 & 0 & \vdots & -\tfrac{4}{7} \\ 0 & 0 & 1 & \vdots & -\tfrac{10}{7} \end{bmatrix}$

| Hence, the solution is $\{(\tfrac{15}{7}, -\tfrac{4}{7}, -\tfrac{10}{7})\}$. | Hence, the solution set can be read from the augmented column as $\{(\tfrac{15}{7}, -\tfrac{4}{7}, -\tfrac{10}{7})\}$. |

Notice that in solving the system of linear equations (A), the system was replaced by an augmented matrix A whose elements are the coefficients and constants occurring in the equations. Notice also that we can work with the augmented matrix instead of the actual equations by performing the following operations on the augmented matrix.

1. Interchange two rows of the matrix ($R_i \leftrightarrow R_j$).

2. Multiply the elements in a row of the matrix by a nonzero number ($R_i \to kR_i, k \neq 0$).

3. Replace a row with the sum of a nonzero multiple of itself and a multiple of another row ($R_i \to kR_j + cR_i, c \neq 0$).

The operations (1), (2), and (3) are called the *elementary row operations*. The resulting matrix I in the above example is called a row-reduced echelon matrix. In general, a matrix is a *row-reduced echelon* matrix if all of the following conditions hold:

SECTION 2 MATRICES AND ROW REDUCTION

i The first nonzero entry in each row is 1; all other entries in that column are zeros.

ii Each row that consists entirely of zeros is below each row which contains a nonzero entry.

iii The first nonzero entry in each row is to the right of the first nonzero entry in the preceding row.

EXAMPLES

Perform the elementary row operations to solve each of the following linear systems, if the system has a unique solution.

1 (A) $\begin{cases} 4x + y = 2 \\ x + 4y = 1 \end{cases}$

SOLUTION. The matrix form of the system is

$$\begin{bmatrix} 4 & 1 & \vdots & 2 \\ 1 & 4 & \vdots & 1 \end{bmatrix}$$

First, multiply row one by $\frac{1}{4}$ to get $R_1 \to \frac{1}{4}R_1$, so that

$$\begin{bmatrix} 1 & \frac{1}{4} & \vdots & \frac{1}{2} \\ 1 & 4 & \vdots & 1 \end{bmatrix}$$

Next, replace row two with the sum of row two and -1 times row one, $(R_2 \to R_2 + (-1)R_1)$, to get

$$\begin{bmatrix} 1 & \frac{1}{4} & \vdots & \frac{1}{2} \\ 0 & \frac{15}{4} & \vdots & \frac{1}{2} \end{bmatrix}$$

Multiply row two by $\frac{4}{15}$ to get $R_2 \to \frac{4}{15}R_2$, so that

$$\begin{bmatrix} 1 & \frac{1}{4} & \vdots & \frac{1}{2} \\ 0 & 1 & \vdots & \frac{2}{15} \end{bmatrix}$$

Finally, replace row one with the sum of row one and $-\frac{1}{4}$ times row two, $(R_1 \to R_1 + (-\frac{1}{4})R_2)$, to get

$$\begin{bmatrix} 1 & 0 & \vdots & \frac{7}{15} \\ 0 & 1 & \vdots & \frac{2}{15} \end{bmatrix}$$

Hence, the solution set is $\{(\frac{7}{15}, \frac{2}{15})\}$.

2 (A) $\begin{cases} x_1 + 2x_2 - 3x_3 = 6 \\ 2x_1 - x_2 + 4x_3 = 2 \\ 4x_1 + 3x_2 - 2x_3 = 14 \end{cases}$

SOLUTION. The matrix form of the system is

$$\begin{bmatrix} 1 & 2 & -3 & \vdots & 6 \\ 2 & -1 & 4 & \vdots & 2 \\ 4 & 3 & -2 & \vdots & 14 \end{bmatrix}$$

First, replace the second row by the sum of -2 times the first row and the second row ($R_2 \to -2R_1 + R_2$) to obtain the equivalent matrix

$$\begin{bmatrix} 1 & 2 & -3 & \vdots & 6 \\ 0 & -5 & 10 & \vdots & -10 \\ 4 & 3 & -2 & \vdots & 14 \end{bmatrix}$$

Next, replace the third row by the sum of -4 times the first row and the third row ($R_3 \to -4R_1 + R_3$) to obtain

$$\begin{bmatrix} 1 & 2 & -3 & \vdots & 6 \\ 0 & -5 & 10 & \vdots & -10 \\ 0 & -5 & 10 & \vdots & -10 \end{bmatrix}$$

Since the second row R_2 and the third row R_3 are identical, the corresponding linear system will have two identical equations

$$-5x_2 + 10x_3 = -10$$

In this case, we say that the system is a *dependent system*, and the system does not have a unique solution.

3 (A) $\begin{cases} 3x_1 - x_2 + x_3 = 1 \\ 7x_1 + x_2 - x_3 = 6 \\ 2x_1 + x_2 - x_3 = 2 \end{cases}$

SOLUTION. The augmented matrix that corresponds to the linear system is given by

$$\begin{bmatrix} 3 & -1 & 1 & \vdots & 1 \\ 7 & 1 & -1 & \vdots & 6 \\ 2 & 1 & -1 & \vdots & 2 \end{bmatrix}$$

which can be reduced by the following row operations.

$$\begin{bmatrix} 3 & -1 & 1 & \vdots & 1 \\ 7 & 1 & -1 & \vdots & 6 \\ 2 & 1 & -1 & \vdots & 2 \end{bmatrix} \xrightarrow{R_3 \to R_1 + R_3} \begin{bmatrix} 3 & -1 & 1 & \vdots & 1 \\ 7 & 1 & -1 & \vdots & 6 \\ 5 & 0 & 0 & \vdots & 3 \end{bmatrix}$$

$$R_2 \to R_1 + R_2 \begin{bmatrix} 3 & -1 & 1 & \vdots & 1 \\ 10 & 0 & 0 & \vdots & 7 \\ 5 & 0 & 0 & \vdots & 3 \end{bmatrix}$$

$$R_2 \to R_2 + (-2)R_3 \begin{bmatrix} 3 & -1 & 1 & \vdots & 1 \\ 0 & 0 & 0 & \vdots & 1 \\ 5 & 0 & 0 & \vdots & 3 \end{bmatrix}$$

But R_2 implies that $0 \cdot x_1 + 0 \cdot x_2 + 0 \cdot x_3 = 1$, that is, $0 = 1$, which, of course, is not possible. Hence, the system has no solution, and we say that the system is *inconsistent*.

In summary, the procedure for solving a linear system is to form the augmented matrix and proceed to reduce it to echelon form:

1. If a resulting augmented echelon matrix has no row with its first non-zero entry in the last column, and has one or more rows consisting entirely of zeros, then the system has more than one solution (one for each arbitrary choice of one or more of the variables) and is called dependent.

2. If a resulting augmented echelon matrix has a row with its first nonzero entry appearing in the last column, then the system has no solution, and the system is said to be inconsistent.

PROBLEM SET 1

1. Solve each of the following systems of linear equations using the elimination method and show the corresponding matrix form of the system in each step.

   a) $3x + y = 14$
      $2x - y = 1$

   b) $4x + 3y = 15$
      $3x + 5y = 14$

   c) $-2x + 3y = 8$
      $2x - y = 5$

   d) $x - 2y = 5$
      $3x - 6y = 4$

   e) $x/2 + y/6 = \frac{2}{3}$
      $3x + y = 4$

   f) $x - 2y - 4 = 0$
      $x + y + 3 = 0$

2. Solve the following systems of linear equations by using the elimination method. In each step of the process, show the corresponding matrix form of the system.

   a) $x + y + z = 6$
      $3x - y + 2z = 7$
      $2x + 3y - z = 5$

   b) $2x + 3y + z = 6$
      $x - 2y + 3z = -3$
      $3x + y - z = 8$

   c) $x + y + 2z = 4$
      $x + y - 2z = 0$
      $x - y = 0$

   d) $x + y + z = 4$
      $x - y + 2z = 8$
      $2x + y - z = 3$

   e) $2x + y - z = 7$
      $y - x = 1$
      $z - y = 1$

   f) $3x + 2y + 2z = 6$
      $x - 5y + 6z = 2$
      $6x - 8z = 12$

3. Suppose that

$$A = \begin{bmatrix} -4 & 0 & 1 \\ 2 & 3 & -1 \\ 5 & 2 & 8 \end{bmatrix}$$

a) What is the size of matrix A?
b) Use subscript notation to identify each of the entries of matrix A.

4. Indicate which of the following augmented matrices are in row-reduced echelon form. Assuming that the matrix represents a linear system, what can be concluded about the solution of the system?

a) $\begin{bmatrix} 1 & 0 & 1 & \vdots & 3 \\ 0 & 1 & 0 & \vdots & 4 \\ 0 & 0 & 0 & \vdots & 0 \end{bmatrix}$
b) $\begin{bmatrix} 1 & 1 & 0 & \vdots & 0 \\ 0 & 1 & 0 & \vdots & 0 \\ 0 & 0 & 1 & \vdots & 2 \end{bmatrix}$

c) $\begin{bmatrix} 1 & 0 & 0 & \vdots & 3 \\ 0 & 0 & 1 & \vdots & 0 \\ 0 & 0 & 0 & \vdots & 5 \end{bmatrix}$
d) $\begin{bmatrix} 1 & 1 & \vdots & 2 \\ 0 & 0 & \vdots & 3 \end{bmatrix}$

e) $\begin{bmatrix} 1 & 0 & \vdots & 1 \\ 0 & 1 & \vdots & 1 \end{bmatrix}$
f) $\begin{bmatrix} 1 & 0 & 0 & 0 & \vdots & 3 \\ 0 & 1 & 0 & 0 & \vdots & -2 \\ 1 & 0 & 0 & 1 & \vdots & 1 \\ 0 & 0 & 1 & 0 & \vdots & 5 \end{bmatrix}$

g) $\begin{bmatrix} 0 & 1 & \vdots & -3 \\ 1 & 0 & \vdots & 4 \end{bmatrix}$

5. For each of the following systems find the augmented matrix, reduce it to a row-reduced echelon matrix, and determine the solution if it is unique. If the solution is not unique, indicate whether the system is dependent or inconsistent.

a) $x - 2y + z = -1$
$3x + y - 2z = 4$
$y - z = 1$

b) $x + y - 2z = 3$
$3x - y + z = 5$
$3x + 3y - 6z = 9$

c) $2x + y + z = 1$
$4x + 2y + 3z = 1$
$-2x - y + z = 2$

d) $x + y + z = 0$
$2x - y - 4z = 15$
$x - 2y - z = 7$

e) $2x - 3y + z = 4$
$x - 4y - z = 3$
$x - 9y - 4z = 5$

f) $2x + 3y - z = -2$
$x - y + 2z = 4$
$-x + 3y + 4z = 3$

3 Determinants

The study of linear functions has led us into an investigation of solving systems of linear equations. Before returning to the main topic of this chapter—solving systems of linear equations—we will study a function called the determinant, which can be used to solve linear systems that have unique solutions. The determinant is a function that associates with each square matrix a real number.

The determinant is quite difficult for the beginning student to understand if it is presented in the precise, formal way that is usually reserved for an advanced presentation of linear algebra. Since it is so difficult to digest the general definition, we will restrict our attention to the manipulative properties of the determinant and its application to solving linear systems.

The *determinant* is a function that has the *set of square matrices as its domain* and the *set of real numbers as its range*. If A is a square matrix, the determinant of A is denoted by $\det A$ or $|A|$.

The determinant of the 2×2 matrix

$$\begin{bmatrix} a_{11} & a_{12} \\ a_{21} & a_{22} \end{bmatrix}$$

is defined to be the number $a_{11}a_{22} - a_{21}a_{12}$, and we write

$$\det \begin{bmatrix} a_{11} & a_{12} \\ a_{21} & a_{22} \end{bmatrix} = \begin{vmatrix} a_{11} & a_{12} \\ a_{21} & a_{22} \end{vmatrix} = a_{11}a_{22} - a_{21}a_{12}$$

Thus,

$$\det \begin{bmatrix} 5 & 3 \\ -3 & -6 \end{bmatrix} = 5(-6) - (-3)(3) = -30 + 9 = -21$$

The determinant of the 3×3 matrix

$$\begin{bmatrix} a_{11} & a_{12} & a_{13} \\ a_{21} & a_{22} & a_{23} \\ a_{31} & a_{32} & a_{33} \end{bmatrix}$$

is defined as follows:

$$\det \begin{bmatrix} a_{11} & a_{12} & a_{13} \\ a_{21} & a_{22} & a_{23} \\ a_{31} & a_{32} & a_{33} \end{bmatrix} = \begin{vmatrix} a_{11} & a_{12} & a_{13} \\ a_{21} & a_{22} & a_{23} \\ a_{31} & a_{32} & a_{33} \end{vmatrix}$$

$$= a_{11} \det \begin{bmatrix} a_{22} & a_{23} \\ a_{32} & a_{33} \end{bmatrix} - a_{12} \det \begin{bmatrix} a_{21} & a_{23} \\ a_{31} & a_{33} \end{bmatrix} + a_{13} \det \begin{bmatrix} a_{21} & a_{22} \\ a_{31} & a_{32} \end{bmatrix}$$

$$= a_{11}(a_{22}a_{33} - a_{32}a_{23}) - a_{12}(a_{21}a_{33} - a_{31}a_{23}) + a_{13}(a_{21}a_{32} - a_{31}a_{22})$$

For example,

$$\det \begin{bmatrix} 3 & 2 & 7 \\ -1 & 5 & 3 \\ 2 & -3 & -6 \end{bmatrix}$$

$$= 3 \det \begin{bmatrix} 5 & 3 \\ -3 & -6 \end{bmatrix} - 2 \det \begin{bmatrix} -1 & 3 \\ 2 & -6 \end{bmatrix} + 7 \det \begin{bmatrix} -1 & 5 \\ 2 & -3 \end{bmatrix}$$

$$= 3(-30 + 9) - 2(6 - 6) + 7(3 - 10)$$
$$= 3(-21) - 2(0) + 7(-7)$$
$$= -63 - 49 = -112$$

(Although this function can be generalized to any square matrix, we will restrict ourselves here to 2 × 2 and 3 × 3 matrices only.)

3.1 Properties of Determinants

The determinant possesses properties that can be used to simplify the task of its evaluation. Although we will be restricting the investigation of these properties to 2 × 2 and 3 × 3 matrices, it is important to realize that the properties hold for evaluating the determinant of *any* square matrix.

THEOREM 1

A common factor that appears in all entries in some "one row" of a determinant can be factored out of the determinant.

For example,

$$\begin{vmatrix} 6 & 9 \\ 1 & 4 \end{vmatrix} = 3 \begin{vmatrix} 2 & 3 \\ 1 & 4 \end{vmatrix}$$

and

$$\begin{vmatrix} 15 & 45 & 60 \\ 1 & 2 & -1 \\ 2 & 4 & 8 \end{vmatrix} = 15 \begin{vmatrix} 1 & 3 & 4 \\ 1 & 2 & -1 \\ 2 & 4 & 8 \end{vmatrix}$$

$$= (15)(2) \begin{vmatrix} 1 & 3 & 4 \\ 1 & 2 & -1 \\ 1 & 2 & 4 \end{vmatrix}$$

PROOF. (for $n = 2$)

$$\begin{vmatrix} a_{11} & a_{12} \\ a_{21} & a_{22} \end{vmatrix} = a_{11}a_{22} - a_{21}a_{12}$$

Hence,

$$\begin{vmatrix} ka_{11} & ka_{12} \\ a_{21} & a_{22} \end{vmatrix} = ka_{11}a_{22} - ka_{21}a_{12}$$

$$= k(a_{11}a_{22} - a_{21}a_{12})$$

$$= k \begin{vmatrix} a_{11} & a_{12} \\ a_{21} & a_{22} \end{vmatrix}$$

THEOREM 2

If two (not necessarily adjacent) rows of a square matrix are interchanged, the values of the determinants of the two matrices differ only in the algebraic sign.

For example,

$$\begin{vmatrix} 2 & 3 \\ 4 & 1 \end{vmatrix} = -10$$

whereas

$$\begin{vmatrix} 4 & 1 \\ 2 & 3 \end{vmatrix} = 10$$

and

$$\begin{vmatrix} 1 & -1 & 0 \\ 3 & 0 & 4 \\ 2 & 1 & 5 \end{vmatrix} = 3 \qquad \text{(Why?)}$$

whereas

$$\begin{vmatrix} 2 & 1 & 5 \\ 3 & 0 & 4 \\ 1 & -1 & 0 \end{vmatrix} = -3 \qquad \text{(Why?)}$$

PROOF. (for $n = 2$)

$$\begin{vmatrix} a_{11} & a_{12} \\ a_{21} & a_{22} \end{vmatrix} = a_{11}a_{22} - a_{21}a_{12}$$

$$= -(a_{21}a_{12} - a_{11}a_{22})$$

$$= - \begin{vmatrix} a_{21} & a_{22} \\ a_{11} & a_{12} \end{vmatrix}$$

THEOREM 3

If any nonzero multiple of one row is added to any other row of a square matrix, the value of the determinant is unaltered.

For example, consider

$$\begin{vmatrix} 1 & 0 & 2 \\ 4 & 6 & 1 \\ -1 & 0 & -1 \end{vmatrix}$$

If we multiply the third row by 2 and add the result to the first row, we get

$$\begin{vmatrix} -1 & 0 & 0 \\ 4 & 6 & 1 \\ -1 & 0 & -1 \end{vmatrix}$$

and we are assured by the theorem that the latter determinant has the same value as the original. Note that this operation affects only the first row, whereas the other two rows remain the same. But we need not stop here; indeed, we can add the third row to the second row in the latter determinant to obtain

$$\begin{vmatrix} -1 & 0 & 0 \\ 3 & 6 & 0 \\ -1 & 0 & -1 \end{vmatrix}$$

Again, this does not change the value of the determinant. Determinants such as the last one, which contain many zero entries, are relatively easy to evaluate; hence, the above theorem simplifies the task of calculating determinants.

For the proof of this theorem see Problem Set 2, Problem 4a.

EXAMPLES

Use the three theorems above to evaluate each of the following determinants.

1. $\begin{vmatrix} 1 & 0 & 2 \\ 4 & 6 & 1 \\ -1 & 0 & -1 \end{vmatrix}$

SOLUTION

$$\begin{vmatrix} 1 & 0 & 2 \\ 4 & 6 & 1 \\ -1 & 0 & -1 \end{vmatrix} = \begin{vmatrix} -1 & 0 & 0 \\ 4 & 6 & 1 \\ -1 & 0 & -1 \end{vmatrix} \qquad \text{(Theorem 3)} \\ (R_1 \to R_1 + 2R_3)$$

$$= \begin{vmatrix} -1 & 0 & 0 \\ 3 & 6 & 0 \\ -1 & 0 & -1 \end{vmatrix} \quad \text{(Theorem 3)} \\ (R_2 \to R_2 + R_3)$$

$$= \begin{vmatrix} -1 & 0 & 0 \\ 3 & 6 & 0 \\ 0 & 0 & -1 \end{vmatrix} \quad \text{(Theorem 3)} \\ (R_3 \to R_3 + (-1)R_1)$$

$$= 3 \begin{vmatrix} -1 & 0 & 0 \\ 1 & 2 & 0 \\ 0 & 0 & -1 \end{vmatrix} \quad \text{(Theorem 1)}$$

$$= (3)(-1) \begin{vmatrix} 1 & 0 & 0 \\ 1 & 2 & 0 \\ 0 & 0 & -1 \end{vmatrix} \quad \text{(Theorem 1)}$$

$$= (3)(-1)(-1) \begin{vmatrix} 1 & 0 & 0 \\ 1 & 2 & 0 \\ 0 & 0 & 1 \end{vmatrix} \quad \text{(Theorem 1)}$$

$$= 3 \begin{vmatrix} 1 & 0 & 0 \\ 0 & 2 & 0 \\ 0 & 0 & 1 \end{vmatrix} \quad \text{(Theorem 3)} \\ (R_2 \to R_2 + (-1)R_1)$$

$$= (3)(2) \begin{vmatrix} 1 & 0 & 0 \\ 0 & 1 & 0 \\ 0 & 0 & 1 \end{vmatrix} \quad \text{(Theorem 1)}$$

$$= 6 \cdot 1 \quad \text{(Why?)}$$

2 $\begin{vmatrix} 3 & 1 & -1 \\ 0 & 2 & 4 \\ -1 & 4 & 2 \end{vmatrix}$

SOLUTION

$$\begin{vmatrix} 3 & 1 & -1 \\ 0 & 2 & 4 \\ -1 & 4 & 2 \end{vmatrix} = 2 \begin{vmatrix} 3 & 1 & -1 \\ 0 & 1 & 2 \\ -1 & 4 & 2 \end{vmatrix} \quad \text{(Theorem 1)}$$

$$= 2 \begin{vmatrix} 0 & 13 & 5 \\ 0 & 1 & 2 \\ -1 & 4 & 2 \end{vmatrix} \quad \text{(Theorem 3)} \\ (R_1 \to R_1 + 3R_3)$$

$$= 2 \begin{vmatrix} 0 & 13 & 5 \\ 0 & 1 & 2 \\ -1 & 3 & 0 \end{vmatrix} \quad \text{(Theorem 3)} \\ (R_3 \to R_3 + (-1)R_2)$$

$$= (2)(-1) \begin{vmatrix} 0 & 13 & 5 \\ 0 & 1 & 2 \\ 1 & -3 & 0 \end{vmatrix} \quad \text{(Theorem 1)}$$

(continued on page 392)

$$= (2)(-1)(-1) \begin{vmatrix} 1 & -3 & 0 \\ 0 & 1 & 2 \\ 0 & 13 & 5 \end{vmatrix} \quad \text{(Theorem 2)}$$

$$= 2 \begin{vmatrix} 1 & -3 & 0 \\ 0 & 1 & 2 \\ 0 & 0 & -21 \end{vmatrix} \quad \begin{array}{l} \text{(Theorem 3)} \\ (R_3 \to R_3 + (-13)R_2) \end{array}$$

$$= (2)(-21) \begin{vmatrix} 1 & -3 & 0 \\ 0 & 1 & 2 \\ 0 & 0 & 1 \end{vmatrix} \quad \text{(Theorem 1)}$$

$$= (-42) \left(1 \begin{vmatrix} 1 & 2 \\ 0 & 1 \end{vmatrix} - (-3) \begin{vmatrix} 0 & 2 \\ 0 & 1 \end{vmatrix} + 0 \begin{vmatrix} 0 & 1 \\ 0 & 0 \end{vmatrix} \right)$$

$$= -42$$

PROBLEM SET 2

1 Evaluate each of the following determinants by using the definition.

a) $\begin{vmatrix} -1 & 3 \\ -7 & 4 \end{vmatrix}$ b) $\begin{vmatrix} 2 & 3 \\ 9 & 4 \end{vmatrix}$

c) $\begin{vmatrix} 2 & -1 & 3 \\ 9 & -7 & 4 \\ 11 & -6 & 2 \end{vmatrix}$ d) $\begin{vmatrix} 3 & -1 & 2 \\ 0 & 1 & -5 \\ 6 & 7 & 4 \end{vmatrix}$

e) $\begin{vmatrix} 2 & 2 & 2 \\ 3 & 3 & 3 \\ 4 & 4 & 4 \end{vmatrix}$ f) $\begin{vmatrix} \frac{1}{2} & 4 & 7 \\ 1 & -1 & 2 \\ 3 & 2 & 5 \end{vmatrix}$

2 Solve the following determinants for x.

a) $\begin{vmatrix} x & -x \\ 5 & 3 \end{vmatrix} = 2$ b) $\begin{vmatrix} x & 4 & 5 \\ 0 & 1 & x \\ 5 & 2 & 1 \end{vmatrix} = 7$

c) $\begin{vmatrix} x & 0 & 0 \\ 3 & 1 & 2 \\ 0 & 4 & 1 \end{vmatrix} = 5$ d) $\begin{vmatrix} x & 0 & 1 \\ 2x & 1 & 2 \\ 3x & 2 & 3 \end{vmatrix} = 0$

3 Show why each of the following is true, not by evaluating each side but by citing the appropriate theorems of Section 3.1 that have been used.

a) $\begin{vmatrix} 4 & 5 \\ 3 & -2 \end{vmatrix} = - \begin{vmatrix} 3 & -2 \\ 4 & 5 \end{vmatrix}$ b) $\begin{vmatrix} 3 & 0 & 1 \\ 1 & 1 & 2 \\ 3 & 0 & 1 \end{vmatrix} = \begin{vmatrix} 0 & 0 & 0 \\ 1 & 1 & 2 \\ 3 & 0 & 1 \end{vmatrix}$

c) $\begin{vmatrix} 3 & -6 & 2 \\ 5 & -3 & 0 \\ 0 & 9 & 18 \end{vmatrix} = 9 \begin{vmatrix} 3 & -6 & 2 \\ 5 & -3 & 0 \\ 0 & 1 & 2 \end{vmatrix}$

d) $\begin{vmatrix} 2 & 4 & 12 \\ -1 & 0 & 3 \\ 1 & 0 & 6 \end{vmatrix} = 18 \begin{vmatrix} 1 & 2 & 6 \\ -1 & 0 & 3 \\ 0 & 0 & 1 \end{vmatrix}$

e) $\begin{vmatrix} 1 & 1 & 1 \\ 3 & 3 & 3 \\ 2 & 2 & 2 \end{vmatrix} = 6 \begin{vmatrix} 0 & 0 & 0 \\ 0 & 0 & 0 \\ 1 & 1 & 1 \end{vmatrix}$

4 a) Prove Theorem 3 of Section 3.1 for $n = 2$. (*Hint:* Compare

$$\begin{vmatrix} a_{11} & a_{12} \\ a_{21} & a_{22} \end{vmatrix} \quad \text{with} \quad \begin{vmatrix} a_{11} + ca_{21} & a_{12} + ca_{22} \\ a_{21} & a_{22} \end{vmatrix} \;)$$

b) Prove, for $n = 2$, that if two rows of a matrix are the same, then the determinant is zero.

c) Prove, for $n = 2$, that if all the entries in one row of a matrix are zeros, then the determinant is zero.

5 Use the theorems given in Section 3.1 to evaluate each of the following determinants.

a) $\begin{vmatrix} -1 & 0 & 2 \\ 0 & 0 & 0 \\ -1 & 5 & 1 \end{vmatrix}$ b) $\begin{vmatrix} 3 & 1 & 1 \\ -1 & 0 & 3 \\ 2 & 1 & 1 \end{vmatrix}$

c) $\begin{vmatrix} 2 & 1 & 3 \\ 1 & 2 & 1 \\ 4 & 0 & 0 \end{vmatrix}$ d) $\begin{vmatrix} 20 & 12 & 8 \\ 5 & 3 & 2 \\ 5 & 7 & 2 \end{vmatrix}$

e) $\begin{vmatrix} 1 & 0 & 2 \\ 1 & -3 & 0 \\ 0 & 3 & 1 \end{vmatrix}$

6 a) *Principle of duality.* Consider Theorems 1, 2, and 3 in Section 3.1. If we replace the word row with the word column, then the theorems are still valid. Rewrite the three theorems with this substitution.

b) Use the theorems of part *a* to evaluate the determinants in Problem 5.

7 Given the triangle of Figure 1, show that the area of the triangle is given by

$$\frac{1}{2} \begin{vmatrix} x_1 & y_1 & 1 \\ 0 & 0 & 1 \\ x_2 & 0 & 1 \end{vmatrix}$$

Figure 1

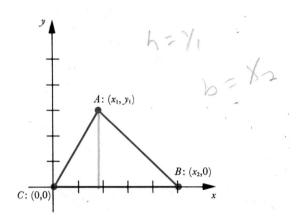

4 Cramer's Rule

We have seen what the determinant function is, and we have investigated methods for computing determinants. *Cramer's rule* provides us with a technique for using determinants to solve systems of linear equations.

Before stating Cramer's rule, let us establish some useful notation. Suppose we are given a linear system (S) containing the same number of equations as unknowns.

$$(S) \cdot \begin{cases} a_{11}x_1 + a_{12}x_2 + \cdots + a_{1n}x_n = c_1 \\ a_{21}x_1 + a_{22}x_2 + \cdots + a_{2n}x_n = c_2 \\ \cdots\cdots\cdots\cdots\cdots\cdots\cdots\cdots\cdots\cdots \\ a_{n1}x_1 + a_{n2}x_2 + \cdots + a_{nn}x_n = c_n \end{cases}$$

The determinant of the matrix of coefficients occurring in the system is called the *determinant of the coefficient matrix* and will be denoted by D. Hence,

$$D = \begin{vmatrix} a_{11} & a_{12} & \cdots & a_{1n} \\ a_{21} & a_{22} & \cdots & a_{2n} \\ \cdots & \cdots & & \cdots \\ a_{n1} & a_{n2} & \cdots & a_{nn} \end{vmatrix}$$

D_j will be used to denote the determinant of the matrix obtained by replacing the jth column in D by the column of constant terms in the system, so that

$$D_j = \begin{vmatrix} a_{11} & a_{12} & \cdots & c_1 & \cdots & a_{1n} \\ a_{21} & a_{22} & \cdots & c_2 & \cdots & a_{2n} \\ \cdots & \cdots & & \vdots & & \cdots \\ a_{n1} & a_{n2} & \cdots & c_n & \cdots & a_{nn} \end{vmatrix}$$

with the jth column indicated.

For example, for

$$\begin{cases} 2x + 3y = 1 \\ x - y = 2 \end{cases}$$

$$D = \begin{vmatrix} 2 & 3 \\ 1 & -1 \end{vmatrix} = -5$$

$$D_1 = \begin{vmatrix} \boxed{1} & 3 \\ \boxed{2} & -1 \end{vmatrix} = -7$$

and

$$D_2 = \begin{vmatrix} 2 & \boxed{1} \\ 1 & \boxed{2} \end{vmatrix} = 3$$

and for

$$\begin{cases} 3x - y + 3z = 1 \\ x + y = 4 \\ -5x + 7y - 2z = -2 \end{cases}$$

$$D = \begin{vmatrix} 3 & -1 & 3 \\ 1 & 1 & 0 \\ -5 & 7 & -2 \end{vmatrix} = 28$$

$$D_1 = \begin{vmatrix} \boxed{1} & -1 & 3 \\ \boxed{4} & 1 & 0 \\ \boxed{-2} & 7 & -2 \end{vmatrix} = 80$$

$$D_2 = \begin{vmatrix} 3 & \boxed{1} & 3 \\ 1 & \boxed{4} & 0 \\ -5 & \boxed{-2} & -2 \end{vmatrix} = 32$$

and

$$D_3 = \begin{vmatrix} 3 & -1 & \boxed{1} \\ 1 & 1 & \boxed{4} \\ -5 & 7 & \boxed{-2} \end{vmatrix} = -60$$

THEOREM (CRAMER'S RULE)

Let (S) be the system of n linear equations in n unknowns described above. Let D be the determinant of the coefficient matrix of (S). If $D \neq 0$, then the system (S) has exactly one solution:

$$x_j = \frac{D_j}{D}, \quad j = 1, 2, \ldots, n$$

where D_j is the determinant defined above.

The proof of Cramer's rule is beyond the scope of this text. However, we will consider its application for cases in which $n = 2$ or 3.

For $n = 2$, Cramer's rule indicates that the solution of

$$\begin{cases} a_{11}x_1 + a_{12}x_2 = c_1 \\ a_{21}x_1 + a_{22}x_2 = c_2 \end{cases}$$

is given by

$$x_1 = \frac{\begin{vmatrix} c_1 & a_{12} \\ c_2 & a_{22} \end{vmatrix}}{\begin{vmatrix} a_{11} & a_{12} \\ a_{21} & a_{22} \end{vmatrix}} \quad \text{and} \quad x_2 = \frac{\begin{vmatrix} a_{11} & c_1 \\ a_{21} & c_2 \end{vmatrix}}{\begin{vmatrix} a_{11} & a_{12} \\ a_{21} & a_{22} \end{vmatrix}}$$

if

$$\begin{vmatrix} a_{11} & a_{12} \\ a_{21} & a_{22} \end{vmatrix} \neq 0$$

For $n = 3$, Cramer's rule can be applied to

$$\begin{cases} a_{11}x_1 + a_{12}x_2 + a_{13}x_3 = c_1 \\ a_{21}x_1 + a_{22}x_2 + a_{23}x_3 = c_2 \\ a_{31}x_1 + a_{32}x_2 + a_{33}x_3 = c_3 \end{cases}$$

as follows.

First, we determine

$$D = \begin{vmatrix} a_{11} & a_{12} & a_{13} \\ a_{21} & a_{22} & a_{23} \\ a_{31} & a_{32} & a_{33} \end{vmatrix} \quad D_1 = \begin{vmatrix} c_1 & a_{12} & a_{13} \\ c_2 & a_{22} & a_{23} \\ c_3 & a_{32} & a_{33} \end{vmatrix}$$

$$D_2 = \begin{vmatrix} a_{11} & c_1 & a_{13} \\ a_{21} & c_2 & a_{23} \\ a_{31} & c_3 & a_{33} \end{vmatrix} \quad \text{and} \quad D_3 = \begin{vmatrix} a_{11} & a_{12} & c_1 \\ a_{21} & a_{22} & c_2 \\ a_{31} & a_{32} & c_3 \end{vmatrix}$$

If $D \neq 0$, then $x_1 = D_1/D$, $x_2 = D_2/D$, and $x_3 = D_3/D$.

Notice that Cramer's rule has nothing to say about the existence of solutions in the case in which $D = 0$. Actually, if $D = 0$, either (1) the system has no solution (inconsistent) or (2) the system has an infinite number of different solutions (dependent).

EXAMPLES

Use Cramer's rule to solve each of the following linear systems if possible.

1. $\begin{cases} 3x + 4y = 12 \\ 3x - 8y = 0 \end{cases}$

SOLUTION. Here

$$D = \begin{vmatrix} 3 & 4 \\ 3 & -8 \end{vmatrix} = -24 - 12 = -36$$

$$D_1 = \begin{vmatrix} 12 & 4 \\ 0 & -8 \end{vmatrix} = -96 \quad \text{and} \quad D_2 = \begin{vmatrix} 3 & 12 \\ 3 & 0 \end{vmatrix} = -36$$

Since $D \neq 0$, this system has one solution $\{(\frac{8}{3}, 1)\}$, since $x = D_1/D = -96/-36 = \frac{8}{3}$ and $y = D_2/D = -36/-36 = 1$ satisfy both equations simultaneously.

2. $\begin{cases} x + 2y = 4 \\ 3x + 6y = -3 \end{cases}$

SOLUTION. Here

$$D = \begin{vmatrix} 1 & 2 \\ 3 & 6 \end{vmatrix} = 0$$

Since $D = 0$, Cramer's rule is not applicable. This situation is consistent with the geometry of the two lines (they are parallel).

3. $\begin{cases} x_1 + x_2 + x_3 = 2 \\ 2x_1 - x_2 + x_3 = 0 \\ x_1 + 2x_2 - x_3 = 4 \end{cases}$

SOLUTION. Here

$$D = \begin{vmatrix} 1 & 1 & 1 \\ 2 & -1 & 1 \\ 1 & 2 & -1 \end{vmatrix} = 7 \quad D_1 = \begin{vmatrix} 2 & 1 & 1 \\ 0 & -1 & 1 \\ 4 & 2 & -1 \end{vmatrix} = 6$$

$$D_2 = \begin{vmatrix} 1 & 2 & 1 \\ 2 & 0 & 1 \\ 1 & 4 & -1 \end{vmatrix} = 10 \quad \text{and} \quad D_3 = \begin{vmatrix} 1 & 1 & 2 \\ 2 & -1 & 0 \\ 1 & 2 & 4 \end{vmatrix} = -2$$

Since $D \neq 0$, this system has exactly one solution $\{(\frac{6}{7}, \frac{10}{7}, -\frac{2}{7})\}$. That is, $x_1 = D_1/D = \frac{6}{7}$, $x_2 = D_2/D = \frac{10}{7}$, and $x_3 = D_3/D = -\frac{2}{7}$ satisfy the system.

PROBLEM SET 3

1. Use Cramer's rule to solve each of the following systems if possible.

a) $2x - y = 0$
   $x + y = 1$

b) $-3x + y = 3$
   $-2x - y = -5$

c) $x + y = 0$
   $x - y = 0$

d) $3x + y = 1$
   $9x + 3y = -4$

e) $2x_1 - x_2 + x_3 = 3$
   $-x_1 + x_2 - x_3 = 1$
   $3x_1 + x_2 + 2x_3 = -1$

f) $3x + 2z = 6 - 2y$
   $x - 5y + 6z = 2$
   $6x - 8z = 12$

g) $x + y + 2z = 4$
   $x + y - 2z = 0$
   $x - y = 0$

h) $2x_1 - 3x_2 = 4$
   $x_1 + x_2 - 2x_3 = 1$
   $x_1 - x_2 - x_3 = 5$

i) $x + y + z = 4$
   $x - y + 2z = 8$
   $2x + y - z = 3$

j) $2x + 3y + z = 6$
   $x - 2y + 3z = -3$
   $3x + y - z = 8$

For each of the following problems, set up a linear system which serves as a model of the situation. Then use determinants to solve the system.

2 a) A watch, chain, and ring together cost $225. The watch costs $50 more than the chain, and the ring costs $25 more than the watch and the chain together. What is the cost of each?

   b) Twice the sum of two numbers is 30, and three times the smaller equals twice the larger. Determine the numbers.

3 A dealer wishes to mix coffee costing him 20 cents a pound with coffee costing 30 cents a pound to make a blend that will cost him 24 cents a pound. How many pounds of each grade must he use to make 50 pounds of the blend?

4 A man invested part of $10,000 at 5 percent and the rest at 8 percent. His annual income from both investments was $608. Find the amount of each investment.

REVIEW PROBLEM SET

1 Use the elimination method to solve each of the following systems of linear equations and show the corresponding matrix form of the system.

   a) $2x - y = 5$
      $x + y = 10$

   b) $2x - 3y = 10$
      $2x + 2y = 5$

   c) $4x - y = 3$
      $-2x + 3y = 1$

   d) $-3x + 5y = 2$
      $2x - 3y = 1$

   e) $4x + 8y + z = 2$
      $x + 7y - 3z = -14$
      $2x - 3y + 2z = 3$

   f) $x + y - z = -2$
      $2x - y + z = -5$
      $x - 2y + 3z = 4$

   g) $2x + 3y - z = 1$
      $x + 2y + 2z = 5$
      $x - y + z = 6$

   h) $x + 3y - 6z = 7$
      $2x - y + z = 1$
      $x + 2y + 2z = -1$

2 For each of the following systems of linear equations find the augmented matrix; then by reducing the matrix to a row-reduced echelon matrix, determine the solution if it is unique. If the solution is not unique, indicate whether the system is dependent or inconsistent.

   a) $3x - 2y = 5$
      $-4x + 5y = 5$

   b) $x - 3y = 4$
      $2x - 6y = 9$

c) $x + y - 2z = 3$
$2x - y + z = 0$
$3x + y - z = 8$

d) $2x - y + z = 1$
$x + 2y - z = 3$
$x + 7y - 4z = 2$

e) $2x - y + z = 1$
$x + 2y - z = 3$
$x + 7y - 4z = 8$

f) $2x - 3y + z = 3$
$x - 4y - z = 4$
$x - 9y - 4z = 5$

3   Use the definitions of determinants to evaluate each of the following determinants.

a) $\begin{vmatrix} 2 & -1 \\ 3 & 4 \end{vmatrix}$

b) $\begin{vmatrix} 7 & 3 \\ 2 & 4 \end{vmatrix}$

c) $\begin{vmatrix} 2 & 3 \\ 5 & 8 \end{vmatrix}$

d) $\begin{vmatrix} 3 & 2 \\ -3 & -1 \end{vmatrix}$

e) $\begin{vmatrix} 4 & 2 & -3 \\ 1 & -4 & 1 \\ -1 & 0 & 2 \end{vmatrix}$

f) $\begin{vmatrix} 2 & -3 & 2 \\ 1 & 2 & 0 \\ 1 & 0 & -3 \end{vmatrix}$

g) $\begin{vmatrix} 1 & 1 & -1 \\ 2 & -3 & 1 \\ -3 & 2 & -3 \end{vmatrix}$

h) $\begin{vmatrix} 3 & 9 & 6 \\ 1 & -1 & -1 \\ 1 & 3 & 2 \end{vmatrix}$

4   Use the properties of determinants to evaluate each of the following determinants.

a) $\begin{vmatrix} 1 & 3 \\ -2 & -6 \end{vmatrix}$

b) $\begin{vmatrix} 2 & 7 \\ 0 & 0 \end{vmatrix}$

c) $\begin{vmatrix} 2 & 0 & 3 \\ -1 & 0 & 3 \\ 5 & 0 & 7 \end{vmatrix}$

d) $\begin{vmatrix} 2 & 6 & 4 \\ 0 & -1 & -2 \\ -2 & -12 & -8 \end{vmatrix}$

e) $\begin{vmatrix} 1 & 3 & 2 \\ 0 & -1 & -2 \\ 2 & 8 & 3 \end{vmatrix}$

f) $\begin{vmatrix} 6 & 2 & 3 \\ 5 & 2 & -1 \\ 3 & 1 & 2 \end{vmatrix}$

5   Use Cramer's rule to solve each of the following systems of linear equations.

a) $2x + y = 3$
$x - 3y = 7$

b) $3x - 2y = 7$
$5x + y = 5$

c) $x - 3y + 4z = 6$
$6x - 9y + 12z = 21$
$x + 2y + 5z = 13$

d) $x - y + 6z = 4$
$3x + 3y - z = 3$
$x + 9y + 2z = 5$

e) $2x + 2y + z = 1$
$x + 3y + 2z = -2$
$x - 3y - z = 7$

f) $3x - y + z = 2$
$3x - y - 6z = -5$
$x + 3y - 2z = 5$

Use systems of equations to solve Problems 6–9.

CHAPTER 8 LINEAR SYSTEMS: MATRICES AND DETERMINANTS

6. A college mailed 120 letters, some requiring 8 cents postage, the rest requiring 10 cents. If the total bill is $11.30 find the number sent at each rate.

7. In a special showing at a certain movie theater, admission tickets are $3.65 cents for adults and $1.25 cents for children. If the receipts from 740 tickets were $1,645.00, how many tickets of each kind were sold?

8. Find values a, b, and c, and sketch the resulting graph for $y = ax^2 + bx + c$ that contains the points $(1, 3)$, $(3, 5)$, and $(-1, 9)$.

9. The specific gravity of an object is defined to be its weight in air, divided by its loss of weight when submerged in water. An object made of gold, of specific gravity 16, and silver, of specific gravity 10.8, weighs 8 ounces in air and 7.30 ounces when submerged in water. How many ounces of gold and how many ounces of silver are there in the object?

CHAPTER 9

Analytic Geometry

CHAPTER 9

Analytic Geometry

1 Introduction

In Chapter 4 we indicated that the graph of the quadratic function defined by the equation $y = ax^2 + bx + c$, $a \neq 0$ is a "parabola." In this chapter we shall give general definitions for the parabola and curves called the *circle*, *ellipse*, and *hyperbola* (Figure 1). We shall derive standard equations for these curves and discuss briefly their most important properties. These standard equations will be of the form $ax^2 + by^2 + cx + dy + f = 0$, where a, b, c, d, and f are constants, with a and b both $\neq 0$. The chapter is concluded with a discussion of systems with second-degree equations.

Figure 1

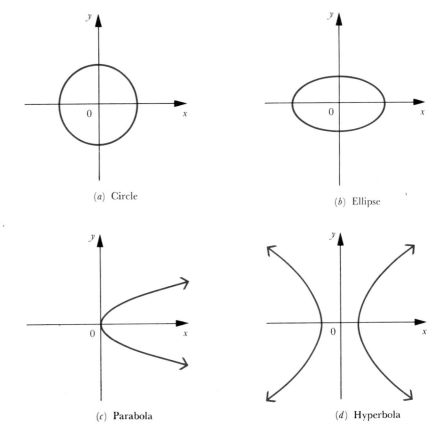

(a) Circle

(b) Ellipse

(c) Parabola

(d) Hyperbola

2 Circle

A *circle* is defined geometrically as a set of all points in the plane that are at a fixed distance r, called the *radius*, from a fixed point C, called the *center*. The length of a diameter of a circle is $2r$ (Figure 1). The distance

Figure 1

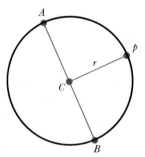

formula can be used to write the equation of a relation whose graph is a circle as follows.

PROPERTY (CIRCLE EQUATION)

Given a circle with center at the point (h, k) and with radius r, the equation of the circle is $(x - h)^2 + (y - k)^2 = r^2$.

PROOF. Let $P = (x, y)$ [for convenience, we will use the notation $P = (x, y)$ to identify points in the plane rather than using the notation $P: (x, y)$] represent any point on the circle whose center C is (h, k). Then, using the distance formula, we have

$$\sqrt{(x - h)^2 + (y - k)^2} = r$$

(Figure 2). Squaring both sides of the equation, we obtain

(1) $\quad (x - h)^2 + (y - k)^2 = r^2$

Figure 2

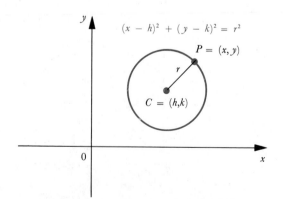

Conversely, if the numbers x and y satisfy Eq. (1), then the point (x, y) belongs to the circle. That is, if $P = (x, y)$ is a point that satisfies Eq. (1), then

$$\sqrt{(x - h)^2 + (y - k)^2} = \sqrt{r^2} = r$$

so that P is r units from the center (h, k) and is a point on the circle.

EXAMPLES

1 Find the equation of a circle if the center C is the point $(-3, 4)$ and the radius r is 3.

SOLUTION. The equation of a circle is $(x - h)^2 + (y - k)^2 = r^2$. Replace (h, k) by $(-3, 4)$ and r by 3 so that $(x + 3)^2 + (y - 4)^2 = 9$ (Figure 3).

Figure 3

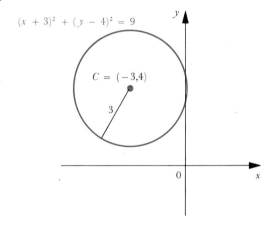

2 Given the equation of a circle $x^2 + y^2 - 6x + 8y - 24 = 0$, find the center and the radius.

SOLUTION. First, we will rewrite the equation in the standard form of the circle equation property. To do this, we "complete the square" as follows:

$$(x^2 - 6x + \quad) + (y^2 + 8y + \quad) = 24$$
$$(x^2 - 6x + 9) + (y^2 + 8y + 16) = 9 + 16 + 24$$
$$(x - 3)^2 + (y + 4)^2 = 49$$

so that the graph is a circle with center at $(3, -4)$ and radius 7 (Figure 4).

3 Find the equation of the circle whose center is $(5, 3)$ and contains the point $(-1, -5)$.

Figure 4

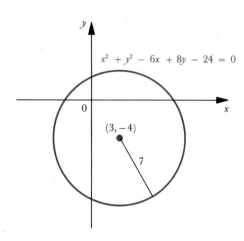

SOLUTION. The radius of the circle is

$$r = \sqrt{(-1-5)^2 + (-5-3)^2} \quad \text{(why?)}$$
$$r = \sqrt{36 + 64}$$
$$= \sqrt{100} = 10$$

so that $(x-5)^2 + (y-3)^2 = 100$ (Figure 5).

Figure 5

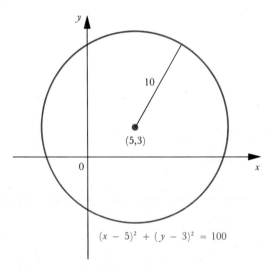

4. Find the equation of the circle which contains the point $(0, 0)$, with radius 13, and whose x coordinate of the center is -12.

SOLUTION. Since the equation contains the point $(0, 0)$, we have $(0-h)^2 + (0-k)^2 = 13^2$, or $h^2 + k^2 = 169$. Here $h = -12$, so that $144 + k^2 = 169$, or $k^2 = 25$, thus $k = \pm 5$. Therefore, the equation.

of the circle is either

$$(x + 12)^2 + (y - 5)^2 = 169$$

or

$$(x + 12)^2 + (y + 5)^2 = 169 \quad \text{(Figure 6)}$$

Figure 6

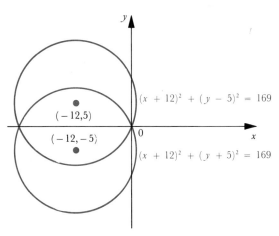

5 From plane geometry, we know that three noncollinear points determine a unique circle. Find the equation of the circle containing the points $(0, 0)$, $(1, 0)$, and $(1, 4)$.

SOLUTION. Substituting the three points into the equation $(x - k)^2 + (y - k)^2 = r^2$, we get

$$(0 - h)^2 + (0 - k)^2 = r^2$$
$$(1 - h)^2 + (0 - k)^2 = r^2$$
$$(1 - h)^2 + (4 - k)^2 = r^2$$

In other words,

$$h^2 + k^2 = r^2$$
$$1 - 2h + h^2 + k^2 = r^2$$
$$1 - 2h + h^2 + 16 - 8k + k^2 = r^2$$

Subtracting the first equation from the second equation and the second equation from the third, we have

$$1 - 2h = 0 \quad \text{and} \quad 16 - 8k = 0$$

Hence $h = \frac{1}{2}$, $k = 2$, and $r = \sqrt{17}/2$, so that the center of this circle is the point $(\frac{1}{2}, 2)$ and the radius is $\sqrt{17}/2$. Thus, the equation of the circle is $(x - \frac{1}{2})^2 + (y - 2)^2 = \frac{17}{4}$ (Figure 7).

Figure 7

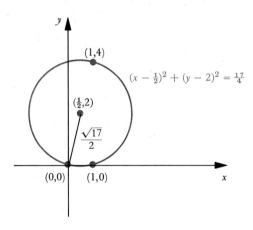

PROBLEM SET 1

1. Write the equation of the circle and sketch the graph in each case.
   a) The circle with center $(6, 4)$ and radius 6.
   b) The circle with center $(4, 3)$ and radius 5.
   c) The circle with center $(5, -12)$ and radius 13.
   d) The circle with center $(1, 2)$ and containing the point $(-2, 3)$.
   e) The circle containing the points $(4, 2)$, $(2, 4)$ and $(-4, 2)$.

2. Find the center and radius of each of the following circles. Also, sketch the graph in each case.
   a) $(x + 2)^2 + (y - 3)^2 = 25$ b) $(x - 1)^2 + (y + 4)^2 = 9$
   c) $(x - 1)^2 + (y + 2)^2 = 16$ d) $(x - 3)^2 + (y - 4)^2 = 25$
   e) $x^2 + y^2 - 8x + 6y + 24 = 0$ f) $x^2 + y^2 - 13x = 0$
   g) $4x^2 + 4y^2 - 12x - 81 = 0$ h) $3x^2 + 3y^2 + 36x - 14y = 0$

3. Sketch the graph of the circle defined by the equation $(x + 4)^2 + (y - 3)^2 = r$, for $r = 16, 9, 4, 1,$ and 0.

4. Find the equation of the circle that contains the points $(2, 5)$ and $(10, 1)$ and has its center on the line $x - y - 4 = 0$. Sketch the graph.

5. We say that a relation R is symmetric with respect to the x axis if $(x, y) \in R$, then $(x, -y) \in R$. Discuss the symmetry of the circle $x^2 + y^2 = r^2$ with respect to the x axis, y axis and the origin.

6. Find the equation of the circle whose center is the point $(4, 4)$ and is tangent to the line $x - y - 4 = 0$.

7. Show that the four points $(0, 6)$, $(4, 8)$, $(12, 0)$, and $(4, -6)$ lie on the same circle.

8 Find the equation of the circle that is tangent to the line $3x - 4y = 2$ at the point $(2, 1)$ and which contains the origin. Sketch the graph.

9 Graph each of the following relations on different coordinate axes: $y = \sqrt{4 - x^2}, y = -\sqrt{4 - x^2}$, and $x^2 + y^2 = 4$. Indicate the domain and range of each relation. Which, if any, of the three relations are functions?

3 Parabola

A *parabola* is defined geometrically to be the set of all points in a plane each of which is equidistant from a fixed point, called the *focus*, and from a fixed line, called the *directrix*. In Figure 1, if P_1, P_2, and P_3 are points on the parabola, F is the focus, and l is the directrix, then $d_1 = c_1$, $d_2 = c_2$, and $d_3 = c_3$.

Figure 1

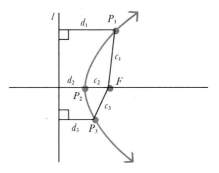

From this geometric description, we can derive an algebraic equation of the parabola.

PROPERTY (PARABOLA EQUATION)

Consider (x, y) to be any point on the parabola with focus $(c, 0)$ and directrix $x = -c$. Then the equation of the parabola is $y^2 = 4cx$, where $c > 0$ (Figure 2).

PROOF. The point $P = (x, y)$ is equidistant from the focus and the directrix, so that $\overline{PF} = \overline{PD}$ or

$$\sqrt{(x - c)^2 + y^2} = |x - (-c)| = |x + c|$$

Squaring both sides of the equation, we have

$$(x - c)^2 + y^2 = (x + c)^2$$

Figure 2

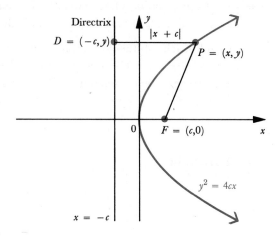

so that

$$x^2 - 2xc + c^2 + y^2 = x^2 + 2xc + c^2$$

or

$$y^2 = 4cx$$

In Figure 3 the line passing through the focus of the parabola perpendicular to the directrix is called the *axis of symmetry;* the point V on the axis midway between the directrix and focus is called the *center* or *vertex* of the parabola; the line segment $\overline{AB}$ with endpoints on the parabola perpendicular to the axis at the focus is called the *focal chord* or *latus rectum* of the parabola.

Figure 3

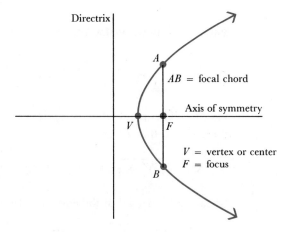

The graph of the parabola $y^2 = 4cx$ has the x axis as its axis of symmetry. Also, since $y^2 > 0$, c and x must have the same sign. The directrix is parallel to the y axis and the focus is a point to its right. This case, which might be described by saying that the parabola "opens to the right," is one of four possible cases which are described as follows (assume that $c > 0$ in all cases).

1 $y^2 = 4cx$. The focus is on the x axis, the directrix is parallel to the y axis, the vertex is $(0, 0)$, and the parabola opens to the right (Figure 4).

Figure 4

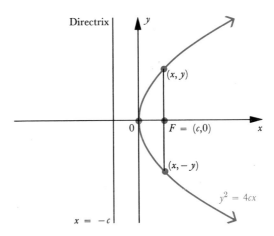

2 $y^2 = -4cx$. The focus is on the x axis, the directrix is parallel to the y axis, the vertex is $(0, 0)$, and the parabola opens to the left (Figure 5).

Figure 5

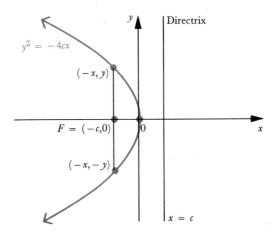

3 $x^2 = 4cy$. The focus is on the y axis, the directrix is parallel to the x axis, the vertex is $(0, 0)$, and the parabola opens upward (Figure 6).

Figure 6

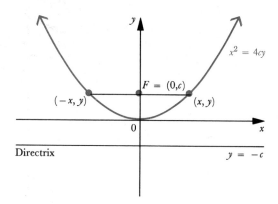

4 $x^2 = -4cy$. The focus is on the y axis, the directrix is parallel to the x axis, the vertex is (0, 0), and the parabola opens downward (Figure 7).

Figure 7

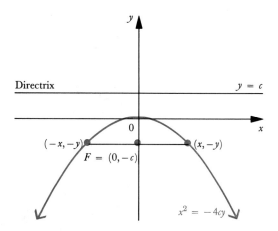

EXAMPLES

In Examples 1 and 2 determine the focus and the directrix and sketch each of the parabolas.

1 $y^2 = 4x$

SOLUTION. This is an example of case 1. The graph of the parabola is symmetric with respect to the x axis. The parabola opens to the right (why?), so that $4c = 4$ or $c = 1$; therefore, the focus is (1, 0) and the equation of the directrix is $x = -1$ (Figure 8).

2 $x^2 = -24y$

SOLUTION. This is an example of case 4. The graph of the parabola is symmetric with respect to the y axis. The parabola opens downward

Figure 8

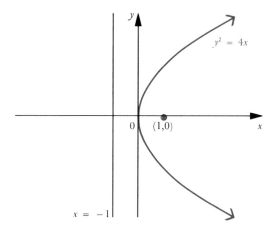

(why?), so that $-4c = -24$ or $c = 6$; therefore, the focus is $(0, -6)$ and the equation of the directrix is $y = 6$ (Figure 9).

Figure 9

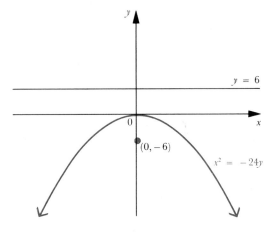

3 A parabola has its focus at $(0, 4)$ and its directrix $y = -4$. Find its equation and sketch the graph.

SOLUTION. The focus is $(0, 4)$ so that $c = 4$, and the directrix is the

Figure 10

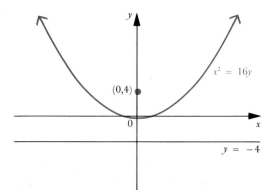

line $y = -4$. The parabola opens upward and the axis of symmetry is the y axis. Hence, the equation is $x^2 = 4cy$, so that $x^2 = 16y$ (Figure 10).

Let us consider the equation of the parabola $y^2 - 6y - 6x + 39 = 0$. In order to write the equation in standard form, we have $y^2 - 6y = 6x - 39$. Now we complete the square in y by adding 9 to both sides of the equation, so that

$$y^2 - 6y + 9 = 6x - 39 + 9$$

This equation may be written in the form $(y - 3)^2 = 6(x - 5)$. The graph of this relation is a parabola with vertex at $(5, 3)$. The value of c is $\frac{3}{2}$ so that the focus is $\frac{3}{2}$ units to the right of the vertex, and the focus is $(\frac{13}{2}, 3)$; also the equation of the directrix is $x = \frac{7}{2}$ (Figure 11).

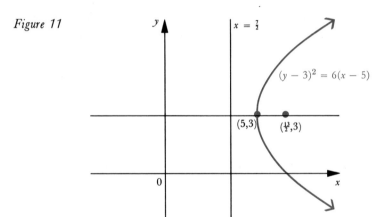

Figure 11

In general, if the vertex of a parabola is at the point (h, k), then its equation can take any one of the following forms (assume that $c > 0$).

1. $(y - k)^2 = 4c(x - h)$. The vertex is (h, k), the directrix is the line $x = h - c$, and the focus is $(h + c, k)$ (Figure 12).

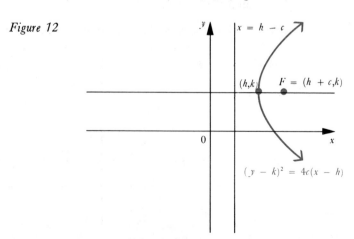

Figure 12

2 $(y - k)^2 = -4c(x - h)$. The vertex is (h, k), the directrix is the line $x = h + c$, and the focus is $(h - c, k)$ (Figure 13).

Figure 13

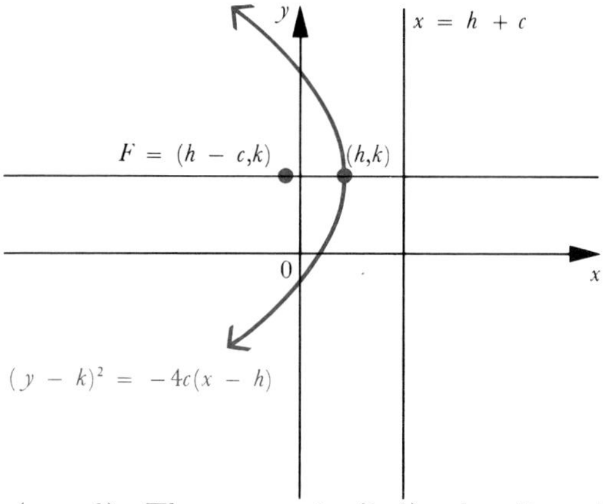

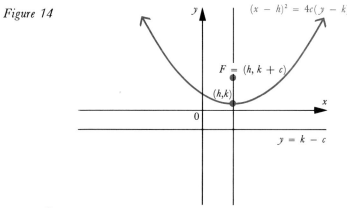

$(y - k)^2 = -4c(x - h)$

3 $(x - h)^2 = 4c(y - k)$. The vertex is (h, k), the directrix is the line $y = k - c$, and the focus is $(h, k + c)$ (Figure 14).

Figure 14

4 $(x - h)^2 = -4c(y - k)$. The vertex is (h, k), the directrix is the line $y = k + c$, and the focus is $(h, k - c)$ (Figure 15).

Figure 15

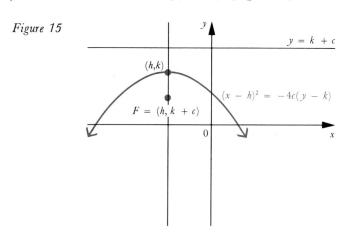

EXAMPLES

Find the vertex, the focus, and the equation of the directrix for each of the given parabolas.

1. $y^2 - 4y - 8x + 28 = 0$

    SOLUTION. First, we will rewrite the equation by completing the square, so that $(y^2 - 4y + \quad) = 8x - 28$, or $y^2 - 4y + 4 = 8x - 28 + 4$, that is, $(y - 2)^2 = 8(x - 3)$. Therefore, the vertex is $(3, 2)$. Since $4c = 8$, it follows that $c = 2$; hence, we have the focus at $(5, 2)$. The equation of the directrix is $x = 3 - 2$ or $x = 1$ (Figure 16).

Figure 16

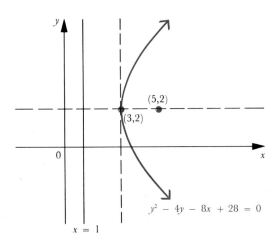

2. $x^2 - 4x - 2y + 10 = 0$

    SOLUTION. After completing the square we have $x^2 - 4x + 4 = 2y - 10 + 4$, so that $(x - 2)^2 = 2(y - 3)$. Therefore, the vertex is $(2, 3)$.

Figure 17

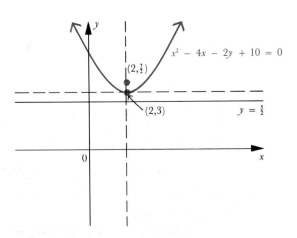

Since $4c = 2$ or $c = \frac{1}{2}$, the focus is $(2, 3 + \frac{1}{2})$ or $(2, \frac{7}{2})$ and the directrix is $y = 3 - \frac{1}{2}$ or $y = \frac{5}{2}$ (Figure 17).

3   $y^2 - 6y + 6x + 15 = 0$

SOLUTION. Completing the square, we have $y^2 - 6y + 9 = -6x - 15 + 9$, or $(y - 3)^2 = -6(x + 1)$. Hence, the vertex is $(-1, 3)$. Since $-4c = -6$, $c = \frac{3}{2}$, so that the focus is $(-1 - \frac{3}{2}, 3)$ or $(-\frac{5}{2}, 3)$ and the directrix is $x = -1 + \frac{3}{2}$ or $x = \frac{1}{2}$ (Figure 18).

Figure 18

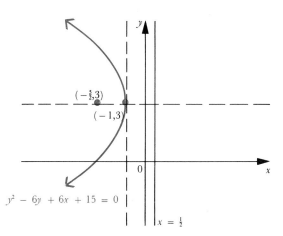

PROBLEM SET 2

1   For each of the following parabolas find the vertex, the focus, and the equation of the directrix, and sketch the graph.
   a) $y^2 = 16x$
   b) $y^2 = -8x$
   c) $x^2 + 8y = 0$
   d) $2x^2 - 9y = 0$
   e) $3y = -4x^2$
   f) $y^2 = 6x$
   g) $15x + 2y^2 = 0$
   h) $(x - 1)^2 = -3(y + 2)$
   i) $(x - 2)^2 = 4(y + 1)$
   j) $(y - 2)^2 = 6(x + 5)$
   k) $y = 4x - x^2$
   l) $4x = 9y^2 - 18y - 2$
   m) $5y = x^2 + 4x - 6$
   n) $x^2 - 2x + 4y + 5 = 0$

2   Show that the length of the latus rectum (focal chord) of a parabola is equal to $4c$. Use this result to find the length of the latus rectums of the parabolas in Problem 1.

3   Find the domain and range of the relations in Problem 1.

4   Find the equation of the parabola in each of the following cases. Also sketch the graph.
   a) Focus at $(0, -4)$ and directrix $y = 4$.

b) Focus at (4, 0) and directrix $x = 3$.
c) Focus at $(-2, 3)$ and directrix $x = 6$.
d) Vertex at $(2, -5)$ and directrix $y = 3$.
e) Vertex at $(4, -3)$ and directrix $y = -6$.

5 Find the equation of the circle that contains the vertex and the endpoints of the latus rectum of the parabola $y^2 = 8x$.

6 Find the equation of the parabola whose axis is parallel to the y axis and that contains the points $(2, 5)$, $(3, 0)$, and $(-7, 0)$.

7 Find the equation of the parabola whose axis is parallel to the x axis and that contains the points $(1, 6)$, $(2, 4)$, and $(4, -2)$.

8 Discuss the symmetry of the parabola $y^2 = 4cx$.

9 How high is a parabolic arch of span 36 feet and 30 feet high at a point 12 feet from one end of the arch (Figure 19)?

Figure 19

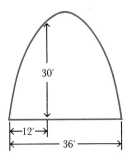

10 Find the equation of the parabola with its vertex on the line $7x + 3y - 4 = 0$ and containing the point $(\frac{3}{2}, 1)$, and its axis parallel to the x axis.

4 Ellipse

An *ellipse* can be defined in a plane as the set of all points each of which has the property that the sum of its distances from two fixed points, called the *foci*, is constant. The definition can be interpreted geometrically as follows: If F_1 and F_2 are the foci, P_1 and P_2 are two points on the ellipse, then $\overline{P_1F_1} + \overline{P_1F_2} = \overline{P_2F_1} + \overline{P_2F_2} = k$; k is a constant and will be denoted by $2a$ (Figure 1). (This constant is written in the form $2a$, so that the equation of the ellipse will have a simpler form.) In constructing the ellipse, we notice that it has two axes of symmetry. One axis is the line through the foci, the other is the perpendicular bisector of the line joining the foci. The point of intersection of these axes is called the *center*

Figure 1

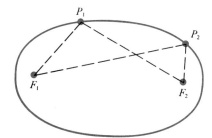

of the ellipse. The four points of intersection of the lines of symmetry and the ellipse are called *vertices* of the ellipse (Figure 2). The longer

Figure 2

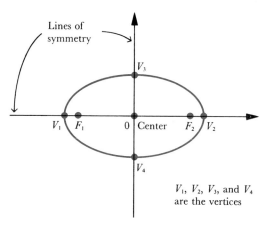

V_1, V_2, V_3, and V_4 are the vertices

line segment determined by the vertices is called the *major axis*, whereas the length of the shorter line segment determined by the vertices is called the *minor axis*. In Figure 2 V_1, V_2, V_3, and V_4 are the vertices; $\overline{V_1V_2}$ is the major axis; $\overline{V_3V_4}$ is the minor axis; and 0 is the center.

This definition of the ellipse is geometric. To express this description in analytic terms, let us assume that $F_1 = (-c, 0)$, $F_2 = (c, 0)$, so that the distance between F_1 and F_2 is $2c$, where $c > 0$. Consider the major axis to be along the x axis, the minor axis to be along the y axis, and the center to be at $(0, 0)$ (Figure 3).

PROPERTY (ELLIPSE EQUATION)

An equation of the ellipse with foci at $F_1 = (-c, 0)$ and $F_2 = (c, 0)$ is

$$\frac{x^2}{a^2} + \frac{y^2}{b^2} = 1 \quad \text{where } b^2 = a^2 - c^2, a > b$$

(Note that $2a$ is the length of the major axis and $2b$ is the length of the minor axis.)

Figure 3

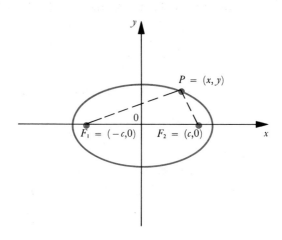

PROOF. Consider $P = (x, y)$ to be a point on the ellipse (Figure 3). Then, by definition of the ellipse, we have $\overline{PF_1} + \overline{PF_2} = 2a$. Note from triangle F_1F_2P that $\overline{PF_1} + \overline{PF_2} > \overline{F_1F_2}$, or $2a > 2c$, so that $a > c$. Hence, by the distance formula

$$\sqrt{(x + c)^2 + y^2} + \sqrt{(x - c)^2 + y^2} = 2a$$

that is,

$$\sqrt{(x + c)^2 + y^2} = 2a - \sqrt{(x - c)^2 + y^2}$$

After squaring both sides of the equation, we obtain

$$(x + c)^2 + y^2 = 4a^2 - 4a\sqrt{(x - c)^2 + y^2} + (x - c)^2 + y^2$$

or

$$x^2 + 2xc + c^2 + y^2 = 4a^2 - 4a\sqrt{(x - c)^2 + y^2} + x^2 - 2xc + c^2 + y^2$$

Simplifying the equation, we get

$$a\sqrt{(x - c)^2 + y^2} = a^2 - cx$$

so we square the latter equation to get

$$a^2[(x - c)^2 + y^2] = a^4 - 2a^2cx + c^2x^2$$

Simplifying again, we have

$$(a^2 - c^2)x^2 + a^2y^2 = a^2(a^2 - c^2)$$

Dividing both sides of the equation by $a^2(a^2 - c^2)$, we get

$$\frac{x^2}{a^2} + \frac{y^2}{a^2 - c^2} = 1$$

so that by substituting b^2 for $a^2 - c^2$, we get

$$\frac{x^2}{a^2} + \frac{y^2}{b^2} = 1 \quad \text{(Figure 4)}$$

Figure 4

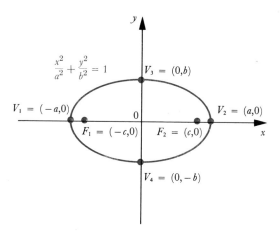

EXAMPLES

1 Given the equation of the ellipse $9x^2 + 25y^2 = 225$, find the vertices and the foci and sketch the graph.

   SOLUTION. Dividing both sides of the equation by 225, we have

   $$\frac{x^2}{25} + \frac{y^2}{9} = 1$$

   This equation is of the form

   $$\frac{x^2}{a^2} + \frac{y^2}{b^2} = 1 \quad \text{with } a^2 = 25 \text{ and } b^2 = 9$$

   so that $a = 5$ and $b = 3$. Thus, the graph is an ellipse with vertices $(-5, 0)$, $(5, 0)$, $(0, -3)$, and $(0, 3)$ and, since $c^2 = a^2 - b^2 = 25 - 9 = 16$, we have $c = 4$, so that the coordinates of the foci are $(-4, 0)$ and $(4, 0)$ (Figure 5).

2 Find the equation of the ellipse with center at the origin, major axis on the x axis, and containing the points $(4, 3)$ and $(6, 2)$. Also sketch the graph.

Figure 5

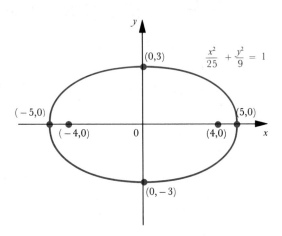

SOLUTION. The equation of the ellipse is of the form $x^2/a^2 + y^2/b^2 = 1$. Since the points (4, 3) and (6, 2) are on the ellipse, we have

$$\frac{16}{a^2} + \frac{9}{b^2} = 1 \quad \text{and} \quad \frac{36}{a^2} + \frac{4}{b^2} = 1$$

so that by solving these two equations simultaneously, we get

$$a^2 = 52 \quad \text{and} \quad b^2 = 13$$

Hence, the equation is

$$\frac{x^2}{52} + \frac{y^2}{13} = 1 \quad \text{(Figure 6)}$$

Figure 6

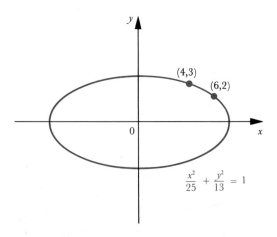

If the foci of the ellipse are $F_1 = (0, -c)$ and $F_2 = (0, c)$ and the center $C = (0, 0)$, the equation of the ellipse will be

$$\frac{x^2}{b^2} + \frac{y^2}{a^2} = 1$$

In this case, the major axis is along the y axis. Again $a > b$, so that we still have $c^2 = a^2 - b^2$.

EXAMPLE

Find the vertices and foci and sketch the graph of the ellipse whose equation is

$$\frac{x^2}{16} + \frac{y^2}{25} = 1$$

SOLUTION. Using the equation

$$\frac{x^2}{b^2} + \frac{y^2}{a^2} = 1$$

we have $b^2 = 16$ and $a^2 = 25$. Then

$$c^2 = a^2 - b^2 = 25 - 16 = 9 \quad \text{or} \quad c = 3$$

The vertices are $(-4, 0)$, $(4, 0)$, $(0, -5)$, and $(0, 5)$. The foci are $(0, -3)$ and $(0, 3)$ (Figure 7).

Figure 7

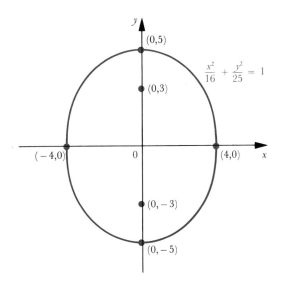

Consider the equation of the ellipse

$$x^2 + 4y^2 - 16x + 16y + 76 = 0$$

This equation can be written in standard form as follows. First write the equation in the form

$$(x^2 - 16x) + (4y^2 + 16y) = -76$$

or

$$(x^2 - 16x) + 4(y^2 + 4y) = -76$$

Now add the numbers necessary to complete the squares of each quadratic, so that

$$(x^2 - 16x + 64) + 4(y^2 + 4y + 4) = -76 + 64 + 16$$

This equation can be written in the form $(x - 8)^2 + 4(y + 2)^2 = 4$. Divide both sides of the equation by 4 to obtain the standard form,

$$\frac{(x - 8)^2}{4} + \frac{(y + 2)^2}{1} = 1$$

The graph of this relation is an ellipse whose center is at the point $(8, -2)$ with the major axis parallel to the x axis and with $a = 2$, $b = 1$, and $c = \sqrt{3}$. The coordinates of the vertices are $(10, -2)$, $(6, -2)$, $(8, -1)$, and $(8, -3)$, and the coordinates of the foci are $(8 - \sqrt{3}, -2)$ and $(8 + \sqrt{3}, -2)$ (Figure 8).

Figure 8

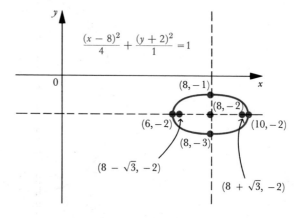

In general, if the center of the ellipse is at the point (h, k), then the equation of the ellipse will be expressed as follows. If the major axis is parallel to the x axis (Figure 9), then

$$\frac{(x-h)^2}{a^2} + \frac{(y-k)^2}{b^2} = 1 \qquad a > b \quad \text{and} \quad c^2 = a^2 - b^2$$

If the major axis is parallel to the y axis (Figure 10), then

$$\frac{(x-h)^2}{b^2} + \frac{(y-k)^2}{a^2} = 1 \qquad a > b \quad \text{and} \quad c^2 = a^2 - b^2$$

EXAMPLES

In Examples 1 and 2 find the coordinates of the center, the vertices, and the foci, and sketch the graph of each of the given ellipses.

1    $9x^2 + 16y^2 - 18x + 32y - 119 = 0$

SOLUTION. Completing the square, we get

$$9(x^2 - 2x + 1) + 16(y^2 + 2y + 1) = 119 + 9 + 16$$

or

$$9(x-1)^2 + 16(y+1)^2 = 144$$

so that

$$\frac{(x-1)^2}{16} + \frac{(y+1)^2}{9} = 1$$

Hence, $h = 1$ and $k = -1$, so that the coordinates of the center are $(1, -1)$. Also, $a^2 = 16$ and $b^2 = 9$ or $a = 4$ and $b = 3$. Since the major axis is parallel to the x axis, the coordinates of the vertices are $(-3, -1)$, $(5, -1)$, $(1, -4)$, and $(1, 2)$. $c^2 = a^2 - b^2 = 16 - 9 = 7$, so that $c = \sqrt{7}$. Hence, the coordinates of the foci are $(1 - \sqrt{7}, -1)$ and $(1 + \sqrt{7}, -1)$ (Figure 9).

Figure 9

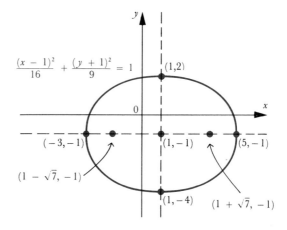

2   $4x^2 + y^2 + 16x - 6y + 21 = 0$

SOLUTION.   Completing the square, we get

$$4(x^2 + 4x + 4) + (y^2 - 6y + 9) = 16 + 9 - 21$$

or

$$4(x + 2)^2 + (y - 3)^2 = 4$$

so that

$$\frac{(x + 2)^2}{1} + \frac{(y - 3)^2}{4} = 1$$

Hence, $h = -2$ and $k = 3$, so that the coordinates of the center are $(-2, 3)$. Also, $a^2 = 4$ and $b^2 = 1$, or $a = 2$ and $b = 1$. Since the major axis is parallel to the y axis, the coordinates of the vertices are $(-3, 3)$, $(-1, 3)$, $(-2, 1)$, and $(-2, 5)$. $c^2 = a^2 - b^2 = 4 - 1 = 3$, so that $c = \sqrt{3}$. Hence, the coordinates of the foci are $(-2, 3 + \sqrt{3})$ and $(-2, 3 - \sqrt{3})$ (Figure 10).

Figure 10

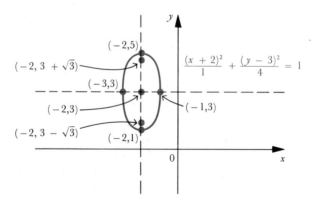

3   Find the equation of the ellipse whose vertices are $(-3, 3)$, $(2, 0)$, $(7, 3)$, and $(2, 6)$.

SOLUTION.   First we plot the vertices (Figure 11). These points are the endpoints of the two axes of the ellipse. Since the distance from $(-3, 3)$ to $(7, 3)$ is 10 units and the distance from $(2, 0)$ to $(2, 6)$ is 6 units, $a = 5$ and $b = 3$, and the major axis is parallel to the x axis. The center is the intersection of the major and minor axes, which in this case is $(2, 3)$. (Why?) Hence, the equation of the ellipse is

$$\frac{(x - 2)^2}{5^2} + \frac{(y - 3)^2}{3^2} = 1 \quad \text{or} \quad \frac{(x - 2)^2}{25} + \frac{(y - 3)^2}{9} = 1$$

Figure 11

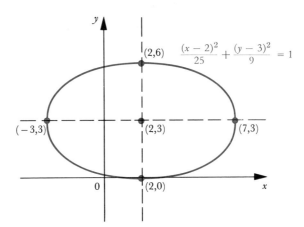

PROBLEM SET 3

1 For each of the following ellipses, find the coordinates of the vertices and the coordinates of the foci, and sketch the graph.

a) $\dfrac{x^2}{16} + \dfrac{y^2}{4} = 1$
b) $\dfrac{x^2}{9} + \dfrac{y^2}{16} = 1$
c) $\dfrac{x^2}{4} + \dfrac{y^2}{64} = 1$
d) $\dfrac{x^2}{3} + \dfrac{y^2}{9} = 1$
e) $5x^2 + y^2 = 10$
f) $9x^2 + 4y^2 = 64$
g) $4x^2 + 25y^2 = 100$
h) $4x^2 + 25y^2 = 1$

2 Indicate the domain and the range of the relations in Problem 1.

3 For each of the following ellipses, find the coordinates of the center, the coordinates of the vertices, and the coordinates of the foci. Also sketch the graph.

a) $(x - 1)^2 + 4(y - 2)^2 = 16$
b) $16(x + 1)^2 + 9(y - 1)^2 = 144$
c) $5(x + 3)^2 + (y - 4)^2 = 10$
d) $(x + 2)^2 + 4(y - 1)^2 = 16$
e) $x^2 - 8x + 4y^2 = 0$
f) $16x^2 + y^2 - 32x + 4y - 44 = 0$
g) $4x^2 + 25y^2 + 100y = 0$
h) $9x^2 + 4y^2 + 36x - 24y - 252 = 0$

4 Find the equation of the ellipse whose vertices are
a) $(-8, 0)$, $(8, 0)$, $(0, 6)$, and $(0, -6)$
b) $(0, -4)$, $(12, -4)$, $(6, 0)$, and $(6, -8)$
c) $(-8, 1)$, $(12, 1)$, $(2, 6)$, and $(2, -4)$

5 Find the equation of the ellipse in each case.
a) Foci at $(-4, 0)$ and $(4, 0)$ and vertices at $(-5, 0)$ and $(5, 0)$.
b) Major axis 20, minor axis 12, vertices on the coordinate axis, and vertical minor axis.

c) With center at $(0, 0)$, axes parallel to coordinate axes, and containing the points $\left(1, \dfrac{3\sqrt{5}}{5}\right)$ and $(\sqrt{5}, 1)$.

d) Center at $(4, -1)$, focus at $(1, -1)$, and containing the point $(8, 0)$.

6 Discuss the symmetry of the ellipse of the form

a) $x^2/a^2 + y^2/b^2 = 1$.

b) $\dfrac{(x - h)^2}{a^2} + \dfrac{(y - k)^2}{b^2} = 1$

7 Sketch the graphs of $y = -\sqrt{1 - x^2/4}$, $y = \sqrt{1 - x^2/4}$, and $y^2 = 1 - x^2/4$. Which of the three equations defines a function?

5 Hyperbola

A *hyperbola* is defined geometrically to be the set of points in a plane such that the absolute value of the difference of the distances from any point to two fixed points, called the *foci*, is a constant.

The hyperbola has two axes of symmetry; the point of intersection of these axes is called the *center* of the hyperbola. One of these axes intersects the hyperbola; it is called the *transverse axis* and its points of intersection with the hyperbola are called the *vertices* of the hyperbola. The foci lie on this axis. The other axis of symmetry is called the *conjugate axis* (Figure 1).

Figure 1

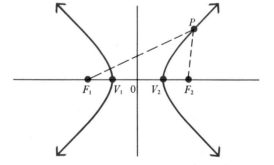

In Figure 1 F_1 and F_2 are the foci, V_1 and V_2 are the vertices, 0 is the center, V_1V_2 is the transverse axis, and $|\overline{PF_1} - \overline{PF_2}| = k$, where k is a constant, as with the ellipse, we call this constant $2a$, for convenience. The equation of the hyperbola will be derived as follows:

PROPERTY (HYPERBOLA EQUATION)

The equation of the hyperbola with foci $F_1 = (-c, 0)$ and $F_2 = (c, 0)$, and with vertices $V_1 = (-a, 0)$ and $V_2 = (a, 0)$ is

$$\dfrac{x^2}{a^2} - \dfrac{y^2}{b^2} = 1 \qquad \text{where } b^2 = c^2 - a^2$$

PROOF. Let $P = (x, y)$ be a point on the hyperbola and let the constant difference k be $2a$; then $|\overline{PF_1} - \overline{PF_2}| = 2a$ (Figure 2). Using the dis-

Figure 2

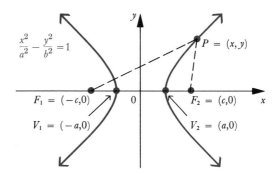

tance formula, we get

$$|\sqrt{(x+c)^2 + y^2} - \sqrt{(x-c)^2 + y^2}| = 2a$$

so that

$$\sqrt{(x+c)^2 + y^2} - \sqrt{(x-c)^2 + y^2} = \pm 2a$$

or, equivalently,

$$\sqrt{(x+c)^2 + y^2} = \sqrt{(x-c)^2 + y^2} \pm 2a$$

After squaring both sides, we have

$$x^2 + 2cx + c^2 + y^2 = 4a^2 \pm 4a\sqrt{(x-c)^2 + y^2} \\ + x^2 - 2cx + c^2 + y^2$$

so that

$$4cx - 4a^2 = \pm 4a\sqrt{(x-c)^2 + y^2}$$

or

$$cx - a^2 = \pm a\sqrt{(x-c)^2 + y^2}$$

Squaring both sides of the equation, we get

$$c^2x^2 - 2a^2cx + a^4 = a^2(x^2 - 2cx + c^2 + y^2)$$

so that

$$(c^2 - a^2)x^2 - a^2y^2 = a^2c^2 - a^4 = a^2(c^2 - a^2)$$

or

$$\frac{x^2}{a^2} - \frac{y^2}{c^2 - a^2} = 1$$

Replacing $c^2 - a^2$ by b^2, we have

$$\frac{x^2}{a^2} - \frac{y^2}{b^2} = 1$$

5.1 Properties of the Hyperbola

The graph of the hyperbola

$$\frac{x^2}{a^2} - \frac{y^2}{b^2} = 1$$

is symmetric with respect to the x axis and the y axis since the points $(-x, y)$ and $(x, -y)$ also satisfy this equation. To find the x intercepts of the hyperbola, let $y = 0$, so that $x^2 = a^2$ or $x = \pm a$. There are no y intercepts since if we substitute $x = 0$ we find that y is not real. Further, by writing the equation

$$\frac{x^2}{a^2} - \frac{y^2}{b^2} = 1$$

in the form

$$y^2 = \frac{b^2}{a^2}(x^2 - a^2)$$

we have

$$y = \pm \frac{b}{a} x \sqrt{1 - \frac{a^2}{x^2}}$$

The expression $1 - a^2/x^2$ approaches 1 as $|x|$ gets very large, so that the larger x becomes in absolute value, the closer the graph of the hyperbola gets to the lines whose equations are $y = \pm(b/a)x$. These lines are called the *asymptotes of the hyperbola*. The asymptotes will serve as a guideline in sketching the graph of the hyperbola (Figure 3).

EXAMPLES

1 Given the equation of the hyperbola

$$\frac{x^2}{4} - \frac{y^2}{1} = 1$$

Figure 3

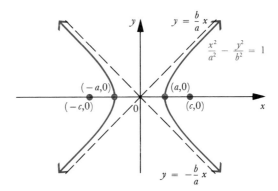

find the coordinates of the foci, the coordinates of the vertices, and the equations of the asymptotes. Also sketch the graph.

SOLUTION. This equation is of the form

$$\frac{x^2}{a^2} - \frac{y^2}{b^2} = 1 \quad \text{with } a^2 = 4 \text{ and } b^2 = 1$$

so that $a = 2$ and $b = 1$. $c = \sqrt{a^2 + b^2} = \sqrt{4 + 1} = \sqrt{5}$. Then the coordinates of the foci are $(-\sqrt{5}, 0)$ and $(\sqrt{5}, 0)$; the coordinates of the vertices are $(-2, 0)$ and $(2, 0)$. The equations of the asymptotes are $y = \pm\frac{1}{2}x$ (Figure 4).

Figure 4

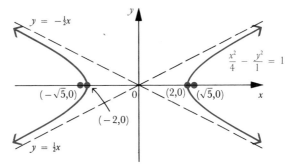

2   Given the equation of the hyperbola

$$\frac{x^2}{9} - \frac{y^2}{16} = 1$$

find the coordinates of the foci, the coordinates of the vertices, and the equations of the asymptotes. Also sketch the graph.

SOLUTION. Here $a^2 = 9$ and $b^2 = 16$, so that $a = 3$ and $b = 4$. $c = \sqrt{a^2 + b^2} = \sqrt{9 + 16} = 5$. Then the coordinates of the foci are

$(-5, 0)$ and $(5, 0)$, the coordinates of the vertices are $(-3, 0)$ and $(3, 0)$, and the equations of the asymptotes are $y = \pm \frac{4}{3}x$ (Figure 5).

Figure 5

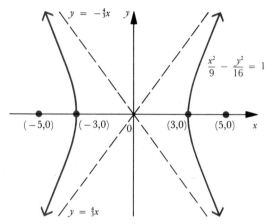

3 Find the equation of the hyperbola with center at the origin, foci at $(-3, 0)$ and $(3, 0)$, and vertices at $(-1, 0)$ and $(1, 0)$. Also sketch the graph.

SOLUTION. Given $a = 1$ and $c = 3$, we have $b = \sqrt{c^2 - a^2} = \sqrt{9 - 1} = \sqrt{8}$. Substituting in the equation

$$\frac{x^2}{a^2} - \frac{y^2}{b^2} = 1$$

so that

$$\frac{x^2}{1} - \frac{y^2}{8} = 1 \quad \text{(Figure 6)}$$

Figure 6

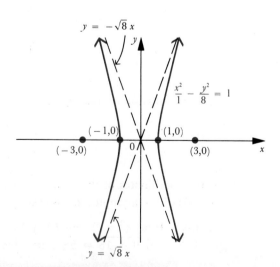

If the foci $F_1 = (0, -c)$ and $F_2 = (0, c)$ and the transverse axis of the hyperbola is on the y axis, then the equation of the hyperbola is written in the form

$$\frac{y^2}{b^2} - \frac{x^2}{a^2} = 1$$

Consider the equation of the hyperbola

$$4y^2 - x^2 + 16y + 2x + 11 = 0$$

This equation can be written in standard form as follows: First write the equation in the form

$$(4y^2 + 16y) - (x^2 - 2x) = -11$$

or

$$4(y^2 + 4y) - (x^2 - 2x) = -11$$

Now add the numbers necessary to complete the squares of the quadratics, so that

$$4(y^2 + 4y + 4) - (x^2 - 2x + 1) = -11 + 16 - 1$$

This equation can be written in the form $4(y + 2)^2 - (x - 1)^2 = 4$. Divide both sides of the equation by 4 to obtain the standard form,

$$\frac{(y + 2)^2}{1} - \frac{(x - 1)^2}{4} = 1$$

The graph of this relation is a hyperbola with center at the point $(1, -2)$,

Figure 7

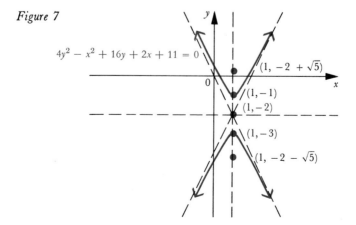

vertices at the points $(1, -1)$ and $(1, -3)$, and foci at $(1, -2 - \sqrt{5})$ and $(1, -2 + \sqrt{5})$. In this case, the transverse axis is parallel to the y axis (Figure 7).

In general, if the center of the hyperbola is $C = (h, k)$ and the transverse axis is parallel to the x axis, the standard form of the hyperbola is

$$\frac{(x - h)^2}{a^2} - \frac{(y - k)^2}{b^2} = 1$$

If the center of the hyperbola is $C = (h, k)$ and the transverse axis is parallel to the y axis, the standard form of the hyperbola is

$$\frac{(y - k)^2}{b^2} - \frac{(x - h)^2}{a^2} = 1$$

The equation of the asymptotes for both of these standard forms is

$$y - k = \pm \frac{b}{a}(x - h)$$

EXAMPLES

In Examples 1 and 2 find the coordinates of the center, the coordinates of the foci, the coordinates of the vertices, and the equations of the asymptotes, and sketch the graph of the given hyperbola.

1   $x^2 - y^2 + 6x + 10y - 2 = 0$

SOLUTION.   First, complete the square to get

$$(x^2 + 6x + 9) - (y^2 - 10y + 25) = 2 + 9 - 25$$

that is,

$$(x + 3)^2 - (y - 5)^2 = -14$$

or

$$(y^2 - 5)^2 - (x + 3)^2 = 14$$

so that

$$\frac{(y - 5)^2}{14} - \frac{(x + 3)^2}{14} = 1$$

The center of the hyperbola $(h, k) = (-3, 5)$. Since $c = \sqrt{a^2 + b^2} = \sqrt{14 + 14} = 2\sqrt{7}$, the coordinates of the foci are $(-3, 5 - 2\sqrt{7})$ and

$(-3, 5 + 2\sqrt{7})$. $b^2 = 14$ or $b = \sqrt{14}$, so that the coordinates of the vertices are $(-3, 5 - \sqrt{14})$ and $(-3, 5 + \sqrt{14})$. Since $a = b$, the equations of the asymptotes are $y - 5 = \pm(x + 3)$ (Figure 8).

Figure 8

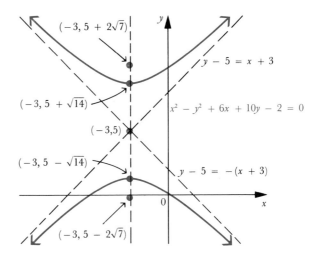

2 $3x^2 - 4y^2 - 18x - 16y - 1 = 0$

SOLUTION. Completing the square, we have

$$3(x^2 - 6x + 9) - 4(y^2 + 4y + 4) = 1 + 27 - 16$$

or

$$3(x - 3)^2 - 4(y + 2)^2 = 12$$

so that

$$\frac{(x - 3)^2}{4} - \frac{(y + 2)^2}{3} = 1$$

$a^2 = 4$ and $b^2 = 3$, so that $a = 2$, $b = \sqrt{3}$, and $c = \sqrt{a^2 + b^2} = \sqrt{4 + 3} = \sqrt{7}$. Since the transverse axis is parallel to the x axis and the center $(h, k) = (3, -2)$, the coordinates of the foci are $(3 - \sqrt{7}, -2)$ and $(3 + \sqrt{7}, -2)$; the coordinates of the vertices are $(3 - 2, -2)$ and $(3 + 2, -2)$ or $(1, -2)$ and $(5, -2)$. The equations of the asymptotes are

$$y + 2 = \pm\frac{\sqrt{3}}{2}(x - 3) \quad \text{(Figure 9)}$$

Figure 9

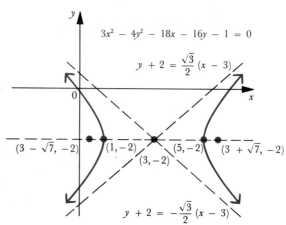

3 Find the equation of the hyperbola whose vertices are the points $(-4, -1)$ and $(0, -1)$ and whose foci are the points $(-5, -1)$ and $(1, -1)$, and sketch its graph.

SOLUTION. By plotting these points, we see that the transverse axis is parallel to the x axis (Figure 10). Hence, the standard form is

$$\frac{(x-h)^2}{a^2} - \frac{(y-k)^2}{b^2} = 1$$

Figure 10

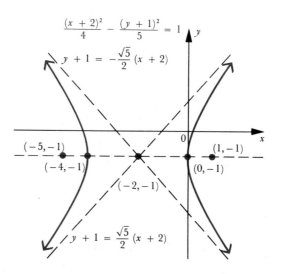

The center is the midpoint of the transverse axis, or $(h, k) = (-2, -1)$. a is the distance from a vertex to the center, or $a = 0 - (-2) = 2$. c is the distance from a focus to the center, or $1 - (-2) = 3$. Since $b^2 = c^2 - a^2$, $b^2 = 3^2 - 2^2 = 5$. Hence, the equation is

$$\frac{(x+2)^2}{4} - \frac{(y+1)^2}{5} = 1$$

PROBLEM SET 4

1. For each of the following hyperbolas, find the coordinates of the vertices, the coordinates of the foci, and the equations of the asymptotes, and sketch the graph.
   a) $4x^2 - 9y^2 = 36$
   b) $x^2 - y^2 = 4$
   c) $\dfrac{x^2}{16} - \dfrac{y^2}{4} = 1$
   d) $\dfrac{x^2}{4} - \dfrac{y^2}{8} = 1$
   e) $x^2 - 2y^2 + 1 = 0$
   f) $3x^2 - y^2 + 9 = 0$
   g) $\dfrac{x^2}{7} - \dfrac{y^2}{5} = 1$
   h) $3x^2 - 8y^2 = 36$

2. Find the domain and range of each of the relations of Problem 1.

3. For each of the following hyperbolas find the coordinates of the center, the coordinates of the vertices, the coordinates of the foci, and the equations of the asymptotes, and sketch the graph.
   a) $\dfrac{(x-1)^2}{16} - \dfrac{(y-3)^2}{9} = 1$
   b) $\dfrac{(x-2)^2}{9} - \dfrac{(y+1)^2}{4} = 1$
   c) $\dfrac{(y+3)^2}{25} - \dfrac{(x-1)^2}{16} = 1$
   d) $\dfrac{(y+1)^2}{16} - \dfrac{(x-3)^2}{25} = 1$
   e) $x^2 - 4y^2 + 4x + 24y - 48 = 0$
   f) $16x^2 - 9y^2 - 96x = 0$
   g) $16x^2 - 9y^2 + 36y + 108 = 0$
   h) $4x^2 - y^2 - 24x + 2y + 35 = 0$

4. Find the equation of the hyperbola with vertices $(-5, 1)$ and $(1, 1)$ and foci $(-7, 1)$ and $(3, 1)$.

5. Find the equation of the hyperbola with center at the origin and foci on the x axis, which contains the points $(6, 2)$ and $(2\sqrt{6}, 1)$.

6. Find the equation of the hyperbola in each of the following cases.
   a) Center at the origin, a vertex at $(3, 0)$, and a focus at $(4, 0)$.
   b) Center at the origin, vertices at $(-16, 0)$ and $(16, 0)$, and asymptotes $y = \pm\frac{5}{4}x$.
   c) Contains the point $(1, 1)$ and with asymptotes $y = \pm 2x$.
   d) Center at $(1, -2)$, vertices at $(-4, -2)$ and $(5, -2)$, and foci at $(6, -2)$ and $(-5, -2)$.

7. Discuss the symmetry of the hyperbola $x^2/a^2 - y^2/b^2 = 1$.

8. Find the equation of the hyperbola such that the product of the slopes of the lines joining (x, y) to $(-2, 1)$ and $(4, 5)$ is 3.

9. Show that the difference of the distances from the point $(8, 8\sqrt{7}/3)$ on the hyperbola $64x^2 - 36y^2 = 2{,}304$ to the foci is equal to the length of the transverse axis.

6 Systems with Second-Degree Equations

One technique for solving systems with second-degree equations is the method of substitution. For example, to solve the system of equations

$$x + y = 7$$
$$x^2 + y^2 = 25$$

first solve the first equation for y and obtain $y = 7 - x$, then substitute in the equation

$$x^2 + y^2 = 25$$

to get

$$x^2 + (7 - x)^2 = 25 \quad \text{or} \quad x^2 - 7x + 12 = 0$$

so that

$$x = 3 \quad \text{or} \quad x = 4$$

Hence, $y = 7 - 3 = 4$ when $x = 3$ and $y = 7 - 4 = 3$ when $x = 4$, so that the solution set is $\{(3, 4), (4, 3)\}$. If we had used the equation $x^2 + y^2 = 25$ to find values of y corresponding to $x = 3$ and $x = 4$, we would have obtained the four ordered pairs $(3, 4)$, $(3, -4)$, $(4, 3)$, and $(4, -3)$, respectively. However, only two ordered pairs, $(3, 4)$ and $(4, 3)$, satisfy both equations of the system. Thus, it is important in solving a system of equations to check a proposed solution in each of the equations in the system.

The above system of equations has two solutions that can be illustrated graphically. The graph of $x^2 + y^2 = 25$ is a circle and the graph of $x + y = 7$ is a straight line. By graphing both equations on the same coordinate axis, we find that they intersect in exactly two points (Figure 1).

EXAMPLES

Solve each of the following systems of equations. Illustrate the system with a graph.

1   $2x + 3y = 8$
    $2x^2 - 3y^2 = -10$

SOLUTION. From the graphs of the two given equations (Figure 2), we see that the solutions lie in the first quadrant. Solving the linear equation

Figure 1

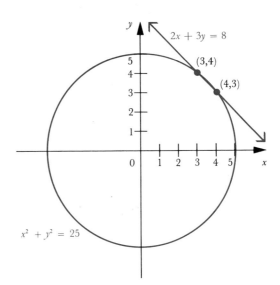

Figure 2

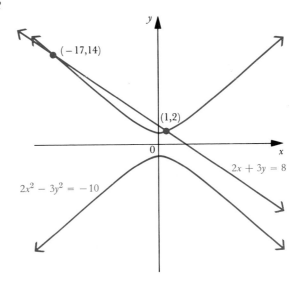

for x, we obtain $x = (8 - 3y)/2$. Then substituting the expression for x into the quadratic equation, we have

$$2\left(\frac{8-3y}{2}\right)^2 - 3y^2 = -10$$

so that

$$3y^2 - 48y + 84 = 0$$

or

$$y^2 - 16y + 28 = 0$$

By factoring, we obtain

$$(y - 2)(y - 14) = 0$$

so that

$$y = 2 \quad \text{or} \quad y = 14$$

When $y = 2$,

$$x = \frac{8 - 3(2)}{2} = 1$$

and when $y = 14$,

$$x = \frac{8 - 3(14)}{2} = -17$$

Hence, the solution set of the system is $\{(1, 2), (-17, 14)\}$.

2 $4x^2 + 7y^2 = 32$
$-3x^2 + 11y^2 = 41$

SOLUTION. From the graphs of the equations $4x^2 + 7y^2 = 12$ and $-3x^2 + 11y^2 = 41$, it is evident that we have four points of intersections (Figure 3). The system

$$4x^2 + 7y^2 = 32$$
$$-3x^2 + 11y^2 = 41$$

can be written

$$12x^2 + 21y^2 = 96$$
$$-12x^2 + 44y^2 = 164$$

which results from multiplying the first equation by 3 and the second equation by 4. Adding these two equations we get

$$65y^2 = 260$$

so that

$$y^2 = 4$$

Figure 3

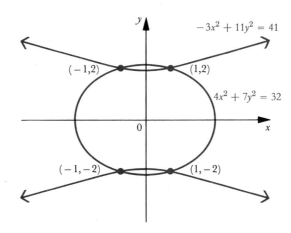

That is,

$$y = -2 \quad \text{or} \quad y = 2$$

When $y = -2$, $4x^2 + 7(-2)^2 = 32$, so that

$$x = -1 \quad \text{or} \quad x = 1$$

When $y = 2$, $4x^2 + 7(2)^2 = 32$, so that

$$x = -1 \quad \text{or} \quad x = 1$$

Hence, the solution set is $\{(1, 2), (-1, 2), (1, -2), (-1, -2)\}$. Examining the graph (Figure 3), we see that the solutions $(1, 2)$, $(-1, 2)$, $(1, -2)$, and $(-1, -2)$, are the points of intersections of the two curves.

3 $x^2 + 2y^2 = 22$
   $2x^2 + y^2 = 17$

SOLUTION. From the graphs of the equations $x^2 + 2y^2 = 22$ and $2x^2 + y^2 = 17$, it is clear that we have four points of intersection (Figure 4). The system

$$x^2 + 2y^2 = 22$$
$$2x^2 + y^2 = 17$$

can be written as

$$2x^2 + 4y^2 = 44$$
$$2x^2 + y^2 = 17$$

Subtracting the second equation from the first equation, we have

$$3y^2 = 27 \quad \text{or} \quad y^2 = 9$$

Figure 4

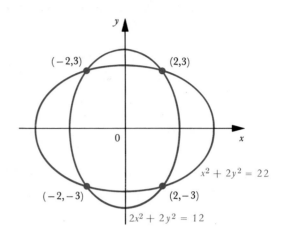

so that

$$y = 3 \quad \text{or} \quad y = -3$$

When $y = 3$, $x^2 + 2(3)^2 = 22$, so that

$$x = 2 \quad \text{or} \quad x = -2$$

When $y = -3$, $x^2 + 2(-3)^2 = 22$, so that

$$x = 2 \quad \text{or} \quad x = -2$$

Hence, the solution set is $\{(2, 3), (-2, 3), (2, -3), (-2, -3)\}$. Examining the graph (Figure 4), we can see that the points of intersection of the two ellipses are $(2, 3)$, $(2, -3)$, $(-2, 3)$, and $(-2, -3)$.

4 A manufacturer of Halloween costumes knows that the total cost C of making x thousand costumes is given by the equation $C = 60x + 600$, where C is in dollars, and that the corresponding sales revenue R is given by $R = 200x - 4x^2$, which is also in dollars. How many costumes will the manufacturer have to produce and sell to break even?

SOLUTION. The manufacturer will break even if the total cost of x thousand costumes is equal to the revenue sale of x thousand costumes, that is, if $C = R$, so that

$$60x + 600 = 200x - 4x^2$$

or, equivalently,

$$4x^2 - 140x + 600 = 0$$

Dividing both sides of the equation by 4, we have

$$x^2 - 35x + 150 = 0 \quad \text{or} \quad (x - 5)(x - 30) = 0, \text{ so that}$$
$$x - 5 = 0 \quad \text{or} \quad x - 30 = 0, \text{ that is,}$$
$$x = 5 \quad \text{or} \quad x = 30$$

So the manufacturer will have to produce either 5,000 or 30,000 costumes to break even.

PROBLEM SET 5

Solve the following systems of equations and check the solution. Illustrate the system with a graph.

1. $x - y = 1$
   $x^2 + y^2 = 5$

2. $x - 2y = 3$
   $x^2 - y^2 = 24$

3. $3x - y = 2$
   $x^2 + y^2 = 20$

4. $x + y = 3$
   $3x^2 - y^2 = \frac{9}{2}$

5. $3x + 2y = 1$
   $3x^2 - y^2 = -4$

6. $x + y = 6$
   $x^2 + y^2 = 20$

7. $5x - 3y = 10$
   $x^2 - y^2 = 6$

8. $2x + y = 10$
   $xy = 12$

9. $2x + 3y = 7$
   $x^2 + y^2 + 4y + 4 = 0$

10. $x - y + 4 = 0$
    $x^2 + 3y^2 = 12$

11. $5x - y = 21$
    $y = x^2 - 5x + 4$

12. $x^2 - 25y^2 = 20$
    $2x^2 + 25y^2 = 88$

13. $x - y^2 = 0$
    $x^2 + 2y^2 = 24$

14. $3x^2 - 8y^2 = 40$
    $5x^2 + y^2 = 81$

15. $2x^2 - 3y^2 = 6$
    $3x^2 + 2y^2 = 35$

16. $x^2 - y^2 = 7$
    $x^2 + y^2 = 25$

17. $x^2 + 9y^2 = 33$
    $x^2 + y^2 = 25$

18. $x^2 + 5y^2 = 70$
    $3x^2 - 5y^2 = 30$

19. $4x^2 - y^2 = 36$
    $4x^2 + \frac{5}{3}y^2 = 36$

20. $x^2 - 2y^2 = 17$
    $2x^2 + y^2 = 54$

21. $2x^2 - 3y^2 = 5$
    $x^2 + 2y^2 = 6$

22. $4x^2 + 3y^2 = 43$
    $3x^2 - y^2 = 3$

23. $x^2 - 2y^2 = 1$
    $x^2 + 4y^2 = 25$

24. $2x^2 - 5y^2 + 8 = 0$
    $x^2 - 7y^2 + 4 = 0$

25. $x^2 + 4y = 8$
    $x^2 + y^2 = 5$

26. $3x - 2y = 9$
    $9x = y^2$

27 $x^2 + y^2 = 16$
    $x^2 - y^2 = -34$

28 $x^2 - 4y^2 = -15$
    $-x^2 + 3y^2 = 11$

29 A manufacturer of toys knows that the total cost C of making x thousand toys of a certain kind is given by the equation $C = 200x - 480$, where C is in dollars, and that the corresponding sales revenue R is given by the equation $R = 266x - 6x^2$, which is also in dollars. How many toys will the manufacturer have to produce and sell to break even?

REVIEW PROBLEM SET

1. Find the equation of the circle in each of the following cases, and sketch the graph.
    a) Center at $(4, -3)$ and radius 5.
    b) Containing the points $(4, 5)$, $(3, -2)$, and $(1, -4)$.
    c) Center at $(1, 2)$ and containing the point $(3, -1)$.
    d) Center at $(-2, 5)$ and tangent to the line $x = 7$.

2. Find the equation of the parabola in each of the following cases, and sketch the graph.
    a) Focus at $(5, 0)$ and directrix $x = -5$.
    b) Focus at $(0, -2)$ and directrix $y = 2$.
    c) Vertex at $(2, 4)$ and focus at $(-3, 4)$.
    d) Vertex at $(3, -2)$, directrix $x = 3$, and length of the latus rectum is 6.

3. Find the equation of the ellipse in each of the following cases, and sketch the graph.
    a) Vertices at $(1, 0)$, $(1, -8)$, $(2, -4)$, and $(0, -4)$.
    b) Vertices at $(3, 0)$ and $(3, 10)$ and focus at $(3, 2)$.
    c) Vertices at $(-1, 8)$ and $(-1, -2)$ and contains the point $(1, 0)$ and whose major axis is parallel to the y axis.
    d) Center at $(-1, 3)$, major axis is 10, and minor axis is 4.

4. Find the equation of the hyperbola in each of the following cases, and sketch the graph.
    a) Vertices at $(-2, 0)$ and $(2, 0)$ and focus at $(-4, 0)$.
    b) Vertices at $(5, 0)$ and $(-5, 0)$ and containing the point $(\frac{9}{5}, -4)$.
    c) Foci at $(-2, 5)$ and $(-2, -3)$ and vertex at $(-2, 2)$.
    d) Center at $(2, 5)$, vertex at $(2, 7)$, and focus $(2, 0)$.
    e) Asymptotes $y = \pm 2x/3$ and vertex at $(6, 0)$.

5. Find the coordinates of the center, the coordinates of the vertices, and the coordinates of the foci in each of the following cases. Also sketch the graph.
    a) $16x^2 - 9y^2 + 54y - 225 = 0$
    b) $4x^2 - y^2 - 2x - y - 16 = 0$

c) $9x^2 - 4y^2 + 36x + 32y + 8 = 0$
d) $25x^2 + 4y^2 + 50x - 24y - 39 = 0$
e) $4x^2 + 9y^2 + 48x - 144y + 684 = 0$
f) $x^2 + 4y^2 - 2x - 3 = 0$
g) $9y^2 - 36x - 12y + 22 = 0$
h) $12x^2 - 12x - 24y + 11 = 0$
i) $25y^2 - 200x - 20y - 116 = 0$

6 Find the equation of the line containing the points on the parabola $y^2 = 8x$ whose y coordinates are 2 and 8, respectively.

7 Consider the equation $(b^2 - k)x^2 + (a^2 - k)y^2 = 1$, where $a \geq b$.
   a) For what choices of k is the graph an ellipse?
   b) For what choices of k is the graph a hyperbola?
   c) For what choices of k is the graph a circle?

8 Solve each of the following systems of equations and check your solutions. Illustrate the system with graphs.
   a) $3x - 4y = 25$
      $x^2 + y^2 = 25$
   b) $2x - y = 2$
      $x^2 + 2y^2 = 12$
   c) $x + y = 6$
      $x^2 + y^2 = 36$
   d) $3x^2 - 2y^2 = 35$
      $7x^2 + 5y^2 = 43$
   e) $x^2 + y^2 = 29$
      $x^2 - y^2 = 21$
   f) $3x^2 - 2y^2 = 180$
      $2x^2 + 5y^2 = 200$

APPENDIX

APPENDIX

approx

TABLE I COMMON LOGARITHMS

mantissas

logs

n	0	1	2	3	4	5	6	7	8	9
10	0000	0043	0086	0128	0170	0212	0253	0294	0334	0374
11	0414	0453	0492	0531	0569	0607	0645	0682	0719	0755
12	0792	0828	0864	0899	0934	0969	1004	1038	1072	1106
13	1139	1173	1206	1239	1271	1303	1335	1367	1399	1430
14	1461	1492	1523	1553	1584	1614	1644	1673	1703	1732
15	1761	1790	1818	1847	1875	1903	1931	1959	1987	2014
16	2041	2068	2095	2122	2148	2175	2201	2227	2253	2279
17	2304	2330	2355	2380	2405	2430	2455	2480	2504	2529
18	2553	2577	2601	2625	2648	2672	2695	2718	2742	2765
19	2788	2810	2833	2856	2878	2900	2923	2945	2967	2989
20	3010	3032	3054	3075	3096	3118	3139	3160	3181	3201
21	3222	3243	3263	3284	3304	3324	3345	3365	3385	3404
22	3424	3444	3464	3483	3502	3522	3541	3560	3579	3598
23	3617	3636	3655	3674	3692	3711	3729	3747	3766	3784
24	3802	3820	3838	3856	3874	3892	3909	3927	3945	3962
25	3979	3997	4014	4031	4048	4065	4082	4099	4116	4133
26	4150	4166	4183	4200	4216	4232	4249	4265	4281	4298
27	4314	4330	4346	4362	4378	4393	4409	4425	4440	4456
28	4472	4487	4502	4518	4533	4548	4564	4579	4594	4609
29	4624	4639	4654	4669	4683	4698	4713	4728	4742	4757
30	4771	4786	4800	4814	4829	4843	4857	4871	4886	4900
31	4914	4928	4942	4955	4969	4983	4997	5011	5024	5038
32	5051	5065	5079	5092	5105	5119	5132	5145	5159	5172
33	5185	5198	5211	5224	5237	5250	5263	5276	5289	5302
34	5315	5328	5340	5353	5366	5378	5391	5403	5416	5428
35	5441	5453	5465	5478	5490	5502	5514	5527	5539	5551
36	5563	5575	5587	5599	5611	5623	5635	5647	5658	5670
37	5682	5694	5705	5717	5729	5740	5752	5763	5775	5786
38	5798	5809	5821	5832	5843	5855	5866	5877	5888	5899
39	5911	5922	5933	5944	5955	5966	5977	5988	5999	6010
40	6021	6031	6042	6053	6064	6075	6085	6096	6107	6117
41	6128	6138	6149	6160	6170	6180	6191	6201	6212	6222
42	6232	6243	6253	6263	6274	6284	6294	6304	6314	6325
43	6335	6345	6355	6365	6375	6385	6395	6405	6415	6425
44	6435	6444	6454	6464	6474	6484	6493	6503	6513	6522
45	6532	6542	6551	6561	6571	6580	6590	6599	6609	6618
46	6628	6637	6646	6656	6665	6675	6684	6693	6702	6712
47	6721	6730	6739	6749	6758	6767	6776	6785	6794	6803
48	6812	6821	6830	6839	6848	6857	6866	6875	6884	6893
49	6902	6911	6920	6928	6937	6946	6955	6964	6972	6981

TABLE I COMMON LOGARITHMS — Cont.

n	0	1	2	3	4	5	6	7	8	9
50	6990	6998	7007	7016	7024	7033	7042	7050	7059	7067
51	7076	7084	7093	7101	7110	7118	7126	7135	7143	7152
52	7160	7168	7177	7185	7193	7202	7210	7218	7226	7235
53	7243	7251	7259	7267	7275	7284	7292	7300	7308	7316
54	7324	7332	7340	7348	7356	7364	7372	7380	7388	7396
55	7404	7412	7419	7427	7435	7443	7451	7459	7466	7474
56	7482	7490	7497	7505	7513	7520	7528	7536	7543	7551
57	7559	7566	7574	7582	7589	7597	7604	7612	7619	7627
58	7634	7642	7649	7657	7664	7672	7679	7686	7694	7701
59	7709	7716	7723	7731	7738	7745	7752	7760	7767	7774
60	7782	7789	7796	7803	7810	7818	7825	7832	7839	7846
61	7853	7860	7868	7875	7882	7889	7896	7903	7910	7917
62	7924	7931	7938	7945	7952	7959	7966	7973	7980	7987
63	7993	8000	8007	8014	8021	8028	8035	8041	8048	8055
64	8062	8069	8075	8082	8089	8096	8102	8109	8116	8122
65	8129	8136	8142	8149	8156	8162	8169	8176	8182	8189
66	8195	8202	8209	8215	8222	8228	8235	8241	8248	8254
67	8261	8267	8274	8280	8287	8293	8299	8306	8312	8319
68	8325	8331	8338	8344	8351	8357	8363	8370	8376	8382
69	8388	8395	8401	8407	8414	8420	8426	8432	8439	8445
70	8451	8457	8463	8470	8476	8482	8488	8494	8500	8506
71	8513	8519	8525	8531	8537	8543	8549	8555	8561	8567
72	8673	8579	8585	8591	8597	8603	8609	8615	8621	8627
73	8633	8639	8645	8651	8657	8663	8669	8675	8681	8686
74	8692	8698	8704	8710	8716	8722	8727	8733	8739	8745
75	8751	8756	8762	8768	8774	8779	8785	8791	8797	8802
76	8808	8814	8820	8825	8831	8837	8842	8848	8854	8859
77	8865	8871	8876	8882	8887	8893	8899	8904	8910	8915
78	8921	8927	8932	8938	8943	8949	8954	8960	8965	8971
79	8976	8982	8987	8993	8998	9004	9009	9015	9020	9025
80	9031	9036	9042	9047	9053	9058	9063	9069	9074	9079
81	9085	9090	9096	9101	9106	9112	9117	9122	9128	9133
82	9138	9143	9149	9154	9159	9165	9170	9175	9180	9186
83	9191	9196	9201	9206	9212	9217	9222	9227	9232	9238
84	9243	9248	9253	9258	9263	9269	9274	9279	9284	9289
85	9294	9299	9304	9309	9315	9320	9325	9330	9335	9340
86	9345	9350	9355	9360	9365	9370	9375	9380	9385	9390
87	9395	9400	9405	9410	9415	9420	9425	9430	9435	9440
88	9445	9450	9455	9460	9465	9469	9474	9479	9484	9489
89	9494	9499	9504	9509	9513	9518	9523	9528	9533	9538
90	9542	9547	9552	9557	9562	9566	9571	9576	9581	9586
91	9590	9595	9600	9605	9609	9614	9619	9624	9628	9633
92	9638	9643	9647	9652	9657	9661	9666	9671	9675	9680
93	9685	9689	9694	9699	9703	9708	9713	9717	9722	9727
94	9731	9736	9741	9745	9750	9754	9759	9763	9768	9773
95	9777	9782	9786	9791	9795	9800	9805	9809	9814	9818
96	9823	9827	9832	9836	9841	9845	9850	9854	9859	9863
97	9868	9872	9877	9881	9886	9890	9894	9899	9903	9908
98	9912	9917	9921	9926	9930	9934	9939	9943	9948	9952
99	9956	9961	9965	9969	9974	9978	9983	9987	9991	9996

TABLE II POWERS AND ROOTS

Number	Square	Square Root	Cube	Cube Root	Number	Square	Square Root	Cube	Cube Root
1	1	1.000	1	1.000	51	2,601	7.141	132,651	3.708
2	4	1.414	8	1.260	52	2,704	7.211	140,608	3.733
3	9	1.732	27	1.442	53	2,809	7.280	148,877	3.756
4	16	2.000	64	1.587	54	2,916	7.348	157,464	3.780
5	25	2.236	125	1.710	55	3,025	7.416	166,375	3.803
6	36	2.449	216	1.817	56	3,136	7.483	175,616	3.826
7	49	2.646	343	1.913	57	3,249	7.550	185,193	3.849
8	64	2.828	512	2.000	58	3,364	7.616	195,112	3.871
9	81	3.000	729	2.080	59	3,481	7.681	205,379	3.893
10	100	3.162	1,000	2.154	60	3,600	7.746	216,000	3.915
11	121	3.317	1,331	2.224	61	3,721	7.810	226,981	3.936
12	144	3.464	1,728	2.289	62	3,844	7.874	238,328	3.958
13	169	3.606	2,197	2.351	63	3,969	7.937	250,047	3.979
14	196	3.742	2,744	2.410	64	4,096	8.000	262,144	4.000
15	225	3.873	3,375	2.466	65	4,225	8.062	274,625	4.021
16	256	4.000	4,096	2.520	66	4,356	8.124	287,496	4.041
17	289	4.123	4,913	2.571	67	4,489	8.185	300,763	4.062
18	324	4.243	5,832	2.621	68	4,624	8.246	314,432	4.082
19	361	4.359	6,859	2.668	69	4,761	8.307	328,509	4.102
20	400	4.472	8,000	2.714	70	4,900	8.367	343,000	4.121
21	441	4.583	9,261	2.759	71	5,041	8.426	357,911	4.141
22	484	4.690	10,648	2.802	72	5,184	8.485	373,248	4.160
23	529	4.796	12,167	2.844	73	5,329	8.544	389,017	4.179
24	576	4.899	13,824	2.884	74	5,476	8.602	405,224	4.198
25	625	5.000	15,625	2.924	75	5,625	8.660	421,875	4.217
26	676	5.099	17,576	2.962	76	5,776	8.718	438,976	4.236
27	729	5.196	19,683	3.000	77	5,929	8.775	456,533	4.254
28	784	5.292	21,952	3.037	78	6,084	8.832	474,552	4.273
29	841	5.385	24,389	3.072	79	6,241	8.888	493,039	4.291
30	900	5.477	27,000	3.107	80	6,400	8.944	512,000	4.309
31	961	5.568	29,791	3.141	81	6,561	9.000	531,441	4.327
32	1,024	5.657	32,768	3.175	82	6,724	9.055	551,368	4.344
33	1,089	5.745	35,937	3.208	83	6,889	9.110	571,787	4.362
34	1,156	5.831	39,304	3.240	84	7,056	9.165	592,704	4.380
35	1,225	5.916	42,875	3.271	85	7,225	9.220	614,125	4.397
36	1,296	6.000	46,656	3.302	86	7,396	9.274	636,056	4.414
37	1,369	6.083	50,653	3.332	87	7,569	9.327	658,503	4.431
38	1,444	6.164	54,872	3.362	88	7,744	9.381	681,472	4.448
39	1,521	6.245	59,319	3.391	89	7,921	9.434	704,969	4.465
40	1,600	6.325	64,000	3.420	90	8,100	9.487	729,000	4.481
41	1,681	6.403	68,921	3.448	91	8,281	9.539	753,571	4.498
42	1,764	6.481	74,088	3.476	92	8,464	9.592	778,688	4.514
43	1,849	6.557	79,507	3.503	93	8,649	9.644	804,357	4.531
44	1,936	6.633	85,184	3.530	94	8,836	9.695	830,584	4.547
45	2,025	6.708	91,125	3.557	95	9,025	9.747	857,375	4.563
46	2,116	6.782	97,336	3.583	96	9,216	9.798	884,736	4.579
47	2,209	6.856	103,823	3.609	97	9,409	9.849	912,673	4.595
48	2,304	6.928	110,592	3.634	98	9,604	9.899	941,192	4.610
49	2,401	7.000	117,649	3.659	99	9,801	9.950	970,299	4.626
50	2,500	7.071	125,000	3.684	100	10,000	10.000	1,000,000	4.642

TABLE III NATURAL LOGARITHMS

t	0.00	0.01	0.02	0.03	0.04	0.05	0.06	0.07	0.08	0.09
1.0	0.0000	0.0100	0.0198	0.0296	0.0392	0.0488	0.0583	0.0677	0.0770	0.0862
1.1	0.0953	0.1044	0.1133	0.1222	0.1310	0.1398	0.1484	0.1570	0.1655	0.1740
1.2	0.1823	0.1906	0.1989	0.2070	0.2151	0.2231	0.2311	0.2390	0.2469	0.2546
1.3	0.2624	0.2700	0.2776	0.2852	0.2927	0.3001	0.3075	0.3148	0.3221	0.3293
1.4	0.3365	0.3436	0.3507	0.3577	0.3646	0.3716	0.3784	0.3853	0.3920	0.3988
1.5	0.4055	0.4121	0.4187	0.4253	0.4318	0.4383	0.4447	0.4511	0.4574	0.4637
1.6	0.4700	0.4762	0.4824	0.4886	0.4947	0.5008	0.5068	0.5128	0.5188	0.5247
1.7	0.5306	0.5365	0.5423	0.5481	0.5539	0.5596	0.5653	0.5710	0.5766	0.5822
1.8	0.5878	0.5933	0.5988	0.6043	0.6098	0.6152	0.6206	0.6259	0.6313	0.6366
1.9	0.6419	0.6471	0.6523	0.6575	0.6627	0.6678	0.6729	0.6780	0.6831	0.6881
2.0	0.6931	0.6981	0.7031	0.7080	0.7130	0.7178	0.7227	0.7275	0.7324	0.7372
2.1	0.7419	0.7467	0.7514	0.7561	0.7608	0.7655	0.7701	0.7747	0.7793	0.7839
2.2	0.7885	0.7930	0.7975	0.8020	0.8065	0.8109	0.8154	0.8198	0.8242	0.8286
2.3	0.8329	0.8372	0.8416	0.8459	0.8502	0.8544	0.8587	0.8629	0.8671	0.8713
2.4	0.8755	0.8796	0.8838	0.8879	0.8920	0.8961	0.9002	0.9042	0.9083	0.9123
2.5	0.9163	0.9203	0.9243	0.9282	0.9322	0.9361	0.9400	0.9439	0.9478	0.9517
2.6	0.9555	0.9594	0.9632	0.9670	0.9708	0.9746	0.9783	0.9821	0.9858	0.9895
2.7	0.9933	0.9969	1.0006	1.0043	1.0080	1.0116	1.0152	0.0188	1.0225	1.0260
2.8	1.0296	1.0332	1.0367	1.0403	1.0438	1.0473	1.0508	1.0543	1.0578	1.0613
2.9	1.0647	1.0682	1.0716	1.0750	1.0784	1.0818	1.0852	1.0886	1.0919	1.0953
3.0	1.0986	1.1019	1.1053	1.1086	1.1119	1.1151	1.1184	1.1217	1.1249	1.1282
3.1	1.1314	1.1346	1.1378	1.1410	1.1442	1.1474	1.1506	1.1537	1.1569	1.1600
3.2	1.1632	1.1663	1.1694	1.1725	1.1756	1.1787	1.1817	1.1848	1.1878	1.1909
3.3	1.1939	1.1970	1.2000	1.2030	1.2060	1.2090	1.2119	1.2149	1.2179	1.2208
3.4	1.2238	1.2267	1.2296	1.2326	1.2355	1.2384	1.2413	1.2442	1.2470	1.2499
3.5	1.2528	1.2556	1.2585	1.2613	1.2641	1.2669	1.2698	1.2726	1.2754	1.2782
3.6	1.2809	1.2837	1.2865	1.2892	1.2920	1.2947	1.2975	1.3002	1.3029	1.3056
3.7	1.3083	1.3110	1.3137	1.3164	1.3191	1.3218	1.3244	1.3271	1.3297	1.3324
3.8	1.3350	1.3376	1.3403	1.3429	1.3455	1.3481	1.3507	1.3533	1.3558	1.3584
3.9	1.3610	1.3635	1.3661	1.3686	1.3712	1.3737	1.3762	1.3788	1.3813	1.3838
4.0	1.3863	1.3888	1.3913	1.3938	1.3962	1.3987	1.4012	1.4036	1.4061	1.4085
4.1	1.4110	1.4134	1.4159	1.4183	1.4207	1.4231	1.4255	1.4279	1.4303	1.4327
4.2	1.4351	1.4375	1.4398	1.4422	1.4446	1.4469	1.4493	1.4516	1.4540	1.4563
4.3	1.4586	1.4609	1.4633	1.4656	1.4679	1.4702	1.4725	1.4748	1.4770	1.4793
4.4	1.4816	1.4839	1.4861	1.4884	1.4907	1.4929	1.4952	1.4974	1.4996	1.5019
4.5	1.5041	1.5063	1.5085	1.5107	1.5129	1.5151	1.5173	1.5195	1.5217	1.5239
4.6	1.5261	1.5282	1.5304	1.5326	1.5347	1.5369	1.5390	1.5412	1.5433	1.5454
4.7	1.5476	1.5497	1.5518	1.5539	1.5560	1.5581	1.5602	1.5623	1.5644	1.5665
4.8	1.5686	1.5707	1.5728	1.5748	1.5769	1.5790	1.5810	1.5831	1.5851	1.5872
4.9	1.5892	1.5913	1.5933	1.5953	1.5974	1.5994	1.6014	1.6034	1.6054	1.6074
5.0	1.6094	1.6114	1.6134	1.6154	1.6174	1.6194	1.6214	1.6233	1.6253	1.6273
5.1	1.6292	1.6312	1.6332	1.6351	1.6371	1.6390	1.6409	1.6429	1.6448	1.6467
5.2	1.6487	1.6506	1.6525	1.6544	1.6563	1.6582	1.6601	1.6620	1.6639	1.6658
5.3	1.6677	1.6696	1.6715	1.6734	1.6752	1.6771	1.6790	1.6808	1.6827	1.6845
5.4	1.6864	1.6882	1.6901	1.6919	1.6938	1.6956	1.6974	1.6993	1.7011	1.7029
5.5	1.7047	1.7066	1.7084	1.7102	1.7120	1.7138	1.7156	1.7174	1.7192	1.7210
5.6	1.7228	1.7246	1.7263	1.7281	1.7299	1.7317	1.7334	1.7352	1.7370	1.7387
5.7	1.7405	1.7422	1.7440	1.7457	1.7475	1.7492	1.7509	1.7527	1.7544	1.7561
5.8	1.7579	1.7596	1.7613	1.7630	1.7647	1.7664	1.7682	1.7699	1.7716	1.7733
5.9	1.7750	1.7766	1.7783	1.7800	1.7817	1.7834	1.7851	1.7867	1.7884	1.7901

TABLE III   NATURAL LOGARITHMS—Cont.

t	0.00	0.01	0.02	0.03	0.04	0.05	0.06	0.07	0.08	0.09
6.0	1.7918	1.7934	1.7951	1.7967	1.7984	1.8001	1.8017	1.8034	1.8050	1.8066
6.1	1.8083	1.8099	1.8116	1.8132	1.8148	1.8165	1.8181	1.8197	1.8213	1.8229
6.2	1.8245	1.8262	1.8278	1.8294	1.8310	1.8326	1.8342	1.8358	1.8374	1.8390
6.3	1.8406	1.8421	1.8437	1.8453	1.8469	1.8485	1.8500	1.8516	1.8532	1.8547
6.4	1.8563	1.8579	1.8594	1.8610	1.8625	1.8641	1.8656	1.8672	1.8687	1.8703
6.5	1.8718	1.8733	1.8749	1.8764	1.8779	1.8795	1.8810	1.8825	1.8840	1.8856
6.6	1.8871	1.8886	1.8901	1.8916	1.8931	1.8946	1.8961	1.8976	1.8991	1.9006
6.7	1.9021	1.9036	1.9051	1.9066	1.9081	1.9095	1.9110	1.9125	1.9140	1.9155
6.8	1.9169	1.9184	1.9199	1.9213	1.9228	1.9242	1.9257	1.9272	1.9286	1.9301
6.9	1.9315	1.9330	1.9344	1.9359	1.9373	1.9387	1.9402	1.9416	1.9430	1.9445
7.0	1.9459	1.9473	1.9488	1.9502	1.9516	1.9530	1.9544	1.9559	1.9573	1.9587
7.1	1.9601	1.9615	1.9629	1.9643	1.9657	1.9671	1.9685	1.9699	1.9713	1.9727
7.2	1.9741	1.9755	1.9769	1.9782	1.9796	1.9810	1.9824	1.9838	1.9851	1.9865
7.3	1.9879	1.9892	1.9906	1.9920	1.9933	1.9947	1.9961	1.9974	1.9988	2.0001
7.4	2.0015	2.0028	2.0042	2.0055	2.0069	2.0082	2.0096	2.0109	2.0122	2.0136
7.5	2.0149	2.0162	2.0176	2.0189	2.0202	2.0215	2.0229	2.0242	2.0255	2.0268
7.6	2.0282	2.0295	2.0308	2.0321	2.0334	2.0347	2.0360	2.0373	2.0386	2.0399
7.7	2.0412	2.0425	2.0438	2.0451	2.0464	2.0477	2.0490	2.0503	2.0516	2.0528
7.8	2.0541	2.0554	2.0567	2.0580	2.0592	2.0605	2.0618	2.0631	2.0643	2.0665
7.9	2.0669	2.0681	2.0694	2.0707	2.0719	2.0732	2.0744	2.0757	2.0769	2.0782
8.0	2.0794	2.0807	2.0819	2.0832	2.0844	2.0857	2.0869	2.0882	2.0894	2.0906
8.1	2.0919	2.0931	2.0943	2.0956	2.0968	2.0980	2.0992	2.1005	2.1017	2.1029
8.2	2.1041	2.1054	2.1066	2.1078	2.1090	2.1102	2.1114	2.1126	2.1138	2.1150
8.3	2.1163	2.1175	2.1187	2.1199	2.1211	2.1223	2.1235	2.1247	2.1258	2.1270
8.4	2.1282	2.1294	2.1306	2.1318	2.1330	2.1342	2.1353	2.1365	2.1377	2.1389
8.5	2.1401	2.1412	2.1424	2.1436	2.1448	2.1459	2.1471	2.1483	2.1494	2.1506
8.6	2.1518	2.1529	2.1541	2.1552	2.1564	2.1576	2.1587	2.1599	2.1610	2.1622
8.7	2.1633	2.1645	2.1656	2.1668	2.1679	2.1691	2.1702	2.1713	2.1725	2.1736
8.8	2.1748	2.1759	2.1770	2.1782	2.1793	2.1804	2.1815	2.1827	2.1838	2.1849
8.9	2.1861	2.1872	2.1883	2.1894	2.1905	2.1917	2.1928	2.1939	2.1950	2.1961
9.0	2.1972	2.1983	2.1994	2.2006	2.2017	2.2028	2.2039	2.2050	2.2061	2.2072
9.1	2.2083	2.2094	2.2105	2.2116	2.2127	2.2138	2.2148	2.2159	2.2170	2.2181
9.2	2.2192	2.2203	2.2214	2.2225	2.2235	2.2246	2.2257	2.2268	2.2279	2.2289
9.3	2.2300	2.2311	2.2322	2.2332	2.2343	2.2354	2.2364	2.2375	2.2386	2.2396
9.4	2.2407	2.2418	2.2428	2.2439	2.2450	2.2460	2.2471	2.2481	2.2492	2.2502
9.5	2.2513	2.2523	2.2534	2.2544	2.2555	2.2565	2.2576	2.2586	2.2597	2.2607
9.6	2.2618	2.2628	2.2638	2.2649	2.2659	2.2670	2.2680	2.2690	2.2701	2.2711
9.7	2.2721	2.2732	2.2742	2.2752	2.2762	2.2773	2.2783	2.2793	2.2803	2.2814
9.8	2.2824	2.2834	2.2844	2.2854	2.2865	2.2875	2.2885	2.2895	2.2905	2.2915
9.9	2.2925	2.2935	2.2946	2.2956	2.2966	2.2976	2.2986	2.2996	2.3006	2.3016

Answers to Selected Problems

Chapter 1

PROBLEM SET 1, Page 12

1. $5 \in A$ $5 \notin B$ $5 \in C$
   $10 \in A$ $10 \in B$ $10 \notin C$
   $12 \notin A$ $12 \notin B$ $12 \in C$
2. a) $\{2, 3, 5, 7, 8\}$ c) $\{5, 7\}$ e) $\{2, 3, 4, 5, 6, 7, 8\}$
3. a) Finite   c) Infinite   e) Infinite   g) Infinite   i) Finite
4. a) $\{x | x \in \{m, a, x, i, u\}\}$, Finite   c) $\{x | x = 5n, n \in N\}$, Infinite
5. a) True   c) True   e) True   g) False
6. a) Disjoint   c) Disjoint   e) Equal
7. a) $\{2\}$, proper subset is $\emptyset$
   c) $\{a, b, c\}$; proper subsets, $\{a\}, \{b\}, \{c\}, \{a, b\}, \{a, c\}, \{b, c\}, \emptyset$
   e) $\{5, 6, 7, 8\}$; proper subsets, $\{5\}, \{6\}, \{7\}, \{8\}, \{5, 6\}, \{5, 7\}, \{5, 8\}, \{6, 7\}, \{6, 8\},$
   $\{7, 8\}, \{5, 6, 7\}, \{5, 6, 8\}, \{5, 7, 8\}, \{6, 7, 8\}, \emptyset$
8. a) 1   c) 4   e) 16
   Yes, if a set has n elements, then there are 2^n subsets.
9. a) $\{2\}$ c) $\{3, 5\}$ e) $\emptyset$ g) $\emptyset$ i) $\{1, 2, 3, 4, 5, 6, 8\}$ k) $\{2\}$
10. c) Yes   e) Yes

PROBLEM SET 2, Page 20

1. a) $N \subset I$ c) $N \cap L = \emptyset$ e) $I \subset Q$ g) $I \subset R$ i) $Q \subset R$
2. a) Q

c) I

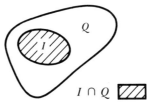

e) R

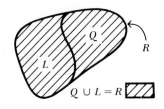

$Q \cup L = R$

3. a) $\frac{3}{8} = 0.375$ c) $\frac{4}{13} = 0.\overline{307692}$ e) $\frac{17}{9} = 1.8\overline{88}$

4. a) Irrational   c) Irrational   e) Irrational

5. a) $0.46\overline{4646} = \frac{46}{99}$ c) $7.3\overline{6262} = \frac{7289}{990}$ e) $0.\overline{777} = \frac{7}{9}$

7.

Polynomial	Degree	Coefficients
a) Trinomial	2	1, −5, 6
c) Binomial	5	4, −21
e) Monomial	2	7
g) Binomial	7	2, −3
i) Binomial	6	4, −13
k) Monomial	0	−3

8. a) Polynomial, 5   c) Polynomial, 3
   e) Not polynomial   g) Polynomial, 2
   i) Polynomial, 2   k) Polynomial, 4
   m) Polynomial, 4

PROBLEM SET 3, Page 31

1. a) Closure property of addition
   c) Commutative property of multiplication
   e) Additive inverse
   g) Multiplicative identity
   i) $a \cdot 0 = 0 \cdot a = 0$, for all a

3. a) $x^2 - 2x + 14$ c) $5x^2 - 4x - 9$
   e) $4x^4 - 2x^3 - 4x^2 + x$ g) $x^3 - 3x + 7$
   i) $-2x^4 + 12x^2 - 7$ k) $4y^4 - 2by^3 - 11b^2y^2$
   m) $-9x^2 + 12x - 17$ o) $-2x^2 + x - 15$

4. a) $2x^3y - 2x^2y^2$
   c) $-x^2y^2 - xy^3$
   e) $35x^2 + 31x + 6$
   g) $12x^2 - 24x + 12$
   i) $6x^2 - 5x - 4$
   k) $2x^2 + xy + 10x + y - y^2 + 12$
   m) $-9x^3 + 18x^2 + x - 2$
   o) $3x^3 - 3x^2y + 3xy + x^4y - x^3y^2 + xy^3 - y^3$
   q) $x^4 - 98x^2 + 2401$
   s) $x^4 - 2x^3y + 2xy^3 - y^4$
   u) $6x^3 + 7x^2 - 8x - 5$
   w) $189x^3 + 186x^2y + 45xy^2 + 63x^2 + 62xy + 15y^2$
   y) $x^3 - 3x + 2$

5. a) $9x^2 + 30xy + 25y^2$
   c) $4x^2 + 16x + 16$
   e) $16 - x^2$
   g) $x^{3n} - 27$
   i) $x^3 - 8y^3$
   k) $27 + x^3$
   m) $x^3 + 15x^2y + 75xy^2 + 125y^3$
   o) $x^3 + y^3 + z^3 + 3x^2y + 3x^2z + 3y^2z + 6xyz + 3xy^2 + 3xz^2 + 3yz^2$
   q) $27x^3 - 135x^2y + 225xy^2 - 125y^3$

PROBLEM SET 4, Page 41

1. a) $x(4x + 7y)$ c) $17x^2(x - 2)$
   e) $(2a + b)(3x + 5y)$ g) $(a^2 + 5)(7x + 13a)$
   i) $xy^2(4z + xz^2 - x^2y)$

2. a) $(x + 2)(x + 3)$ c) $(8 + x)(5 - x)$
   e) $(x - 4)(x + 1)$ g) Not possible
   i) $(x - 8)(x + 5)$ k) $(8 + x)(2 - x)$
   m) $(3x - 2y)(2x - 3y)$ o) $2(2x + 1)(3x - 5)$

3. a) $(x^2 - 3y)(x^2 + 3y)$ c) $(x + y)(1 - x + y)$
   e) $(9x - 2y)(81x^2 + 18xy + 4y^2)$ g) $(1 - 9y)(1 + 9y)$
   i) $(x + y)(x - y - 1)$ k) $(2x - 1)^3$
   m) $(x + 1)^3$ o) $(x - y)(x + y - 9)$

4. a) $(x + 7)(x - 12)$ c) $(y - 4x)(y + 4x)$
   e) $2z(3x - y)(3x + y)$ g) $(5 - x)^2$
   i) $2(y - 3)(y - 4)$ k) $4(12y - 1)(13y + 1)$
   m) $(1 - 5x - 11b)(1 - 5x + 11b)$ o) $(a + 1)(a - 15)$
   q) $(y - 1)(y + 1)(y^2 + 1)(y^4 + 1)(y^8 + 1)$ s) $(x + y - 9z)(x + y + 9z)$

5. c) Equivalent   e) Not equivalent

6. a) $\dfrac{x}{y^2}$ c) $\dfrac{y^2c}{3}$ e) $\dfrac{1}{3x^2y^2z}$

7. a) 9   c) $x^2 + 2xy + y^2$ e) $x^2y - xy^2$ g) $x^2 - 3x$

8. a) $\dfrac{x + 1}{x - 1}$ c) $\dfrac{x - 1}{x + 2}$ e) $\dfrac{3x - y}{2(3x + y)}$

   g) $\dfrac{x^2 - y^2}{x^2 + y^2}$ i) $\dfrac{x + 4}{x + 7}$ k) $\dfrac{x - 6}{3x + 4}$

   m) $x - y$ o) $\dfrac{3}{4x}$ q) $\dfrac{5 - 2x}{x(1 + x)}$

   s) $\dfrac{x - 2}{3x^3 - 1}$ u) $\dfrac{-(x + y)}{2x + y}$

PROBLEM SET 5, Page 48

1. a) $\dfrac{105}{180}, \dfrac{18}{180}, \dfrac{8}{180}$
   c) $\dfrac{5x^2(x - 2)}{25x^3y^3}, \dfrac{y^2(4y + 1)}{25x^3y^3}$

e) $\dfrac{5(x^2-4)}{12x^2(x^2-4)}, \dfrac{14x^2}{12x^2(x^2-4)}, \dfrac{x^2(x+2)}{12x^2(x^2-4)}$

2. a) $\dfrac{8x}{y}$ c) $\tfrac{4}{5}$ e) 7   g) $-2x+9$ i) $\dfrac{2y^3}{x+y}$

3. a) $\dfrac{8+x}{2x^2}$ c) $\dfrac{203}{20x}$ e) $-\tfrac{4}{3}$

   g) $\dfrac{4x-4}{x^2-4}$ i) $\dfrac{y}{x^2-y^2}$ k) $\dfrac{2x-x^2-x^4}{x^2-1}$

   m) $\dfrac{2x^2-3x+4}{6x^2+5x-6}$ o) $\dfrac{x^2-xy-y^2}{x^2-y^2}$

4. a) $\dfrac{2x^3y}{3z^3}$ c) $\dfrac{y(x+4)}{x^2(x-y)}$ e) $\dfrac{1}{x-y}$

   g) $(x-3y)^2$ i) $\dfrac{x+y-z}{x-y}$ k) $\dfrac{1}{2x^2-1}$

5. a) $\tfrac{1}{2}$ c) $\tfrac{113}{32}$ e) $\dfrac{1}{x-1}$

   g) $\dfrac{1}{x+2}$ i) $\dfrac{2}{x^2-x}$ k) $\dfrac{x^2+3x+3}{x^2+3x-1}$

   m) $\dfrac{x^4-x^3-x^2+2x+1}{x-1}$

PROBLEM SET 6, Page 56

1. a) x^{15} c) x^{3n} e) x^{4n}

   g) $\dfrac{x^6}{y^3}$ i) $\dfrac{x^{2n}}{y^{2n}}$ k) $\dfrac{16}{a^8 b^{12}}$

   m) $\dfrac{-z^6}{a^6 x^3 y^3}$ o) x^{6n+6} q) x^{31n}

   s) 3^{4m^2-7m+2} u) $81x^4 y^{12}$ w) $125 a^6 b^9 z^6$

3. a) $\tfrac{1}{45}$ c) $\tfrac{49}{9}$ e) $\dfrac{25}{x^2}$

   g) $\dfrac{3}{x+y}$ i) $\dfrac{y-2}{3x^2}$ k) $\dfrac{1}{x^2 y}$

   m) $\dfrac{x^2+y^2}{x^2 y^2}$ o) $\dfrac{x^6}{4y^8}$ q) $\dfrac{y^2}{x^2}$

4. a) $\tfrac{41}{5}$ c) 1600   e) $\tfrac{4}{5}$

   g) $\dfrac{1+x^4}{x^2}$ i) $\dfrac{x^2 y^2}{y-x}$ k) $\dfrac{x^2+y^2}{xy}$

5. a) x^{-1} c) x^{12-4n} e) $x^{-1} y^{2n-2}$

   g) x^{2n} i) $x^{2n} y^{2n}$

6. a) $-\tfrac{2}{3}$

PROBLEM SET 7, Page 62

1. a) 8   c) $5|x|$ e) $\frac{125}{64}$ g) -5 i) 0.00000081 k) 3   m) $\frac{9}{4}$ o) $\frac{4}{5}$
2. a) $\frac{3}{2}$ c) $\frac{1}{10}$ e) $\frac{125}{4}$ g) $\frac{2}{3}$
   i) x k) $x^{-1/3} - x^{2/3}$ m) $x^{-1/3}$ o) $(x+y)^{1/2}$
   q) $x + \dfrac{1}{x}$ s) $\dfrac{x-y-1}{x-y}$ u) $\dfrac{x^2 y^6}{z^2}$ w) $x^{-1/2} y^{-10/3}$
   y) $\dfrac{x^3 y^{-1/4}}{2}$

3. a) $\frac{5}{3}$ c) $\dfrac{y^5}{x^5}$ e) $-\dfrac{12x^3 y^2}{5}$ g) $\dfrac{x^3 y^4}{3}$

4. a) $\frac{6}{7}$ c) $\dfrac{x}{x^3 - 1}$ e) $\dfrac{x^{2/3}}{1 - x^{2/3} y^{5/3}}$
   g) $\dfrac{1}{x} - \dfrac{1}{y^{4/3}}$ i) $\dfrac{x^2 y^{5/2} - 2 y^{5/2}}{x^{3/2} y^{5/2} - 2x^2}$ k) 8

PROBLEM SET 8, Page 69

1. a) $\sqrt{3}$ c) $\sqrt[4]{16^3}$ e) $\sqrt{(x+y)^{-3}}$ g) $\sqrt[3]{2x}$ i) $-\sqrt[3]{\frac{21}{64}}$
2. a) $8^{1/3}$ c) $7^{1/3}$ e) $-(xy)^{1/3}$ g) $x^{1/2} - y^{1/2}$ i) $(100 x^4 y^2)^{1/3}$
3. a) 3   c) $10\sqrt{3}$ e) $3\sqrt[4]{2}$ g) $2\sqrt[6]{3}$
   i) $3\sqrt{x}$ k) $\dfrac{5x^2}{y}$ m) $\dfrac{9x^3 y^3}{7zw}$ o) $5x^3 y \sqrt[3]{5xy^2}$
   q) 12   s) 2   u) $\sqrt[12]{x^{10} y^{11}}$ w) $y^2 \sqrt[15]{x^{13} y}$
4. a) 3   c) $x - y^2$ e) $x\sqrt{x} + y\sqrt{y}$
5. a) $4\sqrt{3}$ c) $11\sqrt{2}$ e) $7\sqrt[3]{2}$ g) $\dfrac{(-3 + \sqrt{6})\sqrt{x}}{x}$ i) $50x - 6\sqrt[3]{x^2}$
6. a) $\dfrac{2\sqrt{3}}{3}$ c) $\dfrac{8\sqrt[3]{7}}{7}$ e) $\dfrac{\sqrt{5} - 1}{4}$ g) $-5 - 2\sqrt{6}$ i) $\dfrac{x(x - \sqrt{y})}{x^2 - y}$
   k) $19 - 5\sqrt{15}$ m) $-3\sqrt{2} + 2\sqrt{6} - \sqrt{3} + 2$

REVIEW PROBLEM SET, Page 71

1. a) N c) F or Q e) F g) F i) $\emptyset$
2. a) $\{c\}$ c) $\{c, d\}$ e) $\{a, c\}$ g) $\emptyset$ i) $\{a, b, c, d, g\}$
3. a) T   c) F   e) F
4. a) No   c) No   e) $\frac{17}{25} = 0.68$, terminating   g) No
5. a) $\frac{1}{6561}$ c) x^{21} e) $a^{n+1} b^{2n}$ g) 2^{6n+5}
6. a) Degree: 2; coefficients: 4, -39, 100
   c) Degree: 3; coefficients: $\frac{81}{35}$, -2, 25
   e) Degree: 2; coefficients: 4, 36, 14
   g) Degree: 4; coefficients: 1, -3

ANSWERS TO SELECTED PROBLEMS

7. a) $x^3 - 3x^2y - xy^2 + 6x^2$ c) $-6x^7 - x^3 + 63$ e) $3x^2 + 13$

8. a) $-3x^2 + 8x - 5$ c) $x^3 - 2x^2 + 9x - 6$

9. a) $8x^2 + 34x + 21$
c) $-6x^3 + 14x$
e) $39x^2 + 31x - 28$
g) $9x^2 + 36x + 35$
i) $-78x^4 + 31x^2 + 4$
k) $x^2 - 4y^2 - 12yz - 9z^2$
m) $55x^2 - 77xy + 58x - 35y + 15$
o) $4x^2 - 4xy + y^2$
q) $4x^2 + y^2 + z^2 - 4xy + 4xz - 2yz$
s) $4x^2 + 9y^2 + 25z^2 - 12xy - 20xz + 30yz$
u) $8x^3 - 36x^2y + 54xy^2 - 27y^3$
w) $x^4 + 8x^3y + 24x^2y^2 + 32xy^3 + 16y^4$

10. a) $(abc + 4)(abc + 1)$
c) $ab(ab + 4)$
e) $2y(3x^2 + y^2)$
g) $xy(x - 1)(x - 4)$
i) $(x - y)(x + y - a)$
k) $(2x - 5y)(2x + 5y)(4x^2 + 25y^2)$
m) $(x - 1)(x + 1)(y - 1)(y + 1)$
o) $\left(\dfrac{x}{2} + \dfrac{y}{3}\right)\left(\dfrac{x^2}{4} - \dfrac{xy}{6} + \dfrac{y^2}{9}\right)$
q) $2x(x - z)(x + z)(x^2 + 4)$
s) $(2x^2 + 1 - 2x)(2x^2 + 1 + 2x)$

11. a) $\dfrac{2x}{3z}$ c) $\dfrac{2x(a+b)^2}{3y}$ e) $\dfrac{x-y}{x+y}$ g) $\dfrac{3x-2y}{3y-2x}$

12. a) $\dfrac{5}{12x}$ c) $\dfrac{5z - 6x + 7y}{xyz}$ e) $\dfrac{2x+1}{x^2+x-6}$ g) $\dfrac{15x^2 - 22xy - 12y^2}{21x^2 + 34xy + 8y^2}$

i) $\dfrac{3x^2 + 20x - 72}{(x-2)(x+3)(x-4)}$

13. a) $\dfrac{a^3x^4}{z^3}$ c) $\dfrac{y^2 - x^2}{xy}$ e) $\dfrac{45}{32y^{13}}$

g) xy i) $(x+y)(a+b)$ k) $\dfrac{x^2 - xy + y^2}{x^2 + y}$

m) $\dfrac{x-2}{x-7}$ o) $\dfrac{(a+b)(a^2 + b^2 - a - b)}{a-b}$

14. a) $\dfrac{10}{3}$ c) $\dfrac{1}{y^2 + 2y - 1}$ e) $\dfrac{6}{x^2}$ g) $\dfrac{x-1}{x-2}$

15. a) $\dfrac{1}{81}$ c) -4 e) $\dfrac{1}{25}$ g) $-\dfrac{9}{2}$ i) $\dfrac{1}{27}$ k) $\dfrac{5x}{5-x}$

16. a) 2 c) x^4 e) $-2x^2$ g) $4x^2$ i) $8x^{12}y^9$ k) $\dfrac{y^6}{8x^3}$ m) $\dfrac{-8z^9}{x^3y^6}$

17. a) $3\sqrt{2}$ c) $xy\sqrt[3]{15xy^2}$ e) $2xy^2\sqrt[4]{2xy^2}$ g) $3xyz^2\sqrt[3]{2xy^2z}$ i) $\dfrac{5\sqrt{5xy}}{2xy}$

k) $\dfrac{3x\sqrt[4]{5x^3}}{2y^2}$ m) $\sqrt{x}$

18. a) $8\sqrt{2}$ c) $4\sqrt{x}$ e) $5\sqrt{7}$ g) $18\sqrt{2}$ i) $3\sqrt[3]{10x^2}$ k) 2 m) $1 - 4\sqrt{2}$

19. a) $\dfrac{4\sqrt{3}}{3}$ c) $\dfrac{2\sqrt{3x}}{5x}$ e) $\dfrac{2\sqrt{3xy}}{3y}$

g) $5 + \sqrt{15}$ i) $\dfrac{x^2 + x\sqrt{y} - 2y}{x^2 - 4y}$ k) $\dfrac{6\sqrt{2} - 4}{7}$

m) $\dfrac{28 - 3\sqrt{26}}{11}$ o) $\dfrac{7(1 - \sqrt[3]{5} + \sqrt[3]{25})}{6}$

Chapter 2

PROBLEM SET 1, Page 85

1. a) 25   c) 1   e) 0   g) $\frac{2}{5}$ i) $\frac{9}{5}$ k) $\frac{2}{3}$
2. a) 2   c) 15   e) −39   g) $-\frac{1}{3}$ i) −3   k) −3   m) $\frac{15}{22}$ o) $-\frac{1}{15}$ q) 4
3. a) $\dfrac{f-b}{m}$ c) $\dfrac{s-gt^2}{t}$ e) $\dfrac{9bc}{5b+12c}$

   g) $\dfrac{7b^2(bc+3d)}{3(ab^2-7)}$ i) $-\frac{14}{13}$ k) $\dfrac{4-n^2}{n}$

5. 5 and 13
7. $\frac{15}{8}$ hr
9. 105
11. 5 quarters, 7 dimes, 4 pennies
13. $3,000 at 4 percent and $2,000 at 5 percent

PROBLEM SET 2, Page 102

3. a) $8 - 4 = 4 > 0$ c) $-1 - (-2) = 1 > 0$ e) $4 - (-3) = 7 > 0$
5. a) $2 \cdot 2 < 2 \cdot 3$ c) $(-2)(-1) > -2(0)$ e) $-3x > -3y$
6. a) Addition property of inequalities   c) Transitive   e) Trichotomy
7. a) T   c) T   e) T
9. No, $16 < 25$, but $-4 \not< -5$
11. a) $(-\infty, 1]$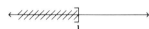

   c) $(-\infty, \infty)$

   e) $[0, 2]$

17. a) $(1, 4) \cup (7, 10)$ c) $(-\infty, -1) \cup (5, 8)$ e) $(-\infty, -1) \cup (1, \infty)$
19. a) $A^c = \{x \mid x \geq 3\}$ c) $C^c = \{x \mid x^2 = 9\} = \{-3, 3\}$
20. a) $(-\infty, 1)$

   c) $(\frac{15}{2}, \infty)$

   e) $(\frac{5}{3}, \infty)$

g) $(-\infty, -\frac{1}{3})$

i) $(-\infty, -1)$

k) $[\frac{9}{4}, \infty)$

m) $(-\infty, \frac{41}{7}]$

o) $[4, \infty)$

q) $(-\infty, \frac{22}{3})$

s) $[-\frac{23}{11}, \infty)$

PROBLEM SET 3, Page 115

1. a) 12   c) 12   e) 35   g) 49   i) 43

3. a) $\{-5, 5\}$ c) $\{-1, 5\}$ e) $\{-11, 1\}$ g) $\{1, 3\}$ i) $\{1, 3\}$

4. a) $\leftrightarrow C$ b) $\leftrightarrow D$ c) $\leftrightarrow E$ d) $\leftrightarrow A$ e) $\leftrightarrow B$

5. a) $(-3, 3)$

c) $[-2, 4]$

e) $(-\infty, 0) \cup (\frac{2}{3}, \infty)$

g) $(-\infty, -3] \cup [2, \infty)$

i) $(-\infty, -3)$

k) $(\frac{1}{6}, \frac{1}{4})$

6. a) 1   c) $x \geq 0$ and $y \geq 0$; $x \leq 0$ and $y \leq 0$
7. a) $x \geq 0$ c) $x \geq 0$ e) $x \neq 0$
9. a) R c) $(\frac{1}{2}, \infty)$ e) R
10. a) $|x| < 3$ c) $|x + 3.05| < 0.05$

PROBLEM SET 4, Page 127

1. a) No   c) Yes   e) Yes
2. a) F   c) T   e) T
4. $A \times B$

B \ A	1	2	3	4
1	(1, 1)	(2, 1)	(3, 1)	(4, 1)
2	(1, 2)	(2, 2)	(3, 2)	(4, 2)

$B \times A$

A \ B	1	2
1	(1, 1)	(2, 1)
2	(1, 2)	(2, 2)
3	(1, 3)	(2, 3)
4	(1, 4)	(2, 4)

5. a) $S \times T = \{(2, a), (2, b), (4, a), (4, b), (6, a), (6, b)\}$
c) $T \times T = \{(a, a), (a, b), (b, a), (b, b)\}$
7. a) $A = \{(-1, -2), (0, 1), (1, 4), (2, 7)\}$
c) $C = \{(1, 2), (2, 4), (3, 6), (4, 8), (5, 10)\}$
8. a) $A \times B = \{(a, x), (a, y), (d, x), (d, y), (c, x), (c, y)\}$
9. 8
11. a) Q_I c) Q_I e) None   g) Q_{II} i) Q_{IV}
16. a)   c)

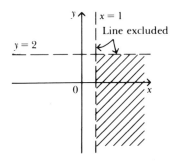

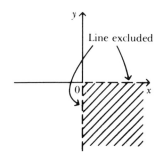

e)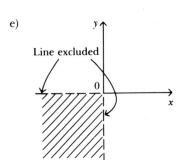

g)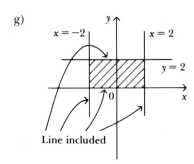

i)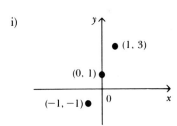

17. a) $\dfrac{\sqrt{41}}{2}$ c) $3\sqrt{17}$ e) $\sqrt{17}$

20. d) i) $(-1, 7)$ ii) $(-\tfrac{7}{2}, \tfrac{7}{2})$ iii) $(\tfrac{11}{2}, \tfrac{11}{2})$
21. a) $7\sqrt{2}, \sqrt{17}, \sqrt{73}$ c) $7, 2\sqrt{17}, \sqrt{89}$

REVIEW PROBLEM SET, Page 130

1. a) $-\tfrac{2}{7}$ c) 16 e) $-\tfrac{150}{103}$ g) $-\dfrac{5a+3}{4}$

   i) $\dfrac{6c-2b}{3a+3b}$ k) 14 m) -24 o) $\tfrac{7}{3}$

3. a) F   c) F   e) T   g) T
4. a) F   c) T   e) T
5. a) $(-\infty, -4)$

   c) $(-\infty, \tfrac{7}{4})$

   e) $(-\infty, 0)$

g) $(-\infty, 4)$

i) $(0, \infty)$

6. a) $\{-\frac{5}{2}, \frac{5}{2}\}$

c) $\{-\frac{16}{3}, \frac{8}{3}\}$

e) $\{-\frac{2}{7}, \frac{8}{7}\}$

g) $\{1, \frac{11}{5}\}$

7. a) $(-7, 7)$

c) $(\frac{7}{2}, \frac{13}{2})$

e) $(-\infty, -3) \cup (\frac{7}{3}, \infty)$

g) $(-\infty, 3) \cup (3, \infty)$

i) $(-\infty, -1) \cup (5, \infty)$

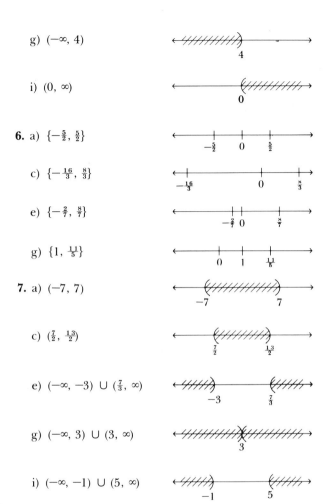

9. 20
11. $2400
12. a) Q_{II} c) Q_{II} e) Q_{II} g) None
13. a) $A \times A = \{(3, 3), (3, -3), (3, 4), (-3, 3), (-3, -3), (-3, 4), (4, 3), (4, -3), (4, 4)\}$
    c) $A \times B = \{(3, -1), (3, 2), (3, 4), (-3, -1), (-3, 2), (-3, 4), (4, -1), (4, 2), (4, 4)\}$
15. a) $\sqrt{53}, \sqrt{53}$ c) 5, 5

Chapter 3

PROBLEM SET 1, Page 141

1. a) Is a relation   c) Is not a relation
2. a) Domain $= \{-1, 2, 3\}$; Range $= \{3, -5, 6\}$

c) Domain = $\{-1\}$; Range = $\{0, 1, 3\}$
e) Domain = $\{-2, -1, 0, 1, 2\}$; Range = $\{-8, -1, 0, 1, 8\}$

3. a) Domain = $\{-1, 0, 1\}$; Range = $\{1, 0\}$

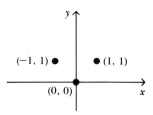

c) Domain = $\{-1, 0, 1, 2\}$; Range = $\{-1, 0, 1, 2\}$

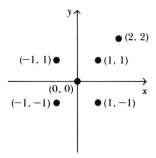

e) Domain = $\{2\}$; Range = $\{0\}$

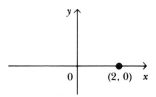

4. a) Domain = R; Range = R

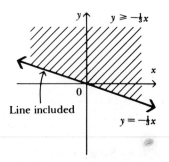

c) Domain = R; Range = R

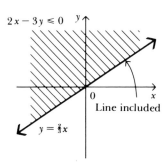

e) Domain = R; Range = R

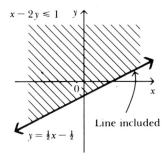

g) Domain = R; Range = $\{-1\}$

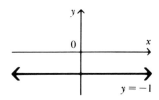

i) Domain = R; Range = $\{y | y \geq 0\}$

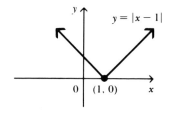

k) Domain = $\{x \mid |x| \geq 1\}$; Range = R

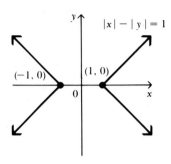

PROBLEM SET 2, Page 157

1. a) Function; Domain = $\{2, 3, 7, 9\}$; Range = $\{4, 6, 2, -3\}$

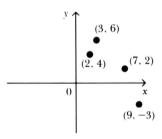

c) Function; Domain = R; Range = R

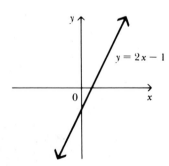

e) Not a function; Domain = $\{x | x \geq 0\}$; Range = R

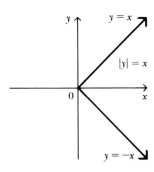

g) Function; Domain = R; Range = R

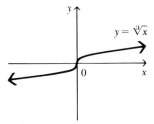

i) Function; Domain = R; Range = R

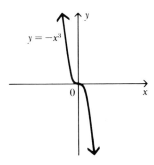

k) Not a function; Domain = R; Range = R

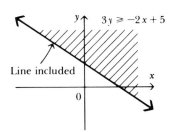

m) Not a function; Domain = $\{x|x \leq 0\}$; Range = R

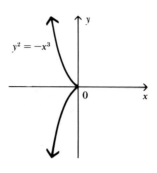

2. a) Function   c) Function
3. a) 4   c) None
4. a) Domain = R, Range = R
   $f(1) = 4$, $f(3) = 10$, $f(5) = 16$

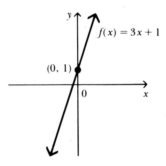

c) Domain = $\{x|x \geq 1\}$, Range = $\{y|y \geq 0\}$
   $f(1) = 0$, $f(3) = \sqrt{2}$, $f(5) = 2$

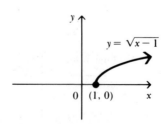

e) Domain = $\{x|x \neq -2\}$, Range = $\{y|y \neq 0\}$
$f(1) = \frac{1}{3}$, $f(3) = \frac{1}{5}$, $f(5) = \frac{1}{7}$

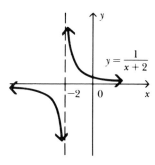

g) Domain = $\{x|x \neq 0\}$, Range = $\{y|y \neq 0\}$
$f(1) = 2$, $f(3) = \frac{2}{3}$, $f(5) = \frac{2}{5}$

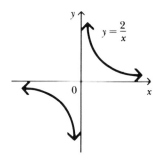

i) Domain = R, Range = $\{y|y \geq 0\}$
$f(1) = 2$, $f(3) = 6$, $f(5) = 10$

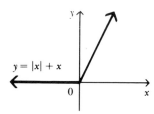

6. a) Domain = R, Range = R
$f(0) = -7$, $f(2) = -1$, $f(3) = 2$, $f(-3) = -16$
$f(a) = 3a - 7$, $f(a+h) = 3(a+h) - 7$
$\dfrac{f(a+h) - f(a)}{h} = 3$

c) Domain = R, Range = $\{y|y \geq 1\}$
$f(0) = 1, f(2) = 13, f(3) = 28, f(-3) = 28$
$f(a) = 3a^2 + 1, f(a+h) = 3(a+h)^2 + 1$
$\dfrac{f(a+h) - f(a)}{h} = 6a + 3h$

e) Domain = R, Range = R
$f(0) = -2, f(2) = 6, f(3) = 25, f(-3) = -29$
$f(a) = a^3 - 2, f(a+h) = (a+h)^3 - 2$
$\dfrac{f(a+h) - f(a)}{h} = 3a^2 + 3ah + h^2$

7. a) $(1, 0), (-2, 4), (-3, 6)$
c) $f(3) = 0, 2f(3) = 0, f(2) = 0, f(3-2) = 0, f(a+1) = |a+1| - a - 1$
8. a) $f[g(x)] = x$ c) $f[f(x)] = 4x + 3$
9. a) $f[g(x)] = |[\![x]\!]|$ c) $f[f(x)] = \big||x|\big| = |x|$
10. a)

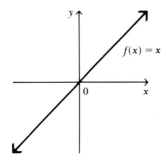
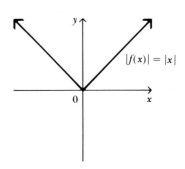

c)

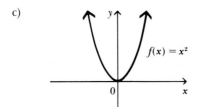
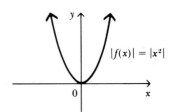

e)

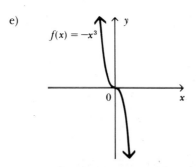
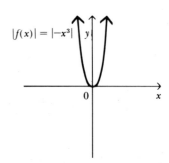

11. $S = 2\pi r^2 + 2\pi rh$, $S = 140\pi$ sq inches

13. $A = \dfrac{300x - x^2}{2}$

14. a) $f + g = \{(-2, -3), (-1, -4), (1, 0)\}$, Domain of $f + g = \{-2, -1, 1\}$
Range of $f + g = \{-3, -4, 0\}$

$f - g = \{(-2, -5), (-1, 0), (1, 4)\}$, Domain of $f - g = \{-2, -1, 1\}$
Range of $f - g = \{-5, 0, 4\}$

$f \cdot g = \{(-2, -4), (-1, 4), (1, -4)\}$, Domain of $f \cdot g = \{-2, -1, 1\}$
Range of $f \cdot g = \{-4, 4\}$

$f/g = \{(-2, -4), (-1, 1), (1, -1)\}$, Domain of $f/g = \{-2, -1, 1\}$
Range of $f/g = \{-4, 1, -1\}$

c) $f + g = \{(x, y) | y = x + 1\}$, Domain of $f + g = R$, Range of $f + g = R$

$f - g = \{(x, y) | y = 3x - 7\}$, Domain of $f - g = R$, Range of $f - g = R$

$f \cdot g = \{(x, y) | y = -2x^2 + 11x - 12\}$, Domain of $f \cdot g = R$
Range of $f \cdot g = \{y | y \leq \frac{25}{8}\}$ (See Chapter 4.)

$f/g = \left\{(x, y) | y = \dfrac{2x - 3}{4 - x}\right\}$, Domain of $f/g = \{x | x \neq 4\}$
Range of $f/g = \{y | y \neq -2\}$ (See Chapter 4.)

15. $f(x) = 2x$ a) $f(2) + f(3) = 10$ c) $\dfrac{f(x + h) - f(x)}{h} = 2$

17. a) $y = \dfrac{x^3}{2}$ c) $y = 3x^3$ e) $y = \frac{14}{1331}x^3$

19. $y = 3$

21. a) $3:2$

23. 2250 newtons

PROBLEM SET 3, Page 168

1. a) $f(x) = -2x$ is symmetric with respect to the origin

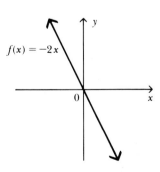

c) $f(x) = 5x^3$ is symmetric with respect to the origin

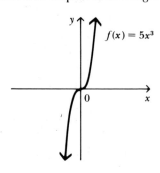

e) $f(x) = -2x^4$ is symmetric with respect to the y axis

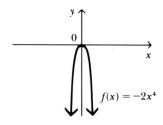

3. a) Odd   c) Neither   e) Even
5. Decreasing

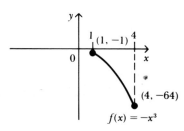

6. a) Decreasing on R

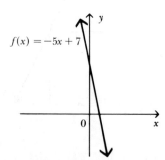

c) Increasing in $(-\infty, 0)$, decreasing on $(0, \infty)$

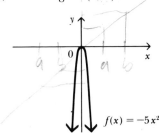

e) Increasing on R

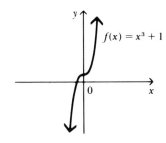

7. a) Even, symmetric with respect to the y axis
 Increasing in $(-\infty, -2) \cup (0, 2)$; decreasing in $(-2, 0) \cup (2, \infty)$
 c) Even, symmetric with respect to the y axis
 Increasing in $(-6, -4) \cup (-2, -1) \cup (0, 1) \cup (2, 4)$;
 decreasing in $(-4, -2) \cup (-1, 0) \cup (1, 2) \cup (4, 6)$
8. a) No, no

REVIEW PROBLEM SET, Page 170

1. a) Function; Domain = $\{4, 3, 5, -1\}$; Range = $\{1, -1\}$
 c) Not a function; Domain = $\{0, 1, 2\}$; Range = $\{1, -1, 3, 4\}$
2. a) Yes   c) No   e) No
3. Domain of $f = R$
 a) $f(-2) = 2$ c) $f(0) = 0$ e) $f(a) = 2a^2 + 3a$
 g) $f(a + b) = 2(a + b)^2 + 3(a + b)$
4. a) Domain = $[-2, 2]$; Range = $[0, 2]$; $f(1) = \sqrt{3}$, $f(0) = 2$, $f(2) = 0$
 c) Domain = R; Range = R; $f(1) = -2$, $f(0) = -1$, $f(2) = -3$
5. a) $V = \frac{4}{3}\pi r^3$ c) $r = \frac{1}{2}\sqrt{\frac{S}{\pi}}$
6. a) $(f + g)(x) = x^2 + 2x - 3$
 c) $(f \cdot g)(x) = 2x^3 - 6x$

e) $(2f + g)(x) = x^2 + 4x - 3$
g) $(3f + f)(x) = 8x$
i)

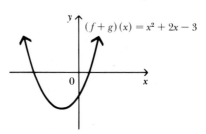

$(f + g)(x) = x^2 + 2x - 3$

7. a) Symmetric with respect to the y axis

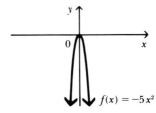

$f(x) = -5x^2$

c) Symmetric with respect to the origin

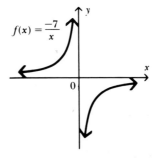

$f(x) = \dfrac{-7}{x}$

e) Symmetric with respect to the origin

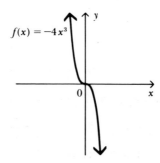

$f(x) = -4x^3$

8. a) Domain = R; Range = R increasing; neither even nor odd; no symmetry

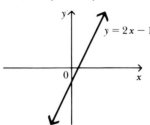

c) Domain = R; Range = $\{y|y \geq 0\}$; decreasing in $(-\infty, 0)$; neither even nor odd; no symmetry

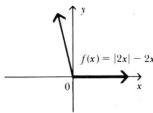

e) Domain = $\{x|x \neq 0\}$; Range = $\{-1, 1\}$; neither increasing nor decreasing; neither even nor odd; no symmetry

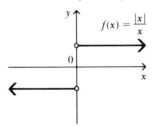

g) Domain = R; Range = $\{y|y \in I\}$; neither increasing nor decreasing; neither even nor odd; no symmetry

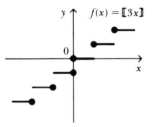

i) Domain = R; Range = R; increasing; neither even nor odd; no symmetry

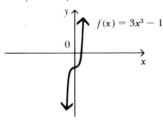

10. a) $5 - 2a$ c) $3 - 2|a|$

11. a) $f(x) = \frac{2}{3}x$ c) $f(x) = 4\sqrt{x}$ e) $f(x) = \frac{20}{x}$

13. a) 3840 rev/min   b) $\frac{5}{4}$ horsepower

Chapter 4

PROBLEM SET 1, Page 193

1. a) Polynomial, degree 2   c) Polynomial, degree 5   e) Polynomial, degree 0

3. $a = b = 0$, c is any real number

4. a) x intercept 0
   y intercept 0

c) x intercept $\frac{7}{2}$
   y intercept 7

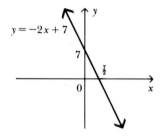

e) x intercept $-\frac{3}{4}$
   y intercept 3

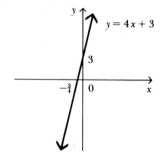

g) x intercept -3
   y intercept -2

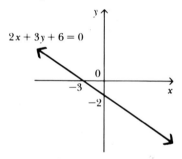

5. In Parts a), c), e), g), i), k) and m): Domain = R, Range = R, neither even nor odd
   a) Decreasing; slope = -3

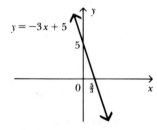

c) Increasing; slope = $\frac{2}{3}$

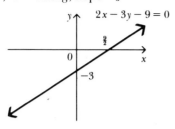

e) Decreasing; slope = -7

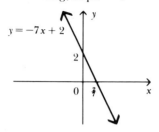

g) Increasing; slope = $\frac{4}{3}$

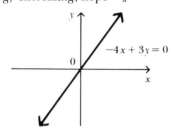

i) Increasing; slope = $\frac{1}{2}$

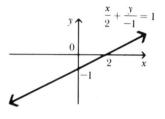

k) Decreasing; slope = -2

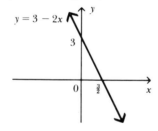

m) Decreasing; slope = $-\frac{3}{4}$

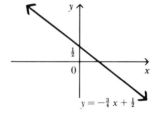

6. a) $f(x) = -3x + 5$ c) $f(x) = -3x - 3$
7. a) -5 c) $\frac{5}{2}$
8. a) -1 c) $\frac{2}{3}$ e) $-\frac{1}{2}$
9. a) $y = -x + 1$ c) $y = \frac{2}{3}x + \frac{13}{3}$ e) $y = -\frac{1}{2}x + 2$
10. a) $f(x) = -\dfrac{x}{2} + \dfrac{5}{2}$ c) $f(x) = x$ (general condition is $f(x) = \left(\dfrac{2b+5}{5}\right)x + b$)
11. b) Slope = $\frac{3}{2}$, y intercept = $\frac{5}{2}$
12. b) $y = -9x - 14$
13. b) $\dfrac{x}{5} + \dfrac{y}{-3} = 1$

14. a) $y - 2 = -3(x + 1)$ c) $y - 4 = \frac{22}{5}(x + 1)$ e) $y - 5 = 3(x - 3)$
15. a) Noncollinear   c) Collinear
16. a) i) -2 ii) $\frac{1}{2}$
    c) i) -5 ii) $\frac{1}{5}$
    e) i) -1 ii) 1
18. $y - 3 = 2(x - 2)$; $y - 1 = -\frac{1}{2}(x + 2)$
20. $v = -55t + 200$, $t = \frac{40}{11}$ sec
23. $236,250

PROBLEM SET 2, Page 207

1. a) $\{(\frac{1}{3}, \frac{2}{3})\}$ c) $\{(\frac{2}{5}, \frac{21}{5})\}$ e) $\{(0, 0)\}$ g) $\emptyset$ i) $\emptyset$ k) $\{(13, 17)\}$
   m) $\{(-2, -3)\}$ o) $\{(7, 9)\}$
2. a) $\{(10, 24)\}$ c) $\{(-\frac{16}{3}, -11)\}$ e) $\{(2, 2, 1)\}$ g) $\{(63\frac{1}{2}, \frac{1}{2}, 16)\}$
   i) $\{(5, 2, 1)\}$ k) $\{(3, -1, 4)\}$ m) $\{(-2, 5, 1)\}$ o) $\{(-1, -2, 7)\}$
   q) $\{(-2, -2, 3)\}$

PROBLEM SET 3, Page 223

1. a) $\{0, \frac{7}{3}\}$ c) $\{-\frac{25}{6}, 5\}$ e) $\{2, 4\}$ g) $\{-2, 2\}$ i) $\{-1, \frac{5}{3}\}$
2. a) $\left\{\frac{4 - \sqrt{10}}{3}, \frac{4 + \sqrt{10}}{3}\right\}$ c) $\left\{\frac{-3 - \sqrt{13}}{2}, \frac{-3 + \sqrt{13}}{2}\right\}$
   e) $\left\{\frac{7 - \sqrt{85}}{6}, \frac{7 + \sqrt{85}}{6}\right\}$ g) $\left\{\frac{13 - \sqrt{157}}{2}, \frac{13 + \sqrt{157}}{2}\right\}$ i) $\{1, 7\}$
3. a) $\left\{\frac{5 - \sqrt{17}}{4}, \frac{5 + \sqrt{17}}{4}\right\}$; $D = 17$ c) $\{2, 10\}$; $D = 64$
   e) No real solution; $D = -3$ g) $\{4, 14\}$; $D = 100$
   i) No real solution; $D = -43$
4. a) Domain = R
   Range = $\{y | y \geq -3\}$
   Extreme point = $(0, -3)$
   x intercepts = $-\frac{\sqrt{6}}{2}, +\frac{\sqrt{6}}{2}$
   y intercept = -3

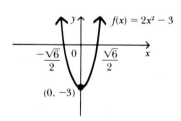

c) Domain = R
   Range = $\{y | y \leq 0\}$
   Extreme point = $(-1, 0)$
   x intercept = -1
   y intercept = -1

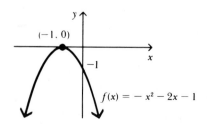

e) Domain = R
   Range = $\{y | y \geq -\frac{1}{4}\}$
   Extreme point = $(-\frac{5}{2}, -\frac{1}{4})$
   x intercepts = $-3, -2$
   y intercept = 6

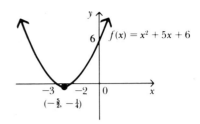

g) Domain = R
   Range = $\{y | y \geq -\frac{9}{8}\}$
   Extreme point = $(\frac{3}{4}, -\frac{9}{8})$
   x intercepts = $0, \frac{3}{2}$
   y intercept = 0

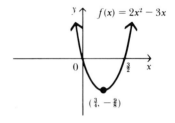

i) Domain = R
   Range = $\{y | y \geq -1\}$
   Extreme point = $(-2, -1)$
   x intercepts = $-1, -3$
   y intercept = 3

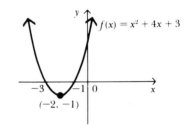

6.

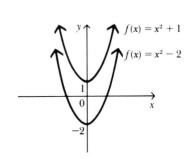

7. a) 128 feet
b) 256 feet

c)
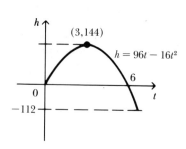

9. $48 - x$ passengers
$80 + 2x$ fare of each passenger
Total receipts: $f(x) = (48 - x)(80 + 2x)$
$f(x) = 3840 + 16x - 2x^2$

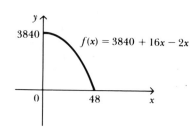

PROBLEM SET 4, Page 230

1. a) $[-\frac{1}{2}, 2]$

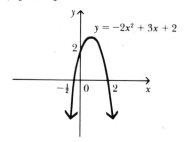

c) $[2, 2]$

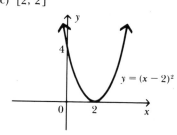

e) $[1, 4]$

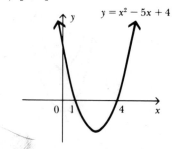

g) $\left(-\infty, \dfrac{9 - \sqrt{33}}{8}\right) \cup \left(\dfrac{9 + \sqrt{33}}{8}, \infty\right)$

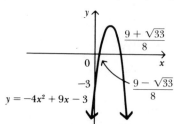

2. a) $-1 < x < \frac{1}{2}$

c) $-5 < x < \frac{1}{2}$

e) $x < -1$ or $x > 2$

g) $x < -3$ or $x > -2$
i) $\varnothing$

PROBLEM SET 5, Page 236

1. $\{-3, -2, 2, 3\}$
3. $\{-\frac{1}{2}, \frac{1}{2}, -1, 1\}$
5. $\{-\frac{1}{4}, 1\}$
7. $\{-3\}$
9. $\{3, -6\}$
11. $\{-\frac{23}{9}\}$
13. $\{2, \frac{17}{4}\}$
15. $\left\{\dfrac{-1-\sqrt{33}}{2}, \dfrac{-1+\sqrt{33}}{2}, \dfrac{-1-\sqrt{29}}{2}, \dfrac{-1+\sqrt{29}}{2}\right\}$
17. $\{-2, -1, \frac{1}{3}, \frac{2}{3}\}$
19. $\{-\frac{1}{4}\}$
21. $\{\frac{11}{2}\}$
23. $\{122\}$ 25. $\{1\}$ 27. $\{-5\}$
29. $\{0, 2\}$ 31. $\{9\}$ 33. $\{5\}$
35. $\{6\}$ 37. $\{\frac{95}{9}\}$ 39. $\{\frac{31}{32}\}$

REVIEW PROBLEM SET, Page 237

1. $f(x) = 2x + 1$
2. a) Domain $= R$
   Range $= R$
   Slope $= -3$
   x intercept $= \frac{5}{3}$
   y intercept $= 5$

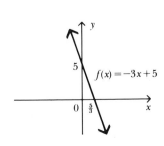

c) Domain = R
   Range = R
   Slope = 2
   x intercept = $\frac{3}{2}$
   y intercept = -3

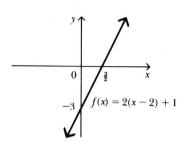

e) Domain = R
   Range = R
   Slope = $-\frac{3}{4}$
   x intercept = $\frac{4}{3}$
   y intercept = 1

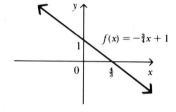

g) Domain = R
   Range = $\{-1\}$
   Slope = 0
   No x intercept
   y intercept = -1

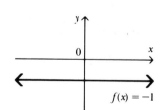

3. a) $f(x) = mx$, $b = 0$
   c) $f(x) = mx$, $b = 0$
   e) $f(x) = 6bx + b$ $\left[\text{or } f(x) = mx + \dfrac{m}{6}\right]$

4. $y = \frac{1}{5}x + \frac{14}{5}$

5. Yes

6. a) $f(2) = -2$ c) $f(2) = -3$

7. a) $m = -3$ c) $m = -1$

8. a) $y = 2x$

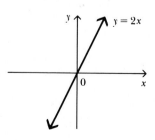

c) $y = -\frac{1}{2}x + 3$

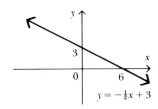

9. a) $\{-\frac{1}{3}, \frac{1}{2}\}$ c) $\{-a-b, b-a\}$
   e) $\{\frac{1}{2}, \frac{3}{2}\}$ g) $\{-1, \frac{2-a}{a}\}$

10. a) $\{a - \sqrt{a^2 - b^2}, a + \sqrt{a^2 - b^2}\}$
    c) No real solution
    e) $\{-5 - 2\sqrt{3}, -5 + 2\sqrt{3}\}$
    g) $\left\{\frac{1-\sqrt{17}}{8}, \frac{1+\sqrt{17}}{8}\right\}$
    i) $\left\{\frac{-5-\sqrt{161}}{4}, \frac{-5+\sqrt{161}}{4}\right\}$

11. a) Domain $= R$
       Range $= \{y|y \geq -\frac{121}{24}\}$
       Extreme point $= (\frac{5}{12}, -\frac{121}{24})$
       x intercepts $= -\frac{1}{2}, \frac{4}{3}$
       y intercept $= -4$

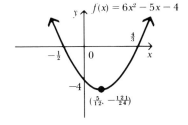

   c) Domain $= R$
      Range $= \{y|y \geq 0\}$
      Extreme point $= (-3, 0)$
      x intercept $= -3$
      y intercept $= 9$

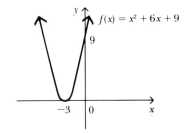

   e) Domain $= R$
      Range $= \{y|y \leq \frac{1}{8}\}$
      Extreme point $= (-\frac{5}{8}, \frac{1}{8})$
      x intercepts $= -\frac{3}{4}, -\frac{1}{2}$
      y intercept $= -3$

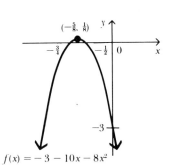

12. a) $(-\infty, -4) \cup (\frac{3}{2}, \infty)$ c) $\emptyset$

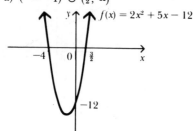
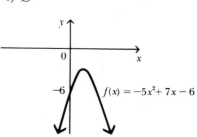

13. a) $\{\frac{1}{81}\}$
    c) $\{-1, 7\}$
    e) $\{-\frac{5}{3}, 3\}$
14. a) $\{(4, 1)\}$ c) $\{(4, 5)\}$ e) $\{(6, -5)\}$ g) $\{(1, 1)\}$
15. a) $\{(10, -2, 8)\}$ c) $\{(0, 1, -1)\}$

Chapter 5

PROBLEM SET 1, Page 250

1. a) $\{\frac{2}{3}\}$ c) $\{1, 2\}$ e) $\{-2, 2\}$ g) $\{0, -3, 3\}$
2. a) $5x^2 + 13x + 42, R = 122$
   c) $5x^4 + 7x^3 + 16x^2 + 33x + 59, R = 121$
   e) $-4x^5 - 4x^4 - 4x^3 - 9x^2 - 6x - 5, R = 2$
4. a. $Q(x) = 3x^2 + 12x + 14, f(2) = 35$
   c) $Q(x) = 2x^2 - 4x + 3, f(\frac{1}{2}) = \frac{25}{2}$
   e) $Q(x) = -3x^3 + 3x^2 - 3x + 8, f(-2) = -20$
5. $f(-5) = 0, f(-4) = 30, f(-3) = 40, f(-1) = 24, f(0) = 10, f(1) = 0, f(2) = 0, f(3) = 16, f(4) = 54, f(5) = 120$
   Factors are $x - 1, x - 2, x + 5$.
7. $k = -5$
8. $k = -10$
9. a) $f(1) = 0$, hence $x - 1$ is a factor of f
10. a) n is an odd positive integer
    c) n is an even positive integer

PROBLEM SET 2, Page 256

1. a) x intercepts $-1, 2$
   y intercept 4

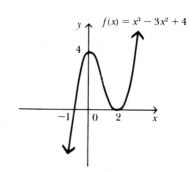

c) x intercepts $0, 1, -2$
y intercept 0

e) x intercept $-\frac{1}{2}$
y intercept 1

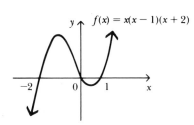
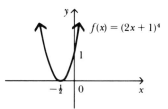

2. a) $(-2, 0) \cup (1, \infty)$

c) $(-\infty, -2) \cup (-1, 1)$

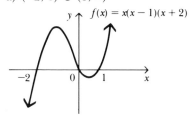
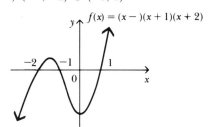

e) $(-1, 0)$

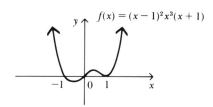

3. a) $\{\pm\frac{1}{3}, \pm\frac{2}{3}, \pm 1, \pm 2\}$; $\frac{1}{3}$ is a root
c) $\{\pm\frac{1}{5}, \pm\frac{2}{5}, \pm 1, \pm 2, \pm 5, \pm 10\}$; 1 is a root
e) $\{\pm\frac{1}{2}, \pm\frac{3}{2}, \pm\frac{5}{2}, \pm\frac{15}{2}, \pm 1, \pm 2, \pm 3, \pm 5, \pm 6, \pm 10, \pm 15, \pm 30\}$; $-\frac{5}{2}$ is a root

PROBLEM SET 3, Page 263

1. Vertical asymptote, $x = 3$
Horizontal asymptote, $y = 0$

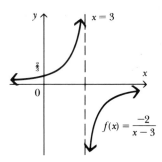

3. Vertical asymptote, $x = \frac{5}{2}$ and $x = 0$
Horizontal asymptote, $y = 0$

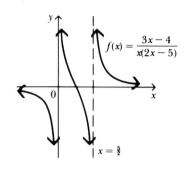

5. Vertical asymptote, $x = 2$
No horizontal asymptote

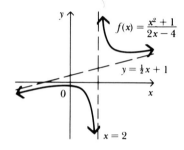

7. Vertical asymptotes, $x = -1$ and $x = 1$
Horizontal asymptote, $y = 0$

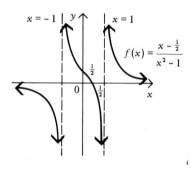

9. Vertical asymptote, $x = 2$
Horizontal asymptote, $y = 5$

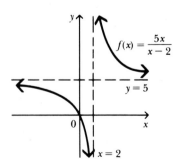

11. Vertical asymptotes, $x = 0$ and $x = 1$
    Horizontal asymptote, $y = 1$
    Yes, at $(-1, 1)$

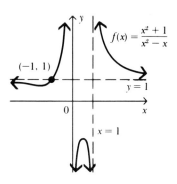

PROBLEM SET 4, Page 271

1. a) $6 - 5i$
    c) $5 - 27i$
    e) $7 - 17i$
    g) $\dfrac{25 + 3\sqrt{2}}{34} + \dfrac{15 - 5\sqrt{2}}{34}i$
    i) $-\frac{6}{7}i$
    k) $\frac{4}{25} - \frac{3}{25}i$
    m) $-\frac{3}{5} - \frac{7}{5}i$
    o) $2 - \frac{35}{4}i$
    q) $-\frac{3}{5}i$
    s) $25 - 8i$
    u) $-\frac{1}{2}i$
    w) $-\frac{3}{25} + \frac{29}{25}i$

2. a) $x = 3, y = 4$
    c) $x = 2, y = \frac{11}{2}$

3. a) $\bar{z} = 2 - \sqrt{3}\,i$, Re $(z) = 2$, Im $(z) = \sqrt{3}$, $\dfrac{1}{z} = \dfrac{2}{7} - \dfrac{\sqrt{3}}{7}i$
    c) $\bar{z} = \dfrac{3 + 2\sqrt{3}}{2} - \dfrac{3\sqrt{3} - 8}{4}i$, Re $(z) = \dfrac{3 + 2\sqrt{3}}{2}$, Im $(z) = \dfrac{3\sqrt{3} - 8}{4}$
    $\dfrac{1}{z} = \dfrac{(24 + 16\sqrt{3}) - (12\sqrt{3} - 32)i}{175}$

7. a) $13 - 8i$ c) $5 - 6i$

9. a) 5 c) $5\sqrt{13}$ e) $\dfrac{5\sqrt{13}}{13}$

10. a) 53 c) 78

PROBLEM SET 5, Page 277

2. a) Yes; 2 is a root of multiplicity 3
    c) Yes; $\frac{2}{3}$ is a root of multiplicity 2

3. a) $f(x) = \left(x - \dfrac{1 - \sqrt{17}}{4}\right)\left(x - \dfrac{1 + \sqrt{17}}{4}\right)$

4. a) $-1 - i, -\sqrt{2}, \sqrt{2}$

5. a) $f(x) = x^3 - 8x^2 + 22x - 20$
    c) $f(x) = x^4 - 4x^3 + 24x^2 - 40x + 100$
    e) $f(x) = x^4 - 3x^3 + x^2 + 4$

REVIEW PROBLEM SET, Page 277

1. a) $\frac{1}{3}, 2$ c) 1, 2, 3 e) $2, -2, 1+\sqrt{3}\,i, 1-\sqrt{3}\,i$
2. a) $Q(x) = 3x^3 + x^2 - 2x - 1, R = -4$
   c) $Q(x) = x^3 + 7x^2 - 15x + 39; R = -47$
   e) $Q(x) = 2x^3 + x^2 + x + 8, R = -23$
3. a) $k = 7$ c) $k = 18$
4. a) $f(-3) = -134, f(2) = 21$ c) $f(2) = -27, f(3) = -1$
5. a) $-8 + 4\sqrt{5}\,i$ c) $4 + 6\sqrt{5}\,i$
   e) $20 + 2i$ g) $14 + 8i$
   i) $8 + 8\sqrt{7}\,i$ k) $-94 + 17i$
   m) 37 o) $-29 + 15i$
   q) $\frac{37}{89} - \frac{101}{89}i$ s) $\frac{82}{125} - \frac{49}{125}i$
6. a) $f(x) = x^3 + x^2 - 4x - 4$ c) $f(x) = x^3 - 5x^2 + 7x - 3$ e) $f(x) = x^3 - x^2 + 2$
7. a) Yes, 1 is a root of multiplicity 1
   c) Yes, -1 is a root of multiplicity 2
   e) Yes, $-2i$ is a root of multiplicity 1
8. a) x intercepts $1, 2, -\frac{1}{2}$
      y intercept 2

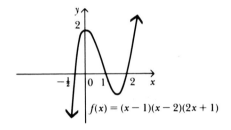

   c) x intercept 2
      y intercept -8

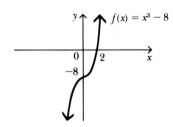

   e) x intercept $-\frac{1}{3}, \frac{3}{2}, -5$
      y intercept -15

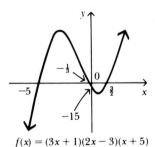

9. a) $(-\infty, -1)$

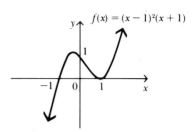

c) $(-3, -1) \cup (1, \infty)$

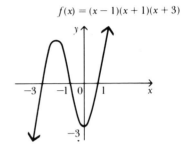

e) $(2, 3)$

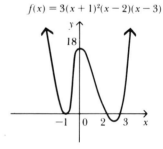

10. a) Vertical asymptote, $x = -1$
Horizontal asymptote, $y = 0$

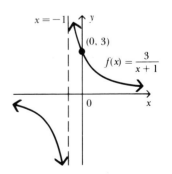

c) Vertical asymptote, $x = 1$
Horizontal asymptote, $y = 0$

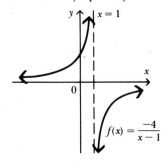

e) Vertical asymptote, $x = 3$
No Horizontal asymptote, $y = \frac{1}{3}x + 1$

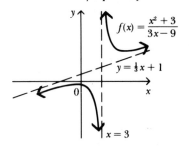

12. a) T   c) T
13. a) $\{-1\}$ c) $\{2, -1, 7\}$ e) $\{-1\}$

Chapter 6

PROBLEM SET 1, Page 290

1. a) $4x^2 + 4x + 1$
c) No

3. a)

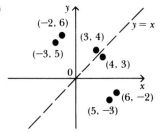

c)

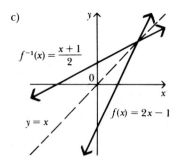

4. a) $f^{-1}(x) = -\frac{1}{3}x + \frac{1}{3}$ **c)** $f^{-1}(x) = \frac{3}{x}$

5. a) Yes   **c)** $f^{-1}(4) = 0$, $f^{-1}(0) = -\sqrt[3]{4}$, $f^{-1}(-4) = -2$

6. a) Yes   **c)** $f^{-1}(0) = \frac{7}{3}$, $f^{-1}(1) = 2$

7. a) Yes   **b)** Yes

9. a) f^{-1} exists

$f^{-1}(x) = \frac{5}{x}$

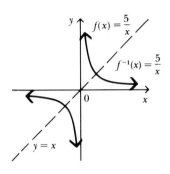

c) No

e) f^{-1} exists

$f^{-1}(x) = \sqrt[3]{\frac{1-x}{2}}$

$f(x) = -2x^3 + 1$

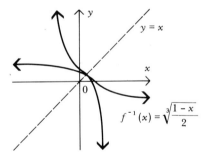

g) f^{-1} exists

$f^{-1}(x) = -\frac{1}{2}x + \frac{3}{2}$

$f(x) = -2x + 3$

PROBLEM SET 2, Page 294

1. a) Domain $= R$
Range $= \{y | y > 0\}$
Increasing

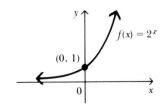

c) Domain = R
Range = $\{y|y>0\}$
Increasing

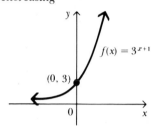

e) Domain = R
Range = $\{y|y<0\}$
Increasing

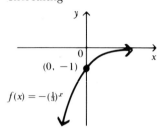

g) Domain = R
Range = $\{y|y>0\}$
Increasing

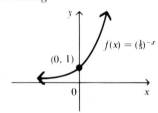

i) Domain = R
Range = $\{y|y>0\}$
Increasing

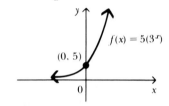

2. a) 1.7   c) 31.5   e) 4.7
3. a) 3   c) 3   e) $b \neq 1, b > 0$
5. Neither
6. a) 2^{-x} c) -8 e) $\frac{1}{4}$
7. a) 2^{x+1} c) $4^x - 4^{-x}$ e) -4
8. a) 1   c) $\frac{5}{2}$ e) $\frac{3}{4}$ g) $-\frac{3}{2}$ i) -1

PROBLEM SET 3, Page 298

1. a) $\log_5 125 = 3$ c) $\log_{32} 2 = \frac{1}{5}$ e) $\log_9 3 = \frac{1}{2}$ g) $\log_x a = 3$ i) $\log_\pi z = t$
2. a) $9^2 = 81$ c) $(\frac{1}{3})^{-2} = 9$ e) $(\sqrt{16})^{1/2} = 2$
g) $x^4 = 2$ i) $x^0 = 1$
3. a) Domain = $\{x|x>0\}$
Range = R
Decreasing

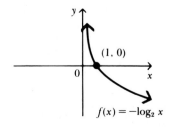

c) Domain = $\{x|x > 0\}$
Range = R
Decreasing

e) Domain = $\{x|x > 0\}$
Range = R
Increasing

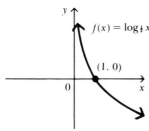

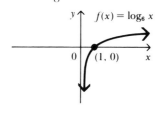

5. a) -1.3 c) 0.25 e) -1
9. $f[f^{-1}(x)] = x$
$f^{-1}[f(x)] = x$
10. a) $x = 1$ c) $x > 1$
11. a) $D = \{x|x < \tfrac{1}{2}\}$ c) $D = \{x|x \neq 0\}$ e) $D = \{x|x \neq -1\}$

PROBLEM SET 4, Page 302

1. a) $-\tfrac{17}{2}$ c) $-\tfrac{43}{222}$
2. a) $x = 11$ c) $x = 9$ e) $x = 3$
4. a) 0.6020 c) 1.2552 e) 3 g) 0.0602 i) 3.4771
6. a) 9 c) $63\tfrac{1}{2}$ e) $\tfrac{8}{3}$ g) 4 or -25
7. a) 7 c) 3 e) 7 g) 3 or 4
8. a) $\log_{10}\dfrac{a}{2}$ c) $\log_{10}\dfrac{x}{3}$ e) $\log_{10}\dfrac{a}{c}$
9. $B - 2A - 3C$

PROBLEM SET 5, Page 318

1.

	Scientific notation	Mantissa	Characteristic	log
a)	1.5×10^{-2}	0.1761	-2	-1.8239
c)	2.304×10^{-3}	0.3625	-3	-2.6375
e)	3.333×10	0.5228	1	1.5228
g)	1.37×10^{-6}	0.1367	-6	-5.8633
i)	5.171×10^{3}	0.7136	3	3.7136
k)	1.753×10	0.2438	1	1.2438

2. a) $x = 1397$ c) $x = 1.413$ e) $x = 3.14 \times 10^{-10}$
3. a) 127 c) 0.0314 e) 5366
4. 1.33

5. a) 16.28 c) 4.029 e) 4.626 g) 9.49 i) 2.029 k) 17.25
6. a) 2.322 c) 1.098 e) −0.2618
7. a) 2.808 c) 0.402 e) −0.00746
8. a) $1961, 20.48 years   c) $1977, 19.88 years
9. a) $574.38   b) 35 years
11. $1,656

REVIEW PROBLEM SET, Page 319

1. a) $f^{-1}(x) = \frac{1}{5} - \frac{1}{5}x$, $g^{-1}(x) = \frac{1}{7}x - \frac{3}{7}$
2. a) $f^{-1}(x) = -\frac{2}{3}x + \frac{2}{3}$

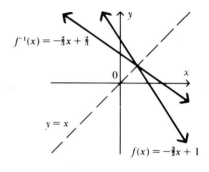

c) $f^{-1}(x) = \sqrt[3]{\frac{3}{2} - \frac{x}{2}}$

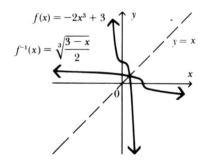

e) No

3. a) Domain $= R$
   Range $= \{y | y > 0\}$
   Increasing
   $f^{-1}(x) = \log_2 x$

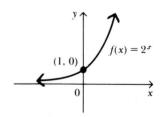

ANSWERS TO SELECTED PROBLEMS   495

c) Domain = R
Range = {y|y > 0}
Increasing
$f^{-1}(x) = \log_2 \left(\dfrac{x}{3}\right)$

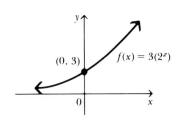

e) Domain = R
Range = {y|y > 0}
Increasing
$f^{-1}(x) = \log_4 x$

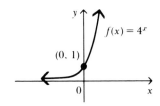

4. a) 2   c) 4   e) $\tfrac{1}{4}$
5. 7th year approx
   $1,275,000

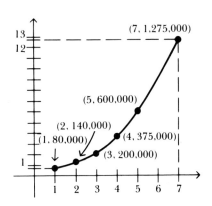

6. a) $2^5 = 32$ c) $8^{5/3} = 32$ e) $27^{2/3} = 9$
7. a) 2   c) 0   e) $-\tfrac{1}{2}$ g) $\tfrac{1}{3}$ i) 3
8. a) 9   c) $\dfrac{\sqrt{2}}{4}$ e) 243   g) 5
9. a) $\log_2 x + 5 \log_2 y$ c) $x \log_a b + y \log_a c$ e) $\tfrac{9}{7} \log_e 5$
10. a) -3 b) -3
11. a) 13.2   c) 2.524   e) 0.92   g) $\sqrt[3]{2.76}$
12. a) 1.7451   c) $0.7275 - 1$ e) -1.2090 g) 1.7875   i) $0.6654 - 1$
13. a) $-\sqrt{5}, \sqrt{5}$ c) $\tfrac{3}{4}$
14. a) $\log 7$ c) $\log x^9$

15. a) Domain = $\{x|x > 0\}$
Range = R
Decreasing
$g^{-1}(x) = (\tfrac{1}{2})^x$

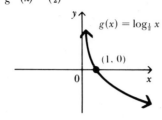

c) Domain = $\{x|x > -2\}$
Range = R
Increasing
$f^{-1}(x) = 3^x - 2$

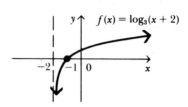

16. $D = \{x|x > 0\}$, Range = R
  a) 2   c) $\tfrac{5}{4}$
17. \$5,073.33

Chapter 7

PROBLEM SET 1, Page 328

1. a) $\tfrac{3}{2}$, 4, $\tfrac{15}{2}$, 12, $\tfrac{35}{2}$ c) $-1, -1, 0, 2, 5$ e) 2, 4, 2, 4, 2
3. a) 15   c) 660   e) 15   g) 15   i) $\tfrac{163}{60}$ k) 500   m) $\tfrac{5}{6}$ o) 148
4. a) $\sum_{k=0}^{4}(3k+1)$ c) $\sum_{k=0}^{3}(\tfrac{3}{5})^{k+1}$
5. a) T   c) F   e) F

PROBLEM SET 2, Page 338

1. a) A.P., $d=3$, $S_{10}=155$ c) A.P., $d=5$, $S_{10}=295$
  e) A.P., $d=-13$, $S_{10}=85$ g) A.P., $d=1.2$, $S_{10}=111$
2. a) $a_{10}=50$, $a_{15}=85$ c) $a_6=16a+4b$, $a_9=25a-8b$
3. $63b-63$
4. a) $a_{10}=33$, $S_{10}=195$ c) $d=13\tfrac{5}{51}$, $a_{18}=239\tfrac{2}{3}$ e) $n=32$, $d=\tfrac{21}{31}$
5. a) $r=3$, $S_5=242$ c) $r=-2$, $S_5=11$ e) $r=\tfrac{2}{3}$, $S_5=211$
  g) $r=-\tfrac{2}{3}$, $S_5=\tfrac{55}{9}$
6. a) $a_{10}=\tfrac{1}{128}$ c) $a_5=2$ e) $a_n=(1+a)^{n-1}$
7. a) $a_6=192$, $a_{10}=3072$, $S_{10}=6138$ c) $a_5=48$
8. a) $r=3$ or -4 c) $r=2$ or -2 e) $a_8=-\tfrac{1}{64}$, $S_8=\tfrac{85}{64}$
  g) $n=10$, $S_n=S_{10}=\tfrac{1023}{16}$ i) $n=5$, $S_5=\tfrac{2882}{5}$
9. a) $\tfrac{2}{27}$, $\tfrac{2}{9}$, $\tfrac{2}{3}$
10. a) $\tfrac{1}{2}$ c) $\tfrac{1}{4}$
11. a) $\tfrac{32}{99}$ c) $\tfrac{46}{99}$ e) $3\tfrac{561}{999}$

13. $500

15. $6,604.50

PROBLEM SET 4, Page 352

1. a) $\frac{1}{30}$ c) $\frac{5}{24}$ e) 0 g) $\dfrac{1}{n(n-1)(n+1)}$ i) $(n+k)(n+k-1)$

2. a) 1 c) 10 e) n g) $\dfrac{n(n-1)}{2}$ i) 1

3. a) 50

5. a) 30 c) No

7. 635,013,559,600

9. a) 42 c) 6720

10. a) $n=9$ c) $n=8$

12. 657,720

13. 120

15. 14,400 (2) = 28,800

17. 126

PROBLEM SET 5, Page 358

1. a) $256z^8 + 1024z^7x + 1792z^6x^2 + 1792z^5x^3 + 1120z^4x^4 + 448z^3x^5 + 112z^2x^6 + 16zx^7 + x^8$
  c) $y^{10} - 10y^8x + 40y^6x^2 - 80y^4x^3 + 80y^2x^4 - 32x^5$

2. a) $x^{20} - 20x^{18}a + 180x^{16}a^2 - 960x^{14}a^3$
  c) $\left(\dfrac{x}{2}\right)^{7/2} + 14\left(\dfrac{x}{2}\right)^3 y + 84\left(\dfrac{x}{2}\right)^{5/2} y^2 + 280\left(\dfrac{x}{2}\right)^2 y^3$
  e) $a^{12} - 16a^{21/2}x^2 + 112a^9x^4 - 448a^{15/2}x^6$

3. a) $x^{16} + 16x^{15}y + 120x^{14}y^2 + 560x^{13}y^3 + 1820x^{12}y^4$
  c) $a^{11} - 22a^{10}b^2 + 220a^9b^4 - 1320a^8b^6 + 5280a^7b^8$
  e) $x^7 - 14x^6y + 84x^5y^2 - 280x^4y^3 + 560x^3y^4$
  g) $a^{27} - 9a^{26} + 36a^{25} - 84a^{24} + 126a^{23}$

4. a) $\frac{455}{4096}x^{24}a^3$ c) $\frac{672}{729}x^6a^{12}$ e) $\frac{14}{9}a^5x^{12}$

PROBLEM SET 6, Page 370

1. a) $\{(H,H,H),(H,H,T),(H,T,H),(T,H,H),(H,T,T),(T,H,T),(T,T,H),(T,T,T)\}$
  c) $\{(H,H,H,H),(H,H,H,T),(H,H,T,H),(H,T,H,H),(T,H,H,H),(H,H,T,T),$
  $(H,T,H,T),(T,H,H,T),(H,T,T,H),(T,H,T,H),(T,T,H,H),(H,T,T,T),$
  $(T,H,T,T),(T,T,H,T),(T,T,T,H),(T,T,T,T)\}$

2. a) $\{1,2,3,4,5,6\}$ b) $\frac{1}{6}$

3. a) $\{(B,B,B),(B,B,G),(B,G,B),(G,B,B),(B,G,G),(G,B,G),(G,G,B),(G,G,G)\}$
  b) $\{(G,G,B),(G,B,G)\}$ c) $\frac{1}{4}$

4. a) $\{1,2,3,4,5,6\}$ b) $\frac{2}{3}$

5. a) $\frac{1}{2}$ c) $\frac{3}{4}$ e) $\frac{3}{26}$

6. a) $\frac{1}{3}$ c) $\frac{1}{4}$ e) $\frac{5}{12}$ g) $\frac{7}{12}$ i) $\frac{2}{3}$ k) $\frac{1}{4}$

7. a) $\frac{1}{6}$ c) $\frac{5}{6}$ e) $\frac{7}{18}$
8. a) $\frac{5}{12}$ c) $\frac{1}{6}$
9. $\{(H,H,H),(H,H,T),(H,T,H),(T,H,H),(H,T,T),(T,H,T),(T,T,H),(T,T,T)\}$
    a) $\frac{3}{4}$ c) $\frac{1}{2}$
11. $\frac{1}{6}$
13. $\frac{1}{2}$
15. a) $\frac{1}{4}$ c) $\frac{1}{8}$
17. $\frac{13}{42}$
19. $\frac{9}{20}$

REVIEW PROBLEM SET, Page 372

1. a) 0, 2, 0, 2   c) 2, $\frac{7}{2}$, 4, $\frac{17}{4}$ e) 1, $\frac{18}{5}$, $\frac{39}{5}$, $\frac{68}{5}$
2. a) $a_9 = 44$, $S_9 = 216$ c) $a_{11} = 12$, $S_{11} = 297$ e) $a_{24} = 4$, $S_{24} = 50$
    g) $a_{16} = \frac{65}{8}$, $S_{16} = 70$
3. a) $x = 3$ c) $x = 3$ or $x = -2$
4. a) $a_8 = 49{,}152$; $S_8 = 65{,}535$ c) $a_6 = -\frac{1}{3}$; $S_6 = 60\frac{2}{3}$
    e) $a_{18} = -768\sqrt{2}$, $S_{18} = -1533(\sqrt{2}-1)$ g) $a_{13} = \frac{1}{81}$, $S_{13} = \dfrac{3^7 + \sqrt{3}}{3^5 + 3^4\sqrt{3}}$
5. a) $x = -2$ or $x = 24$ c) $x = \frac{1}{7}$ or 10
6. a) 95   c) 20   e) 160
7. a) 5005   c) n   e) 21
8. b) 2^n
9. a) $x^4 + 8x^3y + 24x^2y^2 + 32xy^3 + 16y^4$
    c) $1 + 5x + 10x^2 + 10x^3 + 5x^4 + x^5$
    e) $1 - 12x + 60x^2 - 160x^3 + 240x^4 - 192x^5 + 64x^6$
    g) $81x^4 + 108x^3y + 54x^2y^2 + 12xy^3 + y^4$
    i) $243x^5 + 405x^{9/2} + 270x^4 + 90x^{7/2} + 15x^3 + x^{5/2}$
    k) $8x^3 + \dfrac{12x^2}{y} + \dfrac{6x}{y^2} + \dfrac{1}{y^3}$
10. a) $210x^6y^4$ c) $13{,}440x^6y^4$ e) $1{,}082{,}565x^8y^3$
11. 31
12. 210
13. a) $\frac{1}{2}$ c) $\frac{5}{26}$
14. a) $\frac{2}{9}$ c) $\frac{4}{9}$ e) $\frac{2}{3}$
15. 120
16. 5040
17. a) 40,320   b) 720
18. $\frac{1}{32}$
19. a) $\frac{1}{3}$ c) $\frac{3}{5}$ e) $\frac{11}{15}$ g) $\frac{1}{3}$
20. $\frac{11}{30}$
21. 0.62
22. $\frac{1}{4}$

Chapter 8

PROBLEM SET 1, Page 385

1. a) $\{(3, 5)\}$ c) $\{(\frac{23}{4}, \frac{13}{2})\}$ e) Dependent
2. a) $\{(1, 2, 3)\}$ c) $\{(1, 1, 1)\}$ e) $\{(4, 5, 6)\}$
3. a) A is 3×3 matrix
4. a) Echelon, dependent   c) Not echelon, inconsistent   e) Echelon, $\{(1, 1)\}$
   g) Not echelon, $\{(4, -3)\}$
5. a) $\{(1, 1, 0)\}$ c) Inconsistent   e) Dependent

PROBLEM SET 2, Page 392

1. a) 17   c) 63   e) 0
2. a) $x = \frac{1}{4}$ c) $x = -\frac{5}{7}$
3. a) Theorem 2   c) Theorem 1   e) Theorems 1 and 3
5. a) 0   c) -20 e) 3

PROBLEM SET 3, Page 397

1. a) $\{(\frac{1}{3}, \frac{2}{3})\}$ c) $\{(0, 0)\}$ e) $\{(4, -1, -6)\}$ g) $\{(1, 1, 1)\}$ i) $\{(3, -1, 2)\}$
2. a) Watch = \$75, chain = \$25, ring = \$125
3. 30 pounds at 20¢ and 20 pounds at 30¢
4. \$3600 at 8%, \$6400 at 5%

REVIEW PROBLEM SET, Page 398

1. a) $\{(5, 5)\}$ c) $\{(1, 1)\}$ e) $\{(-3, 1, 6)\}$ g) $\{(3, -1, 2)\}$
2. a) $\{(5, 5)\}$ c) $\{(\frac{8}{5}, 5, \frac{9}{5})\}$ e) Dependent
3. a) 11   c) 1   e) -26 g) 15
4. a) 0   c) 0   e) 5
5. a) $\{(\frac{16}{7}, -\frac{11}{7})\}$ c) $\{(1, 1, 2)\}$ e) $\{(2, -2, 1)\}$
7. 300 at \$3.65 and 440 at \$1.25
9. 1.35 gold, 6.65 silver

Chapter 9

PROBLEM SET 1, Page 408

1. a) $(x-6)^2 + (y-4)^2 = 36$ c) $(x-5)^2 + (y+12)^2 = 169$ e) $x^2 + y^2 = 20$
2. a) Center $= (-2, 3)$, radius $= 5$ c) Center $= (1, -2)$, radius $= 4$
   e) Center $= (4, -3)$, radius $= 1$ g) Center $= (\frac{3}{2}, 0)$, radius $= \dfrac{3\sqrt{10}}{2}$

4. $(x-5)^2 + (y-1)^2 = 25$

5. $x^2 + y^2 = r^2$ is symmetric with respect to the x axis, y axis, and the origin

6. $(x-4)^2 + (y-4)^2 = 8$

8. $(x + \frac{7}{4})^2 + (y-6)^2 = \frac{625}{16}$

PROBLEM SET 2, Page 417

1. a) Vertex $= (0, 0)$
Focus $= (4, 0)$
Directrix: $x = -4$

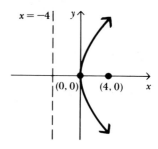

c) Vertex $= (0, 0)$
Focus $= (0, -2)$
Directrix: $y = 2$

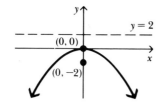

e) Vertex $= (0, 0)$
Focus $= (0, -\frac{3}{16})$
Directrix: $y = \frac{3}{16}$

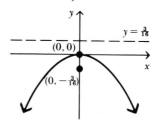

g) Vertex $= (0, 0)$
Focus $= (-\frac{15}{8}, 0)$
Directrix: $x = \frac{15}{8}$

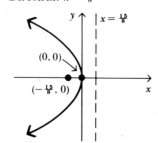

i) Vertex $= (2, -1)$
Focus $= (2, 0)$
Directrix: $y = -2$

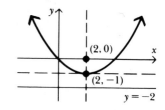

ANSWERS TO SELECTED PROBLEMS 501

k) Vertex = (2, 4)
Focus = (2, $3\frac{3}{4}$)
Directrix: $y = 4\frac{1}{4}$

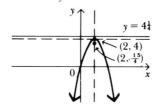

m) Vertex = (−2, −2)
Focus = (−2, $-\frac{3}{4}$)
Directrix: $y = -3\frac{1}{4}$

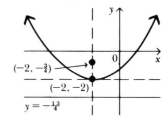

2. a) 16   c) 8   e) $\frac{3}{4}$
g) $\frac{15}{2}$ i) 4   k) 1   m) 5

3. a) Domain = $\{x|x \geq 0\}$, Range = R
c) Domain = R, Range = $\{y|y \leq 0\}$
e) Domain = R, Range = $\{y|y \leq 0\}$
g) Domain = $\{x|x \leq 0\}$, Range = R
i) Domain = R, Range = $\{y|y \geq -1\}$
k) Domain = R, Range = $\{y|y \leq 4\}$
m) Domain = R, Range = $\{y|y \geq -2\}$

4. a) $x^2 = -16y$

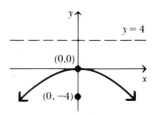

c) $(y - 3)^2 = -16(x - 2)$

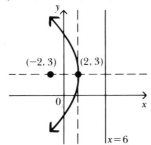

e) $(x - 4)^2 = 12(y + 3)$

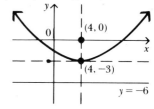

5. $(x-5)^2 + y^2 = 25$

7. $(y+7)^2 = -48(x - \frac{217}{48})$

9. $33\frac{3}{4}$ ft

PROBLEM SET 3, Page 427

1. a) Vertices: $(-4, 0)$, $(4, 0)$, $(0, -2)$, $(0, 2)$
Foci: $(-2\sqrt{3}, 0)$, $(2\sqrt{3}, 0)$

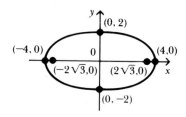

c) Vertices: $(-2, 0)$, $(2, 0)$, $(0, -8)$, $(0, 8)$
Foci: $(0, -2\sqrt{15})$, $(0, 2\sqrt{15})$

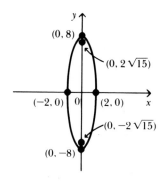

e) Vertices: $(-\sqrt{2}, 0)$, $(\sqrt{2}, 0)$, $(0, -\sqrt{10})$, $(0, \sqrt{10})$
Foci: $(0, -2\sqrt{2})$, $(0, 2\sqrt{2})$

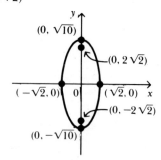

g) Vertices: $(-5, 0)$, $(5, 0)$, $(0, -2)$, $(0, 2)$
   Foci: $(-\sqrt{21}, 0)$, $(\sqrt{21}, 0)$

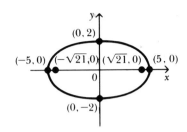

2. a) Domain = $[-4, 4]$, Range = $[-2, 2]$
   c) Domain = $[-2, 2]$, Range = $[-8, 8]$
   e) Domain = $[-\sqrt{2}, \sqrt{2}]$, Range = $[-\sqrt{10}, \sqrt{10}]$
   g) Domain = $[-5, 5]$, Range = $[-2, 2]$

3. a) Center: $(1, 2)$
   Vertices: $(5, 2)$, $(-3, 2)$, $(1, 4)$, $(1, 0)$
   Foci: $(1 - 2\sqrt{3}, 2)$, $(1 + 2\sqrt{3}, 2)$

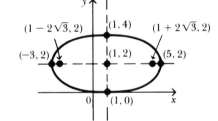

c) Center: $(-3, 4)$
   Vertices: $(-3 - \sqrt{2}, 4)$, $(-3 + \sqrt{2}, 4)$, $(-3, 4 - \sqrt{10})$, $(-3, 4 + \sqrt{10})$
   Foci: $(-3, 4 + 2\sqrt{2})$, $(-3, 4 - 2\sqrt{2})$

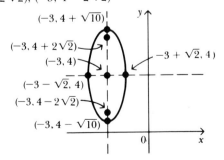

e) Center: (4, 0)
   Vertices: (8, 0), (0, 0), (4, −2), (4, 2)
   Foci: $(4 - 2\sqrt{3}, 0), (4 + 2\sqrt{3}, 0)$

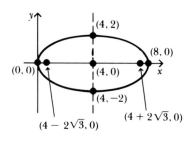

g) Center: (0, −2)
   Vertices: (5, −2), (−5, −2), (0, 0), (0, −4)
   Foci: $(-\sqrt{21}, -2), (\sqrt{21}, -2)$

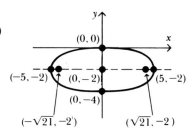

4. a) $\dfrac{x^2}{64} + \dfrac{y^2}{36} = 1$ c) $\dfrac{(x-2)^2}{100} + \dfrac{(y-1)^2}{25} = 1$

5. a) $\dfrac{x^2}{25} + \dfrac{y^2}{9} = 1$ c) $\dfrac{x^2}{10} + \dfrac{y^2}{2} = 1$

7. $y = -\sqrt{1 - \dfrac{x^2}{4}},\ y = \sqrt{1 - \dfrac{x^2}{4}}$

PROBLEM SET 4, Page 437

1. a) Vertices: (−3, 0), (3, 0)
      Foci: $(-\sqrt{13}, 0), (\sqrt{13}, 0)$
      Asymptotes: $y = \pm \tfrac{2}{3} x$

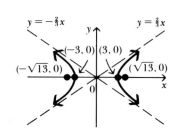

   c) Vertices: (−4, 0), (4, 0)
      Foci: $(-2\sqrt{5}, 0), (2\sqrt{5}, 0)$
      Asymptotes: $y = \pm \tfrac{1}{2} x$

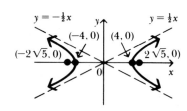

e) Vertices: $\left(0, \dfrac{\sqrt{2}}{2}\right)$, $\left(0, -\dfrac{\sqrt{2}}{2}\right)$

Foci: $\left(0, \dfrac{\sqrt{6}}{2}\right)$, $\left(0, -\dfrac{\sqrt{6}}{2}\right)$

Asymptotes: $y = \pm\dfrac{\sqrt{2}}{2}x$

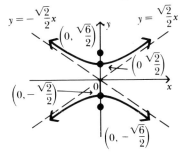

g) Vertices: $(-\sqrt{7}, 0)$, $(\sqrt{7}, 0)$

Foci: $(-2\sqrt{3}, 0)$, $(2\sqrt{3}, 0)$

Asymptotes: $y = \pm\dfrac{\sqrt{35}}{7}x$

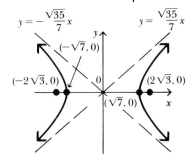

2. a) Domain $= (-\infty, -3] \cup [3, \infty)$, Range $= R$
   c) Domain $= (-\infty, -4] \cup [4, \infty)$, Range $= R$
   e) Domain $= R$, Range $= \left(-\infty, -\dfrac{1}{\sqrt{2}}\right] \cup \left[\dfrac{1}{\sqrt{2}}, \infty\right)$
   g) Domain $= (-\infty, -\sqrt{7}) \cup (\sqrt{7}, \infty)$, Range $= R$

3. a) Center: $(1, 3)$
   Vertices: $(5, 3)$, $(-3, 3)$
   Foci: $(6, 3)$, $(-4, 3)$
   Asymptotes: $y - 3 = \pm\tfrac{3}{4}(x - 1)$

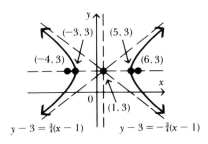

c) Center: $(1, -3)$
   Vertices: $(1, 2)$, $(1, -8)$
   Foci: $(1, -3 - \sqrt{41})$, $(1, -3 + \sqrt{41})$
   Asymptotes: $y + 3 = \pm\tfrac{5}{4}(x - 1)$

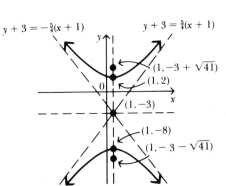

e) Center: $(-2, 3)$
   Vertices: $(-6, 3), (2, 3)$
   Foci: $(-2 + 2\sqrt{5}, 3), (-2 - 2\sqrt{5}, 3)$
   Asymptotes: $y - 3 = \pm\frac{1}{2}(x + 2)$

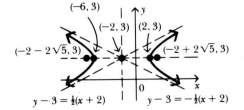

g) Center: $(0, 2)$
   Vertices: $(0, 6), (0, -2)$
   Foci: $(0, 7), (0, -3)$
   Asymptotes: $y - 2 = \pm\frac{4}{3}x$

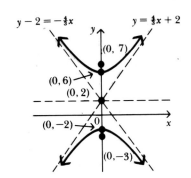

4. $\dfrac{(x + 2)^2}{9} - \dfrac{(y - 1)^2}{16} = 1$

5. $\dfrac{x^2}{20} - \dfrac{y^2}{5} = 1$

6. a) $\dfrac{x^2}{9} - \dfrac{y^2}{7} = 1$ c) $\dfrac{4x^2}{3} - \dfrac{y^2}{3} = 1$

8. $y^2 - 3x^2 - 6y + 6x + 29 = 0$

PROBLEM SET 5, Page 443

1. $\{(-1, -2), (2, 1)\}$
3. $\{(2, 4), (-\frac{4}{5}, -\frac{22}{5})\}$
5. $\{(-1 + 2i, 2 - 3i), (-1 - 2i, 2 + 3i)\}$
7. $\{(\frac{7}{2}, \frac{5}{2}), (\frac{11}{4}, \frac{5}{4})\}$
9. $\{(2 + 3i, 1 - 2i), (2 - 3i, 1 + 2i)\}$
11. $\{(5, 4)\}$
13. $\{(4, 2), (4, -2), (-6, \sqrt{6}i), (-6, -\sqrt{6}i)\}$
15. $\{(3, 2), (-3, 2), (3, -2), (-3, -2)\}$
17. $\{(-2\sqrt{6}, 1), (-2\sqrt{6}, -1), (2\sqrt{6}, 1), (2\sqrt{6}, -1)\}$
19. $\{(3, 0), (-3, 0)\}$
21. $\{(2, 1), (2, -1), (-2, 1), (-2, -1)\}$
23. $\{(3, 2), (-3, 2), (-3, -2), (3, -2)\}$
25. $\{(2i, 3), (-2i, 3), (-2, 1), (2, 1)\}$
27. $\{(3i, 5), (3i, -5), (-3i, 5), (-3i, -5)\}$
29. 16,000

REVIEW PROBLEM SET, Page 444

1. a) $(x-4)^2 + (y+3)^2 = 25$
   c) $(x-1)^2 + (y-2)^2 = 13$
2. a) $y^2 = 20x$
   c) $(y-4)^2 = -20(x-2)$
3. a) $\dfrac{(x-1)^2}{1} + \dfrac{(y+4)^2}{16} = 1$

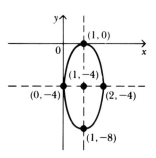

   c) $\dfrac{(x+1)^2}{25/4} + \dfrac{(y-3)^2}{25} = 1$

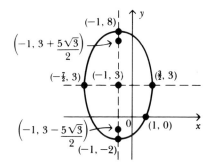

4. a) $\dfrac{x^2}{4} - \dfrac{y^2}{12} = 1$

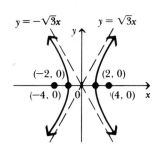

c) $\dfrac{(y-1)^2}{1} - \dfrac{(x+2)^2}{15} = 1$

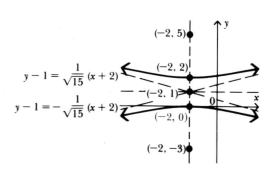

e) $\dfrac{x^2}{36} - \dfrac{y^2}{16} = 1$

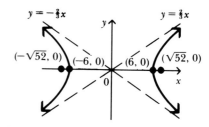

5. a) Center: $(0, 3)$
Vertices: $(-3, 3), (3, 3)$
Foci: $(-5, 3), (5, 3)$

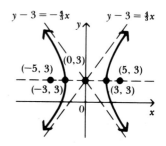

c) Center: $(-2, 4)$
Vertices: $(-2, 7), (-2, 1)$
Foci: $(-2, 4 + \sqrt{13}), (-2, 4 - \sqrt{13})$

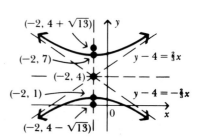

e) Center: (−6, 8)
   Vertices: (−9, 8), (−3, 8), (−6, 10), (−6, 6)
   Foci: (−6 − √5, 8), (−6 + √5, 8)

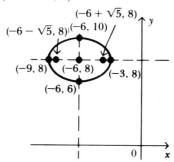

g) Vertex: $(\frac{1}{2}, \frac{2}{3})$
   Focus: $(\frac{3}{2}, \frac{2}{3})$

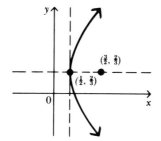

i) Vertex: $(-\frac{3}{5}, \frac{2}{5})$
   Focus: $(\frac{7}{5}, \frac{2}{5})$

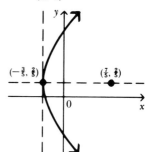

6. $y - 8 = \frac{4}{5}(x - 8)$

7. a) $k < b^2$, $k < a^2$ and $a > b$ c) $k < b^2$ and $a^2 = b^2$

8. a) $\{(3, -4)\}$ c) $\{(6, 0), (0, 6)\}$ e) $\{(-5, 2), (-5, -2), (5, 2), (5, -2)\}$

Index

Abscissa, 120
Absolute inequality, 110
Absolute value, 105, 269
    complex number, 269
    definition of, 106
    equations involving, 109
    function, 150
    properties, 107, 110
Addition
    associative law, 22
    commutative law, 22
    of complex numbers, 264
    of fractions, 43
    of polynomials, 26
    of radical expressions, 66
    of real numbers, 22
    of two functions, 155
Additive inverse, 23
Algebra
    expressions, 18
    fundamental theorem, 274
    of functions, 155
    of polynomials, 26
Algebraic expressions, 18
Analytic geometry, 403
Antilogarithm, 306
Approximation of logarithms, 305, 308
Asymptotes
    of a hyperbola, 430
    of a rational function, 257
Augmented matrix of a system of linear equations, 379
Arithmetic progression, 330
    common difference, 330
    definition of, 330
    nth term of an, 330
    sum of n terms of an, 331
Associative law
    of addition, 22
    of multiplication, 22
Axiom(s)
    for inequalities, 90
    for real numbers, 22

Axis of a coordinate system, 120
    conjugate, 428
    major, 419
    minor, 419
    of symmetry of parabola, 410
    transverse, 428

Base
    of a logarithm, 304
    of a power, 18
    of an exponential function, 291
Binomial
    coefficients, 354
    definition of a, 19
    expansion of a, 355
    factors, 35
    products, 29
    theorem, 355

Cancellation law
    for addition, 25
    for multiplication, 25
Cartesian coordinate system, 119
Cartesian product, 118
Center
    of a circle, 404
    of an ellipse, 419
Characteristic of a logarithm, 304
Circle, 404
    equation of a, 404
Closure
    for addition, 22
    for multiplication, 22
Coefficient of polynomials, 19
Combinations, 347
Collinear, 195
Common factors, 34
Common logarithm, 304
Commutative law, 22
    of addition, 22
    of multiplication, 22
Completely factored form
    of a polynomial, 34
    use of known products, 35

Completing the square method, 209
Complex fraction, 47
Complex numbers, 264
   addition, 264
   conjugate, 267
   division, 267
   equality, 264
   form, 264
   geometric representation, 268
   imaginary part, 264
   multiplication, 264
   real part, 264
   roots, 273
   subtraction, 267
Complex zeros, 273
Composite functions, 283
Compound interest, 317
Conjugate of a complex number, 267
Conjugate root theorem, 275
Constant function, 177
Constant of proportionality, 153
Coordinate(s) of points on a number line, 88
Coordinate system, 119
Cramer's rule, 394

Decreasing functions, 166
Degree of a polynomial, 19
Denominator
   least common, 45
   rationalizing a, 67
Dependent
   event, 369
   system of linear equations, 385
Determinant
   definition, 387
   expansion, 387
   function, 387
   properties, 388
   use in solutions of systems, 395
   value of, 387
Difference
   common, 330
   definition of, 24
   of two squares, 35
   of two complex numbers, 267
   quotient, 156
Direct variation, 153
Directrix of a parabola, 409
Discriminant, 222
Disjoint sets, 6

Distance
   between two points, 124
   formula, 126
Distributive law, 23
Division
   of complex number, 267
   of polynomials, 243
   of real numbers, 24
   of two functions, 156
   synthetic, 245
Domain
   of an absolute value function, 150
   of an exponential function, 291
   of a function, 142, 143
   of a logarithmic function, 296
   of a polynomial function, 175
   of a relation, 136
   of a sequence, 325

e, 304
Echelon form of a matrix, 382
Element of a set, 3
Elimination method, 204
Ellipse
   center, 419
   definition, 418
   foci, 418
   vertices, 419
Empty set, 3
Entries of a matrix, 378
Equality
   addition law of, 80
   of complex numbers, 264
   multiplication law of, 80
   of sets, 5
Equation
   absolute value in, 107
   dependent, 385
   first-degree in one variable, 79
   first-degree in two variables, 197
   first-degree in three variables, 204
   inconsistent, 385
   linear, 79
   quadratic, 208
   quadratic in form, 231
   radical, 231
   second-degree in one variable, 208
   second-degree in two variables, 438
   solution of an, 79
   solution set of an, 79

systems of linear, 196, 379
systems of quadratic, 438
Equivalent equations, 79
Even functions, 161
Events, 360
Expansion
  binomial, 355
  of a determinant, 387
Experiment, 359
Exponent(s)
  laws of, 51, 61
  negative, 51
  positive integers, 51
  rational, 61
  real numbers, 292
  zero, 54
Exponential function, 291
Expression
  graph of, 287
  inverse, 287
Extreme point, 217

Factorial notation, 347
Factoring
  common, 34
  complete, 34
  differences of cubes, 35
  differences of squares, 35
  polynomials, 34
  second-degree polynomials, 37
  solving equations by, 209
  sums of cubes, 35
Factorization theorem, 274
Finite set, 4
First-degree equation, 79
Focal chord of a parabola, 410
Focus (foci)
  of an ellipse, 418
  of a hyperbola, 428
  of a parabola, 409
Fraction(s)
  addition of, 43
  complex, 47
  definition of a, 33
  division of, 47
  equality of, 39
  fundamental principle of, 40
  least common denominator of, 45
  multiplication of, 46
  reduced, 33
  subtraction of, 44

Functions
  absolute value, 150
  algebra, 155
  composite, 149
  constant, 177
  decreasing, 166
  definition, 143
  determinant, 387
  exponential, 291
  even, 161
  greatest integer, 151
  identity, 150
  image, 149
  increasing, 166
  inverse, 283
  linear, 175
  logarithmic, 296
  mapping, 149
  notation, 147
  odd, 161
  one-to-one, 285
  polynomial, 175
  quadratic, 208
  rational, 257
  sequence, 325
Fundamental Theorem of Algebra, 274
Fundamental Principle of Fractions, 40

Geometric progression
  definition of a, 334
  nth term of a, 334
  ratio of, 334
  sum of n terms, 335
Geometric series, 337
Geometry of lines, 87, 185
Graph
  of absolute value function, 151
  of exponential function, 292
  of greatest integer function, 151
  of inverse function, 287
  of a linear function, 176
  of a number line, 88
  of an ordered pair, 120
  of polynomial function of degree greater than two, 251
  of quadratic function, 215
  of rational function, 258
  of sequence function, 326
Graphical solutions of inequalities, 224, 253

Horizontal axis, 120

Hyperbola, 428
　asymptotes, 430
　center, 428
　definition, 428
　foci, 428
　transverse axis, 428
　vertices, 428

Identity
　element for addition, 23
　element for multiplication, 22
　function, 150
Imaginary part of complex numbers, 264
Inclination of a line, 179
Inconsistent system, 202, 385
Independent event, 368
Index
　of a radical, 63
　of summation, 327
Induction, mathematical, 340
Inequalities
　absolute, 110
　equivalent, 101
　first-degree in one variable, 101
　graphs of, 102
　properties, 92
　quadratic, 224
　solutions of, 101
　solution sets of, 101
　symbolism for, 90
Infinite series, 336
Infinite sets, 4
Integers, set of, 14
Intercept of a graph, 187, 208
Interest, compound, 317
Interpolation, linear, 308
Intersection of sets, 9
Interval
　bounded, 97
　unbounded, 98
Inverse
　additive, 23
　additive of a complex number, 266
　multiplicative, 23
　of an exponential function, 296
　of a function, 283
Inverse variation, 154
Irrational numbers, 17
　as exponents, 291
　sets of, 17

Joint variation, 155

Law(s)
　axioms of real numbers, 22
　cancellation, for addition, 25
　cancellation, for multiplication, 25
　of exponents, 60
　of logarithms, 300
　zero-factorial, 348
Latus rectum of a parabola, 410
Least common denominator, 45
Less than, 90
Line(s)
　equations, 187, 188, 195
　graph, 176
　number, 88
　parallel, 186
　perpendicular, 186
　slope, 180
Linear equation(s), 79
　definition of, 79
　general form for, 195
　intercept form for, 195
　point-slope form for, 188
　slope-intercept form for, 187
　systems of, 196, 377
Linear function
　definition of a, 175
　as direct variation, 153
　graph of a, 176
　intercepts of the graph of a, 187
　slope of the graph of a, 180
Linear interpolation, 308
Linear system(s), 196, 377
　solution of by:
　　Cramers rule, 395
　　substitution, 198
　dependent, 202, 385
　elimination method, 203, 379
　inconsistent, 202, 385
Logarithm(s)
　base, 296
　changing the base of, 316
　characteristic of a, 304
　common, 304
　computations using, 311
　mantissa of a, 304
　properties of, 299
Logarithmic equations, 302
Logarithmic function, 295
Lowest terms, 39

Major axis of an ellipse, 419
Mantissa, 304
Mapping, 149
Mathematical induction, 340
Matrix (Matrices)
  augmented, 379
  column, 378
  determinant, 386
  entry, 378
  row, 378
  row-reduced echelon form, 382
  size, 378
Maximum value of quadratic function, 217
Midpoint, 129
Minimum value of quadratic function, 217
Minor axis of an ellipse, 419
Modulus of complex number, 269
Monomial, 19
  definition of a, 19
  degree of a, 19
Multiplication
  Associative Law, 22
  Commutative Law, 22
  of complex numbers, 264
  of fractions, 47
  of functions, 155
  of polynomials, 28
  of real numbers, 22
  using logarithms, 311
Multiplicative-inverse axiom, 23

n factorial, 347
nth root, 63
Natural numbers, 14
  set of, 14
Negative number
  as exponent, 53
  on line graph, 89
Notation
  factorial, 347
  function, 147
  logarithmic, 296
  scientific, 304
  set, 3
  set-builder, 4
  sigma, 326
  summation, 326
Null set, 3
Number(s)
  absolute value, 105
  complex, 264
  counting, 14
  graph of a, 89
  imaginary, 264
  integers, 14
  irrational, 17
  line, 88
  natural, 14
  negative, 90
  ordered pairs of, 117
  positive, 90
  rational, 14
  real, 17
  roots of, 58

Odd function, 160
One to one
  correspondence, 87
  function, 285
One, multiplicative property, 23
Order, 90
  axioms, 92
  symbols, 90
Ordered pairs, 117
  as functions, 143
  as relations, 135
  in Cartesian product, 119
  as solutions of equations, 198
Order relation
  definition, 91
  notation, 90
  positive number axiom, 90
  properties, 92
  transitive, 92
  trichotomy, 90
Ordinate, 120
Origin
  of Cartesian coordinate system, 119
  on a number line, 88

Parabola
  axis of symmetry, 410
  definition, 409
  directrix, 409
  focal chord, 410
  focus, 409
  latus rectum, 410
  vertex, 410
Parallel lines, 136
Partial sums, 336
Pascal's triangle, 354

Perpendicular lines, 186
Permutations, 350
Point slope form of a line, 188
Polynomials
   addition of, 26
   coefficients of, 19
   completely factored form, 34
   definition of, 19
   degree of, 19
   division, 243
   factoring, 34
   multiplication of polynomials, 25
   subtraction, 26
Positive numbers, 90
Positive number axiom, 90
Powers, definition of, 18
Prime factors, 34
Probability
   definition of, 360
   of an event, 360
   of more than one event, 364
Product(s)
   Cartesian, 119
   of complex numbers, 264
   of fractions, 46
   involving radicals, 64
   of polynomials, 26
   of real numbers, 22
   with zero factors, 25, 209
Progression(s)
   arithmetic, 330
   geometric, 333
Proper subset, 5
Property
   of order relation, 92
   of real numbers, 22
   reflexive, 22
   symmetric, 22
   transitive, 22, 92

Quadrant, 121
Quadratic equation(s)
   definition of, 208
   discriminant of a, 222
   in form, 231
   formula, 213
   kind of roots, 222
   solution of:
      by completing the square, 209
      by extraction of roots, 210
      by factoring, 209
      by formula, 212
Quadratic function
   definition of a, 208
   graph of a, 215
Quadratic inequalities, 224
Quotient(s)
   of complex numbers, 267
   definition of a, 24
   of fractions, 47
   of polynomials, 243

Radical(s), 63
   addition of, 66
   definition of, 63
   equations, 231
   index of a, 63
   properties of, 63
   similar, 66
   subtraction of, 66
Radicand, 63
Radius of a circle, 404
Range
   of an exponential function, 292
   of a function, 143
   of a logarithmic function, 296
   of a relation, 136
Ratio of a geometric progression, 334
Rational exponents, 58
Rational expression, 33
Rational functions, 257
Rational Root Theorem, 254
Rationalizing the denominator, 67
Real line, 88
Real numbers, 17
Real part of a complex number, 264
Reciprocal, 23
Reducing fractions, 39
Reflection across $y = x$, 288
Reflexive law, 22
Relation
   domain of a, 136
   graph of a, 136
   range of a, 136
   as set of ordered pairs, 136
Remainder theorem, 248
Repeating decimal, 16
Root(s), 58
   cube, 58
   of equations, 243

nth, 63
    kind of, for quadratic equations, 222
    of numbers, 58, 63
    square, 58
Row operations, 382
Row reduced echelon matrix, 382

Sample space, 359
Scientific notation, 304
Second-degree equations, 438
Second-degree systems
    solution of, 438
    solution by substitution, 438
Sequences, 325
Series
    geometric, 336
    infinite, 336
Set(s), 3
    equality of, 5
    elements of a, 3
    empty, 3
    enumeration, 3
    finite, 4
    infinite, 4
    of integers, 14
    intersection of, 9
    of irrational numbers, 17
    of natural (counting) numbers, 14
    null, 3
    of rational numbers, 14
    of real numbers, 17
    solution, 79
    union of, 9
    universal, 6
Set builder notation, 4
Sigma notation, 326
Size of a matrix, 378
Slope of a line, 179
Slope-intercept form, 187
Solution(s)
    of equations, 79
    extraneous, 84
    of inequalities, 101
    ordered pairs as, 197
    of systems, 197
Solution set
    of an equation, 79
    of an inequality, 101
    of a system, 198, 382, 438
Standard form of logarithm, 304

Subset, 5
Substitution
    property, 22
    solution by, 198, 438
Subtraction
    of complex numbers, 267
    of fractions, 44
    of polynomials, 26
    of real numbers, 24
Sum
    of an arithmetic progression, 331
    of complex numbers, 266
    finite, 326
    of a geometric progression, 335
    of two real numbers, 22
Summation
    index of, 327
Summation notation, 327
Symmetry of the graph, 160
Synthetic division, 245
System(s)
    of linear equations:
        in two variables, 198, 377
        in three variables, 202, 380
    with second-degree equations, 438
    solution set of a, 198

Term(s), 19
    addition of similar, 26, 66
    of an arithmetic progression, 330
    of an expression, 19
    of a geometric progression, 334
    of a sequence, 325
    similar, 66
Terminating decimals, 16
Transitive property of order, 92
Transverse axis, 428
Triangle inequality, 114
Trichotomy, 90
Trinomial
    definition of, 19
    factoring, 37

Union of sets, 9
Universal set, 6

Value, absolute, 105
Variable(s), 18, 79
Variation
    constant of, 153

Variation—*Continued*
  direct, 153
  as a function, 153
  inverse, 154
  joint, 155
Venn diagram, 7
Vertex
  of a hyperbola, 428
  of a parabola, 410
  of an ellipse, 419

Vertical
  asymptote, 259
  axis, 120

Zero, 23
  additive property, 23
  division by, 39
  factorial, 348
  of a polynomial function, 254
  rational, 254